T0291707

Marine Hydrodynamics

Marine Hydrodynamics

40th anniversary edition

J. N. Newman
foreword by John Grue

The MIT Press
Cambridge, Massachusetts
London, England

This book was set in ITC Stone Serif Std and ITC Stone Sans Std by Toppan Best-set Premedia Limited.

Library of Congress Cataloging-in-Publication Data

Names: Newman, J. N. (John Nicholas), 1935– author.
Title: Marine hydrodynamics / J. N. Newman ; foreword by John Grue.
Description: 40th anniversary edition. | Cambridge, MA : The MIT Press, [2017] | Includes bibliographical references and index.
Identifiers: LCCN 2017023272 | ISBN 9780262534826 (pbk. : alk. paper)
Subjects: LCSH: Ships--Hydrodynamics. | Hydrodynamics.
Classification: LCC VM156 .N48 2017 | DDC 623.8/12--dc23 LC record available at https://lccn.loc.gov/2017023272

To my family and friends

Contents

Foreword

I became acquainted with this book in 1980 while completing my master's degree in mechanics at the University of Oslo. After the North Sea oil boom of the 1970s, it was apparent that there was a need for improved higher education and research related to the oil industry. The University of Oslo developed a curriculum in marine hydrodynamics, and interest in the field grew rapidly as students signed up for the program. I was one of them.

The course in marine hydrodynamics was offered for the first time during the 1980–1981 academic year, and has been lectured approximately once a year since its inception. J. N. Newman's textbook felt like a gift to our cohort and to our professor, Enok Palm. The entire book comprised a new curriculum with the exception of chapter 3, as viscous flow was covered in other courses. Like us, Palm was new to the subject, so the course was taught as a seminar. I was assigned the duty of giving lectures based on the book because our professor believed that having the youngest (and presumably the least experienced) student give the lectures would slow the pace of the course. I presented two hours of lectures each week for the entire academic year. During the lectures, Palm conducted a detailed examination of all statements and deductions I wrote on the blackboard, and I fielded questions from my classmates as well. By the end of the spring term, I knew the book by heart. The following year, the entire research division of Det Norske Veritas (now DNV GL) attended the course. This time, Enok Palm lectured the course himself. "I just can't ask a master's student to teach such an important course to the scientific leaders of Veritas," he told me. I was given the duty of leading the exercises instead.

Reading this book—a real classic—has been the most defining experience of my scientific career. I use it frequently in my teaching and scientific work. This text is a continuous font of inspiration for a master's or PhD project.

How does one define marine hydrodynamics? First, it's important to understand the meaning of offshore engineering. Put simply, offshore engineering deals with the design, construction, and safety considerations, including insurance, of both stationary and moving structures at sea. Regarding the first question, marine hydrodynamics defines the theoretical framework of offshore engineering. Marine hydrodynamics is an extensive subject as it treats the hydrodynamics in relation to all conceivable geometries exposed to the forces in the ocean environment. This includes the forces from the waves as well as those acting on lifting surfaces, either in water or in air. Sailors, rowers, canoeists, and kayakers are all faced with the same challenges: how does one reduce wave or frictional resistance, and optimize sail, rudder shape, or paddle motion? These are all examples of marine hydrodynamics. Calculating extreme loads and thereby ensuring the survival of structures exposed to extreme conditions—such as the record high Draupner wave, detected in the North Sea on New Year's Day 1995 with a crest height of more than 18 meters, a wave height of more than 25 meters, and a wavelength slightly exceeding 200 meters—is the ultimate goal. (Note that the design waves of the platforms in the North Sea are stronger than the Draupner wave.)

Stationary bodies are either fixed to the seafloor or floating on the surface with a stiff or slack mooring maintaining their position. There is a variety of such bodies, including all kinds of ships, barges, offshore platforms in transit from one position to another, and other objects towed by ships. Offshore wind turbines are more recently studied geometries. In most cases these are organized in large wind farms that are mounted on the seafloor as at Horns Rev 1 and 2 in Danish waters, and at Dudgeon on the coast of the United Kingdom. Hywind, developed by the Norwegian oil and gas company Statoil, is an offshore wind turbine prototype that floats on a spar buoy. The current target wing span of these geometries is twice that of an Airbus A380! Other geometries in offshore engineering include cages for aquaculture, which may be either closed or open. Making them larger and sufficiently strong enough to be placed in harsh ocean environments is a new development in the fish farming industry. New development of coastal infrastructure would require road crossings of fjords. Proposed bridges would be supported by floating, moored pontoons;

submerged floating tunnels are another alternative. Wave power devices are another example of structures where the subject of marine hydrodynamics is relevant, and where the design of successful structures must be based on this subject.

This textbook is ideal for a masters- or PhD-level course. Students using this book should also have completed courses in introductory mathematics, classical physics, and field theory. Additionally, a course in fluid mechanics provides a foundation to build upon with the material contained herein. Most introductory courses in mathematics have some components of complex theory and provide a sufficient background for the more advanced exercises in the book; a full course in complex analysis may help the reader but is not required for a good learning outcome.

The highlight of the book is the second half of chapter 6, which introduces the subject of wave effects. The primary goal is to calculate the responses of a floating body exposed to incoming waves in the six degrees of freedom: heave, sway, surge, roll, pitch, and yaw. Chapter 6, much like the rest of the book, is self-consistent, with the required background of wave theory—including concepts like superposition, group velocity, mass and energy flux, and third-order Stokes wave theory—given in the first half of the chapter. On the basis of potential theory, the matrix equation of the body responses includes the forces and moments due to added mass, wave damping, hydrostatics, and wave excitation. Perturbation-wise, the wave-body interaction problem is linearized and the forces and motions are modeled at each spectral frequency, a common assumption and decomposition that is used in marine hydrodynamics. The method provides a platform for a rich mathematical analysis in which the wave effects and responses are analyzed at the fundamental frequency. The resonance frequencies in the vertical modes of motion where the hydrostatic forces balance the inertia forces are defined. The linear analysis corresponds to the basic calculation method used in offshore engineering. Quadratic and higher-order wave effects may be studied using other texts.

The introduction of Morison's equation—a coefficient-based force formulation, particularly suitable for slender geometries—is motivated by the dimensional analysis in chapter 2. It is directly related to the long-wave analysis of the wave effects part of chapter 6. The viscous drag term included in Morison's equation is not modeled by the potential theory analysis of chapter 6, however. Chapter 2 introduces the concept of added mass—an inertia effect due to the accelerated mass of the surrounding fluid. It contributes in a similar way to the mass of a body in classical mechanics. In chapter 4—also directly related to the wave effects part of chapter 6—the forces and moments on a geometry moving in the six degrees of freedom, in unbounded fluid, are derived on exact form, expressed in terms of the six-by-six added mass matrix. This elegant, exact analysis is directly suitable as a model of the motion of a remotely operated underwater vehicle (ROV).

The ways in which animals swim and fly, airplanes and ski jumpers control their flight, and a yacht sails across the water are the result of flow at thin streamlined bodies. Chapter 5 introduces the fundamentals of the flow and force on a lifting surface. Moreover, it describes the propulsive force created by a wing's flapping motion. The lift problem of a two-dimensional foil section is analyzed in the beginning of the chapter—a starting point in the training of candidates for the growing wind power industry. The sharp trailing edge of the wing enforces the flow velocity to be finite, imposed mathematically by the Kutta condition, controlling the initial flow separation.

The mathematical modeling in this book is represented by partial differential equations (PDEs). The solution methods of field equations in marine hydrodynamics—commonly the Laplace equation—differ fundamentally from the common solution methods of PDEs. The main reason is this: The cause of the motion (the geometry) is local whereas the motion is global (the infinite extension of the wave field). Popular methods for solving partial differential equations are often based on volume methods, either difference methods or variants of the finite element method. The special method of separation of variables employs sets of orthogonal functions particularly fitted to geometries of special shape. However, for floating or submerged large-volume geometries of arbitrary shape, use of integral equations is unsurpassed and is a standard method in marine hydrodynamics. This method is extensively used in the offshore industry. The formulation, in terms of integral equations, is directly related to the scattering and radiation problems in theoretical physics, although the wave Green functions differ. The application of multipole expansions including the surface wave effects is an efficient, accurate, and convergent method that has found a lesser application in the industry. Thus, chapter 4 involves solving PDEs for the flow in an unbounded fluid by use of integral equations expressed in terms of source and dipole distributions. In chapter 5, the lifting and flapping problems are modeled by vortex and source distributions. The resulting integral equation becomes a singular Fredholm integral equation of the first kind. The equation is directly invertable—a superior, analytical, and generic method directly transferable to other related formulations and problems. The nonsingular integral equation formulated in section 11 of chapter 4 is a Fredholm equation of the second kind. This formulation is robust and easy to invert. Variants of this equation are widely used in the offshore industry.

How do I teach this book? I start with the integral representation of the potential flow at a geometry in unbounded fluid found in section 4.11, and subsequently introduce the forces and moments in an unbounded fluid expressed in terms of the added mass matrix. The first assignment is the calculation of the two-dimensional added mass of a circle, an ellipse, and a square, obtaining the various potentials through integral equations and Python or MATLAB scripting. Problems 4, 6, 13, and 15 are mandatory.

I continue with chapter 6. I usually give one lecture on linear wave theory and a second lecture on energy flux and group velocity. The wave effects part of the chapter comes next, introducing the decomposition of the diffraction and radiation potentials, the latter according to the six modes of motion, the added mass and damping, the exciting forces, and the restoring forces. Sometimes, depending on the student group, I will assign section 6.16 (hydrostatics) separately. The students then derive the mathematical formulas in the chapter for subsequent presentation in class. The theories are reworked by completing problems 10–17 on wave effects. Problems 1–8 cover the fundamentals of wave theory.

In chapter 5, my lecture covers the introduction and sections 5.1–5.6, as well as problems 8 and 9. The results are put in context of the dimensional analysis of the foil in section 7 of chapter 2.

One of my final lectures covers the dimensional analysis of chapter 2, with highlights including the low-level introduction of the concept of added mass and the coefficient-based force formulation included in Morison's equation. Problem 14 illustrates how the contributions from the inertia term and the drag term depend on the cylinder diameter. Froude's hypothesis for the drag on a ship hull is one of my favorites; I frequently use the ITTC-line in figure 2.12 to estimate the frictional resistance on ships. I used this curve as well as equation (23) of chapter 2 when advising the kayaker Eirik Verås Larsen as he prepared for the 2012 London Olympics. His question was how to reduce frictional and wave resistance. I suggested that the only variable to play with was the wet area of the kayak—the portion touching the water, which should be minimal. Larsen and his team eventually discovered that he could reduce his weight—which he eventually reduced by 10 percent—and won the gold medal on the K1 1000m event.

John Grue
University of Oslo
March 16, 2017

Preface to the 40th Anniversary Edition

The field of marine hydrodynamics has broadened greatly over the past 40 years, with applications to a wide variety of vessels and structures. These include systems for converting energy from the wind, waves, and currents; yachts, high-speed vessels, aquaculture facilities, and various types of submerged vessels; and traditional applications to ships and offshore platforms. The support for education and research has grown accordingly; it is gratifying that the term "marine hydrodynamics" has become ubiquitous for university departments, research laboratories, conferences, and publications.

The basic topics of the field are unchanged, corresponding broadly to the chapters of this text. Numerical methods that extend the applications of the theory have been developed. Practicing engineers and naval architects are now making routine use of well-established programs to optimize their designs and predict performance. These programs include Navier–Stokes solvers, which analyze viscous effects including turbulence, and "panel" programs based on boundary-integral equations to solve potential-flow problems including lifting and wave effects. This evolution has been accelerated by the universal access to computers of increasing capacity and convenience. Nevertheless, it is essential to understand the underlying principles covered in this text, and to compare the results of computations with simpler approximations to be confident of their validity.

I am grateful to the MIT Press for suggesting this special edition and for making it available economically, both as a paperback and as an open access e-book. I am especially grateful to Professor John Grue for the foreword, which reflects his long experience using this text.

Preface to the First Edition

The applications of hydrodynamics to naval architecture and ocean engineering have expanded dramatically in recent years. Ship design has been related increasingly to the results of scientific research, and a new field of ocean engineering has emerged from the utilization of offshore resources. The number of technical symposia and journals has increased in proportion to this expansion, but the publication of textbooks has not kept pace.

This volume has been prepared to satisfy the need for a textbook on the applications of hydrodynamics to marine problems. These pages have evolved from lecture notes prepared for a first-year graduate subject in the Department of Ocean Engineering at MIT, and used subsequently for undergraduate and graduate courses at several other universities. Most of the students involved have taken an introductory course in fluid mechanics, but the necessary fundamentals are presented in a self-contained manner. A knowledge of advanced calculus is assumed, including vector analysis and complex-variable theory.

The subject matter has been chosen primarily for its practical importance, tempered by the limitations of space and complexity that can be tolerated in a textbook. Notably absent are topics from the field of numerical hydrodynamics such as three-dimensional boundary-layer computations, lifting-surface techniques including propeller theory, and various numerical solutions of wave-body problems. A textbook on these subjects would be a valuable companion to this volume.

Since most countries of the world have adopted the rationalized metric *Système International d'Unités* (SI), this is used here except for occasional references to the "knot" as a unit of speed. Conversion factors for English units of measure are given in the appendix, together with short tables of the relevant physical properties for water and air. A unified notation has been adopted, despite the specialized conventions of some fields. Cartesian coordinates are chosen with the y-axis directed upward. Forces, moments, and body velocities are defined by an indicial notation that differs from the standard convention of ship maneuvering. The symbol L is reserved for the lift force, and D for drag. Thus length is denoted by l and diameter by d. Vessels with a preferred direction of forward motion are oriented toward the positive x-axis, following the practice of naval architecture but contrary to the usual convention of aerodynamics; a fortunate consequence is that a hydrofoil with upward lift force will possess a positive circulation as defined in the counterclockwise sense.

This text was initiated with the enthusiastic encouragement of Alfred H. Keil, Dean of Engineering at MIT, and Ira Dyer, Head of the Department of Ocean Engineering. Financial support has been provided by the Office of Naval Research Fluid Mechanics Program, which for the past thirty years has fulfilled an invaluable role in the development of this field. Additional thanks are due to the National Science Foundation and the David Taylor Naval Ship Research and Development Center for their support of the research activities that have filtered down into these pages.

Many colleagues and former students have helped significantly with encouragement, advice, and assistance. John V. Wehausen of the University of California, Berkeley, pioneered in applying the discipline of contemporary fluid mechanics on a broad front to the teaching of naval architecture; he has been generous with his advice as well as his own extensive lecture notes. Justin E. Kerwin of MIT shared in developing the course from which this text has evolved, and he has been particularly helpful in discussing a broad range of topics. Other colleagues to whom I am indebted include Chryssostomos Chryssostomidis, Edward C. Kern, Patrick Leehey, Chiang C. Mei, Jerome H. Milgram, Owen H. Oakley, Jr., and Ronald W Yeung of MIT; Keith P. Kerney, Choung M. Lee, and Nils Salvesen of the David Taylor Naval Ship Research and Development Center; Robert F. Beck and T. Francis Ogilvie of the University of Michigan; T. Yao-tsu Wu of the California Institute of Technology; Odd Faltinsen of the Norwegian Technical University; P. Thomas Fink of the University of New South Wales; and Ernest O. Tuck of the University of Adelaide. Former students who have been particularly helpful in a variety of ways include Elwyn S. Baker, Charles N. Flagg, George S. Hazen, Ki-Han Kim, James H. Mays, and Paul J. Shapiro.

The original illustrations are from the talented pen of Lessel Mansour. The manuscript was typed with proficiency by Jan M. Klimmek, Jacqueline A. Sciacca, and Kathy C. Barrington. My wife Kathleen helped with many editorial tasks and has patiently endured the diversion of my time.

1 Introduction

The applications of hydrodynamics to naval architecture and ocean engineering cover many separate topics and range over a broad level of sophistication. The topics are as diverse as the propulsion and steering of ships and the behavior in waves of a moored buoy or oil-drilling platform. The former are classical problems of naval architecture, predated only by Archimedean hydrostatics. The buoy and platform problems are more recent from the standpoint of scientific and engineering analyses. The degree of sophistication varies from empirical design methodology to theoretical research activities whose justification is based on long-range hopes of application.

The fields of technology are also diverse, and to solve these problems requires a knowledge not only of fluid mechanics but also of solid mechanics (to describe the mooring system especially), control theory (to represent the mechanical and human systems involved), as well as statistics and random processes (to deal with the highly irregular environment of the ocean). Here we shall focus our attention specifically on the hydrodynamic aspects, emphasizing those unique to this field as opposed to the other engineering disciplines where fluid mechanics is applicable.

Faced with the choice between empirical design information and esoteric theory, we will follow a middle course to provide the necessary background for an intelligent evaluation and application of empirical procedures and also serve as an introduction to more specialized study on the research end of the spectrum. This approach has the advantage of being a compromise between two viewpoints, which sometimes appear to conflict; it also unifies the seemingly diverse problems of marine hydrodynamics by examining them not as separate problems but instead as related applications of the general field of hydrodynamics. For example, propellers, rudders, antirolling fins, yacht keels, and sails are all fundamentally related to wings and hydrofoils, or *lifting surfaces,* and can be treated and understood together. Similarly, the unsteady ship, buoy, or platform motions in waves and the maneuvering of ships or submarines in nonstraight paths can be analyzed, to some extent, from the same basic equations of motion. In fact, however, the maneuvering problem generally involves separation and lifting effects, whereas the motions of bodies in waves are not as significantly affected by viscosity or vorticity.

The dynamics of fluid motions, like the dynamics of rigid bodies, are governed by the opposing actions of different forces, and moments, which are implied when not explicitly included. In fluid dynamics, these forces can no longer be considered as acting at a single point or discrete points of the system; instead they must be distributed in a relatively smooth or continuous manner throughout the mass of fluid particles. The force distribution and the kinematic description of the fluid motion are in fact continuous if and only if we assume that the discrete molecules of fluid can be analyzed as a *continuum.*

Typically, we can anticipate force mechanisms associated with the fluid inertia, its weight, viscous stresses, and secondary effects such as surface tension. In general, the three principal force mechanisms—inertial, gravitational, and viscous—are of comparable importance. With very few exceptions, it is not possible to analyze such a complicated situation, either theoretically or experimentally, and we can either give up or try to assess intelligently the role of each force mechanism, in the hope of subsequently treating them in pairs. As we shall see, this relatively simple state of affairs is still fraught with difficulties.

It is useful first to estimate the orders of magnitude of the inertial, gravitational, and viscous forces. We shall suppose that the problem at hand can be characterized by a physical length l, velocity U, fluid density ρ, gravitational acceleration g, and a coefficient of fluid viscosity μ. We then can estimate the three forces:

Type of Force	Order of Magnitude
Inertial	$\rho U^2 l^2$
Gravitational	$\rho g l^3$
Viscous	$\mu U l$

These estimates should not be interpreted too strictly. For example, it could be argued, from Bernoulli's equation, that a factor of $\frac{1}{2}$ should be associated with the magnitude of the inertial force. Nevertheless, the estimates are valid in the sense that changes in the magnitudes of any of the physical parameters l, U, ρ, g, or μ, will affect the forces as indicated. Thus, suppose the length scale is doubled, as might follow from attempts to compare directly the forces acting on a 100 m ship and on a 200 m ship, moving with the same speed; then the corresponding changes in the inertial, gravitational, and viscous forces will be multiplicative factors of 2^2, 2^3, and 2, respectively. Therefore, the fundamental balance among the three types of force will change as the length scale changes, and the effects will be more pronounced if we anticipate the relatively large changes of length scale (on the order of ten or one hundred) associated with a comparison of small-scale models and full-scale vessels.

To predict full-scale phenomena from tests with a scale-model, the absolute magnitude of any one force acting alone could be corrected by a suitable multiplicative factor. The principal concern is that all three forces act simultaneously and that their relative magnitudes be preserved so that the resulting flow is dynamically similar. For this reason, it is illuminating to form the ratios of the three forces, which yield a set of three nondimensional parameters to describe the fluid flow:

$$\frac{\text{Inertial Force}}{\text{Gravitational Force}} = \frac{\rho U^2 l^2}{\rho g l^3} = U^2/gl,$$

$$\frac{\text{Inertial Force}}{\text{Viscous Force}} = \frac{\rho U^2 l^2}{\mu U l} = \rho U l/\mu,$$

$$\frac{\text{Gravitational Force}}{\text{Viscous Force}} = \frac{\rho g l^3}{\mu U l} = \rho g l^2/\mu U.$$

Any two of these three ratios are sufficient to define the third and hence to determine the balance of forces in the fluid motion. Customarily, the first two are employed in the forms

F = Froude Number $= U/(gl)^{1/2}$,

R = Reynolds Number $= \rho U l/\mu = Ul/\nu$,

where $\nu = \mu/\rho$ is the kinematic viscosity coefficient of the fluid. A short table of the density and viscosity coefficients for water and air is given in the appendix. Typical values for the kinematic viscosity ν are 10^{-6} m^2/s (10^{-5}

ft^2/s) for water and 1.5×10^{-5} m^2/s (1.5×10^{-4} ft^2/s) for air. That this coefficient is small when expressed in terms of conventional units implies that the Reynolds number R will be large; hence viscous forces will be negligible relative to inertial forces. This is generally true, but one must be more cautious before concluding that viscosity can be completely ignored. In fact, viscosity can be neglected for the bulk of the fluid but must be included in singular regions such as the boundary layer very close to a body.

In this analysis we have tacitly assumed steady motion and hence constant characteristic velocity U. If instead the motion is oscillatory in time, as in the case of a buoy oscillating in a seaway, then the characteristic velocity U should be replaced by the combination ωl, where ω is the frequency of oscillations in radians per unit time. The counterpart of the Froude number for such motions is the nondimensional frequency parameter $\omega(l/g)^{1/2}$. Alternatively, we can use the ratio λ/l, where λ is the wavelength, or distance between successive wave crests, since λ can be explicitly related to ω ($\lambda = 2\pi g/\omega^2$ for waves in deep water).

Other important parameters arise when we examine the kinematic and thermodynamic aspects of the fluid. The most familiar of these is the Mach number of aerodynamics, which is the ratio between the velocity U and the speed of sound in the fluid medium; it arises from considerations of the elasticity or compressibility of the fluid. These effects are not important in the motions and behavior of ocean vehicles because water is not significantly compressible at the speeds of interest; in terms of the Mach number we note that the speed of sound in water is on the order of 1,500 m/s or 3,000 knots, which implies a negligibly small Mach number for water—based craft and thus insignificant compressibility effects. On the other hand we must consider in certain circumstances the possibility of cavitation, for when the pressure in the fluid is reduced below the *vapor pressure* p_v, the physical state of the fluid abruptly changes to that of a gas. Thus, while the fluid is extremely inelastic in compression, it cannot normally sustain significant tension. Dimensional considerations suggest the cavitation number

$$\sigma = \frac{p_0 - p_v}{\frac{1}{2}\rho U^2},$$

which is a measure of the likelihood of cavitation and parametrically describes the subsequent details. Here p_0 is a characteristic pressure level in the fluid, such as the hydrostatic pressure at the depth in question, and the vapor pressure p_v depends on the properties of the fluid and its temperature. If the cavitation number is large, cavitation will not occur, and the precise value of σ is immaterial to the description and analysis of the flow. If the cavitation number is sufficiently small so that cavitation occurs in the flow field, dynamic similarity between two flows will exist only if the corresponding cavitation numbers are equal. At normal temperatures p_v is substantially less than the atmospheric pressure, and thus cavitation is significant only at very high speeds.

Fluid motions that have similar geometries but different values of the relevant physical parameters are said to be dynamically similar if the relevant nondimensional parameters such as the Froude and Reynolds numbers are equal. It follows that the relative balance between the inertial, viscous, and gravitational forces is identical, and the resulting hydrodynamic effects, including fluid velocity, pressure, and forces acting on the boundaries of the fluid can all be analyzed by means of relationships between the two different flows. Dynamic similitude is clearly desirable if one is conducting small-scale model tests that can be used to design large-scale vessels. However, the simultaneous scaling of both Froude and Reynolds numbers is not possible, at least for reasonable changes of length scale with water as the full-scale fluid. To see this simply note that the ratio of the Reynolds and Froude numbers, $(g^{1/2}l^{3/2}/\nu)$, must stay constant. For a model length substantially smaller than the full-scale vessel, either the gravitational acceleration must be increased or the viscosity coefficient decreased, by an order of magnitude. The former suggests a centrifuge and the latter a superfluid, but neither is amenable to exploitation in this context.

This discussion of dynamic similitude demonstrates that assumptions or simplifications are often necessary to apply experimental methods with models to the hydrodynamics of ocean vessels. This is equally true, if not more so, if one wishes to pursue a strictly analytical prediction based on rational mechanics. Ultimately, therefore, we must expect both experiments and theory to be used and to be supplemented by full-scale observations to verify the original predictions.

In the theoretical approach the motion of the fluid is defined at each point in time and space by a kinematic description, usually of the vectorial velocity of the fluid particles or alternatively of the three scalar components of the velocity. This unknown velocity function is related by means of Newton's equations to the forces that act upon the fluid; this yields a system of partial differential equations. For a fluid with conventional stress relations, the resulting system of governing equations is the Navier-Stokes equations, supplemented by the continuity equation expressing conservation of fluid mass. In principle, one can solve these equations, subject to boundary conditions on the boundary surfaces of the fluid. If this procedure could actually be carried out, it would be possible to calculate desired answers for arbitrary values of the Reynolds number and Froude number, and the scaling dilemma of model testing would be circumvented. In practice, however, it has not been possible to solve the Navier-Stokes equations exactly, except for a few cases involving very simple geometries that at first glance have no relation to the shape of marine vessels.

Considerably more analytical progress can be made if the viscous forces are ignored in the Navier-Stokes equations and the fluid is assumed to be inviscid or *ideal*. It is then feasible to construct mathematical solutions for the flow past bodies of realistic form and even to include the effects of wave motions on the free surface, albeit only after further idealizations. Clearly, however, neglecting viscosity will lead to results of only academic interest, unless the justification is more relevant than the mathematical desire to simplify a system of partial differential equations. For predictions of ship resistance, this justification was initially provided by Froude's hypothesis that the resistance could be composed of two separate components, frictional and residual. The frictional component is related to a much simpler geometry, a deeply submerged flat plate, and thus depends only on the Reynolds number. The residual component is assumed to depend only on the Froude number. This hypothesis was essentially an empirical one; its principal justification was that it led to a workable procedure for making model tests and obtaining full-scale predictions not totally at variance with the actual full-scale results. The neglect of viscosity in treating certain aspects of the flow, and Froude's assumption that the frictional resistance of a ship hull could be related to that of a flat plate of the same length and area, found a more rational justification after Prandtl developed the boundary-layer theory. Thus it became evident that at the relatively large Reynolds numbers of interest to naval architects and aerodynamicists, viscous stresses are significant only within the very thin layer of the fluid adjoining the rigid surfaces such as the ship's hull or the airplane. Outside this layer the fluid is essentially inviscid, not because its viscosity has suddenly changed but because viscous stresses are a consequence of large gradients in the fluid velocity and these large gradients are restricted to the immediate vicinity of the boundary. Moreover, the flow within the boundary layer is relatively insensitive to the form of the boundary, provided the body's radii of curvature are large compared to the thickness of the boundary layer. The consequences of Prandtl's boundary-layer theory would appear to be fundamental to Froude's hypothesis, and it is remarkable that Froude predated Prandtl by thirty years.

In the chapters to follow, model testing will be discussed first, with emphasis on the use of dimensional analysis to preserve dynamic similarity between the model and full-scale flows. This glimpse of the "real-world" will serve to introduce the theories that follow in subsequent chapters. The theoretical approach will commence with a study of viscous flows so that the importance of viscosity can be set in proper perspective.

Each chapter begins with a general treatment of the essentials and progresses to more specialized and advanced material. The order in which this is studied can be varied to suit one's interests and background. Readers anxious to proceed to their favorite subject, but lacking the prerequisites to do so directly, should find most of their needs met in sections 3.1 to 3.9 and 4.1 to 4.5. In the references listed at the end of each chapter, preference has been given to recent surveys and papers with comprehensive bibliographies where additional information on each topic may be sought.

2 Model Testing

For most of us, toy boats and models are a useful childhood introduction to the subject of hydrodynamics. Given a block of wood, a sharp knife, and the plans of a ship or small boat, few would resist the impulse to produce a model. Our choice of scale ratio would be determined by the dimensions of the block of wood. Perhaps future engineering interests are indicated if a ruler is used to ensure a "true scale model." If the model is subsequently operated in its natural medium, the builder must determine the proper weight and, in the most refined examples, an appropriate power system to drive the model at the "true" speed of the real ship. The weight may be determined by an application of Archimedes' principle in the bathtub or sink, but the questions pertaining to speed and power are more complex and, strictly speaking, impossible to answer. Superficial bathtub observations might reveal that the flow of water around the model is reasonably like that of a large ship, meaning that the observed waves look realistic, if the model speed is on the order of half a meter per second, or one knot. On the other hand, if we could observe the viscous flow very close to the surface of the hull and in the wake immediately behind the stern, gross differences would be apparent between the ship and its scale model.[1]

This situation can be better understood by a dimensional analysis. Denote the physical quantity of interest by Q (for example, the resistance of a body moving through the water, the thickness of the boundary layer, or the acceleration of a buoy in waves). First it is necessary to list, from a qualitative knowledge of the underlying physical mechanisms involved, the significant parameters that affect the value of Q (length, velocity, density, viscosity, and so forth). In general, some or all of these parameters can be used to make Q nondimensional, and any not needed for this purpose can themselves be rendered nondimensional. We then assert that when expressed in nondimensional form, the quantity of interest must depend only on the remaining nondimensional parameters. This assertion is equivalent to the statement that the event in question is not affected by the choice of units of measurement.

We are concerned exclusively with problems where the three fundamental units are mass (**M**), length (**L**), and time (**T**); hence the unknown Q and the significant parameters upon which it depends can be expressed in terms of these units. If Q depends on $N - 1$ significant parameters, there will be a total of N interrelated dimensional quantities, including Q. The number of interrelated nondimensional quantities is always smaller, hence the resulting simplification of the problem. Since there are three fundamental units (**M**, **L**, **T**), the number of independent nondimensional parameters will be reduced by the same number. Hence a total of $N - 3$ nondimensional quantities must be interrelated.

This statement is essentially an intuitive one. A careful mathematical proof is possible; the end result, known as the *Pi theorem,* is derived by Birkhoff (1955) and Sedov (1959). Our approach here will be more pragmatic; it will illustrate the procedure of dimensional analysis and the conclusions that follow from the Pi theorem by considering a sequence of progressively more complicated physical problems.

2.1 Falling Body in a Vacuum

Let us consider first the problem of a body falling freely in a vacuum. The unknown quantity Q will be taken as the vertical position y, with dimension **L**. The unknown y might depend on time t, body mass m, and the gravitational acceleration g which has units $\mathbf{L/T^2}$. Since the body is falling in a vacuum, its size and shape cannot affect the descent. The vertical position y can be nondimensionalized by dividing by the quantity gt^2 to form the unknown y/gt^2. However, there is no way of forming the significant parameters t, m, g into nondimensional parameters, since neither g nor t contains the units of mass. Thus we conclude that

$$y / gt^2 = C, \tag{1}$$

where C is a pure constant, a nondimensional function of nonexistent, nondimensional parameters involved in the problem. The constant C is known from Newtonian mechanics to be equal to 1/2; but if we did not know this, we could determine C from a single experimental observation, and dimensional analysis would relate the results of this one experiment to all other possible values of the physical parameters g and t. Since there are no nondimensional parameters other than the unknown y/gt^2, dimensional analysis has almost provided the complete solution without any knowledge of dynamics. Our choice of position and time as the dependent and independent parameters of the problem was entirely arbitrary. We might also have considered, for example, the velocity of the body as a function of the distance it has fallen.

2.2 Pendulum

Next, we consider the period T of a simple pendulum, under the usual assumptions of a plane motion about a frictionless pivot in a vacuum. The physical parameters of significance are the pendulum length l, mass m, and maximum angle of its swinging motion θ_0, together with the gravitational acceleration g. A suitable multiplicative combination with the desired units of time is $(l/g)^{1/2}$, and the only nondimensional parameter that can be formed from the four physical parameters is θ_0, which is itself nondimensional. It follows that

$$T(g/l)^{1/2} = f(\theta_0), \tag{2}$$

where f is a nondimensional function of the maximum angle. If we also assume that this angle is very small, and if we observe that under these circumstances the period T tends to a finite limit, then

$$T(g/l)^{1/2} = f(0). \tag{3}$$

The constant $f(0)$ is known from mechanics to be 2π and could be measured from a single experiment.

Rather than using the period T, we could employ the frequency $1/T$ or the radian frequency $\omega = 2\pi/T$. In terms of the latter parameter, the above relations take the form

$$\omega(l/g)^{1/2} = 2\pi/f(\theta_0) \simeq 2\pi/f(0). \tag{4}$$

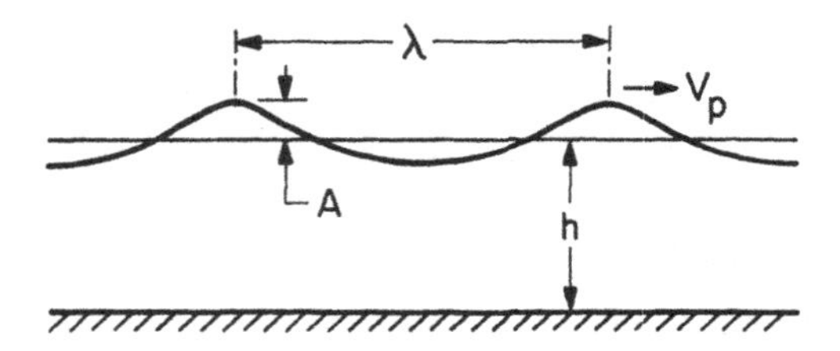

Figure 2.1
Sketch of a periodic progressive wave in a fluid of mean depth h. Note that λ is the wavelength, A the wave amplitude, and the wave translates with phase velocity V_p in the direction shown by this vector.

2.3 Water Waves

As our first application of dimensional analysis to a problem in hydrodynamics, we shall consider the motion of an oscillatory wave system on the free surface. In some respects this is a complicated problem with which to commence our discussion of fluid motions, but the analogy with the motion of a simple pendulum is useful. Moreover, the remaining problems will involve the interactions of the fluid with a rigid body of some sort, and it is convenient to discuss wave motions before inserting the body into the fluid.

Figure 2.1 shows a periodic progressive wave system which is characterized by its amplitude A, wavelength λ, and period T. As in section 2.2, the period T can be replaced by the radian frequency $\omega = 2\pi/T$. The wave motion propagates through the fluid with a *phase* velocity V_p such that any prescribed point of constant phase, such as a crest or trough of the wave, will progress a distance λ in time T. Thus

$$V_P = \lambda / T = \omega\lambda / 2\pi. \tag{5}$$

Like the simple pendulum, this wave motion is the result of a balance between kinetic energy and potential energy. The most important physical parameters are the fluid density ρ, gravity g, and the depth h. Thus, the wave period T must be a function only of the wavelength λ, amplitude A, density ρ, gravitational acceleration g, and depth h. Note that the frequency ω and phase velocity V_p are not included here, since these are not independent and can be substituted for, or replaced by, T and λ.

As in the simpler problems of the falling mass and pendulum, only one physical parameter, the density of the fluid, contains the units of mass. Since it is impossible to nondimensionalize ρ in combination with the other parameters, it can be deleted from the analysis. A total of five dimensional quantities (T, λ, A, g, h) remain, and a nondimensional form for the period is

$$T(g/\lambda)^{1/2} = f(A/\lambda, h/\lambda). \tag{6}$$

Comparing this result with equation (2), we find that the nondimensional amplitude ratio A/λ is analogous to the maximum pendulum angle θ_0, but an additional depth parameter h/λ must now be considered.

Two possible simplifications of (6) can be pursued. First, as in (3), if the wave amplitude is sufficiently small compared to the wavelength, a *linearized* result follows from (6),

$$T(g/\lambda)^{1/2} \simeq f(0, h/\lambda), \qquad A/\lambda \ll 1. \tag{7}$$

Alternatively, if the fluid depth is very large compared to the wavelength, as in the deep ocean, (6) can be replaced by

$$T(g/\lambda)^{1/2} \simeq f(A/\lambda, \infty), \qquad \lambda/h \ll 1. \tag{8}$$

If both inequalities are satisfied simultaneously,

$$T(g/\lambda)^{1/2} \simeq f(0, \infty) = \text{constant}. \tag{9}$$

Thus, for small amplitude waves in deep water, the period is proportional to $\lambda^{1/2}$ and, from (5), the phase velocity likewise is proportional to the square root of the wavelength.

As in the simpler problems of the falling mass and pendulum, the constant in (9) cannot be determined from dimensional analysis. A more complete solution of the linearized problem in chapter 6 will show that this constant is equal to $(2\pi)^{1/2}$ and will provide the more general functional relation (7) for finite depth.

Since the phase velocity V_p depends upon the wavelength, it follows that water waves of different wavelengths or periods will propagate at different phase velocities and are thus *dispersive*. Long waves will travel faster than short waves, and a spectrum of waves will constantly change its appearance in a manner obvious from observation of ocean waves. The monochromatic wave system shown in figure 2.1 is an exception, where the free-surface profile can propagate without change of form because only one wavelength is present.

Other physical properties of secondary importance include the viscosity of the water, the surface tension of the air-water interface, and the properties of the air. The fluid viscosity exerts a small dissipative effect, comparable to the friction of the pendulum bearing, and typical water waves can travel for hundreds or thousands of wavelengths without significant attenuation. Qualitatively this could be anticipated by forming a nondimensional viscosity ratio $\nu g^{-1/2}\lambda^{-3/2}$, which is equal to 3×10^{-4} for $\lambda = 1$ m and is even smaller for longer wavelengths.

The surface tension of the air-water interface exerts a tensile force, per unit width of the free surface, proportional to the curvature of the free surface and analogous to the restoring force of waves on a string or membrane. A suitable nondimensional parameter is $T_s/\rho g\lambda^2$ where the surface tension T_s for an air-water interface is about 0.07 N/m. Thus $T_s/\rho g\lambda^2$ is on the order of 10^{-5} for $\lambda = 1$ m, and smaller for longer waves. It follows that surface tension effects are negligible for all but the shortest waves or ripples.

Generally one can ignore the effects of the air on water waves since the air density is on the order of 10^{-3} times that of water. An important exception is the process by which ocean waves are generated by the wind, but that is a complicated problem beyond the scope of this text.

2.4 Drag Force on a Sphere

Dimensional analysis is particularly useful in studying the forces exerted by the fluid on a moving body. As our first example we shall consider the drag force acting upon a sphere moving with constant velocity U through a viscous fluid of unbounded extent. Of course, no fluid is truly unbounded, and this restriction implies that the sphere is very small compared to the distance to the nearest boundary or other body within the fluid.

If the surface of the sphere is smooth, the only length scale of the problem is the sphere diameter. The drag force D must be a unique function[2] of the diameter d, the sphere velocity U, the fluid density ρ, and the kinematic viscosity coefficient υ. In the dimensional form

$$D = f(d, U, \rho, \nu). \tag{10}$$

Nondimensionalizing these five parameters yields two nondimensional quantities, which can be expressed in the form

$$\frac{D}{\rho U^2 d^2} = f(Ud / \nu), \tag{11}$$

where $R = Ud/\nu$ is the Reynolds number based on the sphere diameter.

In this analysis the body weight W and buoyancy force $\rho g \forall$, where $\forall$ = displaced volume, are not involved because we are considering only the hydrodynamic drag force associated with steady translation of the body. The translation may be in the horizontal plane or in the vertical direction with the static weight and buoyancy forces subtracted from the total force. On the other hand, if the body is freely falling (or rising), the excess weight $(W - \rho g \forall)$ is an important physical parameter. In the steady case, where a freely falling body has reached its *terminal velocity*, the excess weight is balanced by the steady-state drag force.

Writing (11) in a more conventional form, we have

$$\frac{D}{\frac{1}{2}\rho U^2 S} = C_D(R), \tag{12}$$

where $S = \pi d^2/4$ is the frontal area of the sphere and C_D is the drag coefficient. Nothing more can be said about the drag, unless experimental observations of the drag coefficient are available for various Reynolds numbers, but the information now required has been reduced to a bare minimum. To predict the drag of any sphere in any fluid, it is sufficient to perform experiments with a single sphere over a range of velocities and in a single fluid. This is illustrated in figure 2.2, which shows experimental measurements of the drag coefficient for several fluids, including water and air, and for a variety of sphere diameters.

An explanation of the results shown in figure 2.2 requires a description of fluid mechanics beyond the scope of dimensional analysis. Nevertheless, a brief qualitative discussion will provide an understanding of this and other drag problems to be studied subsequently in this chapter. For moderate Reynolds numbers, the dominant contribution to the drag force is due to *separation*, which occurs near the midplane of the sphere. Upstream of the separation point viscous effects are confined to a thin boundary layer adjacent to the sphere surface, and the vicinity of the stagnation point at the front of the sphere will be a region of relatively high

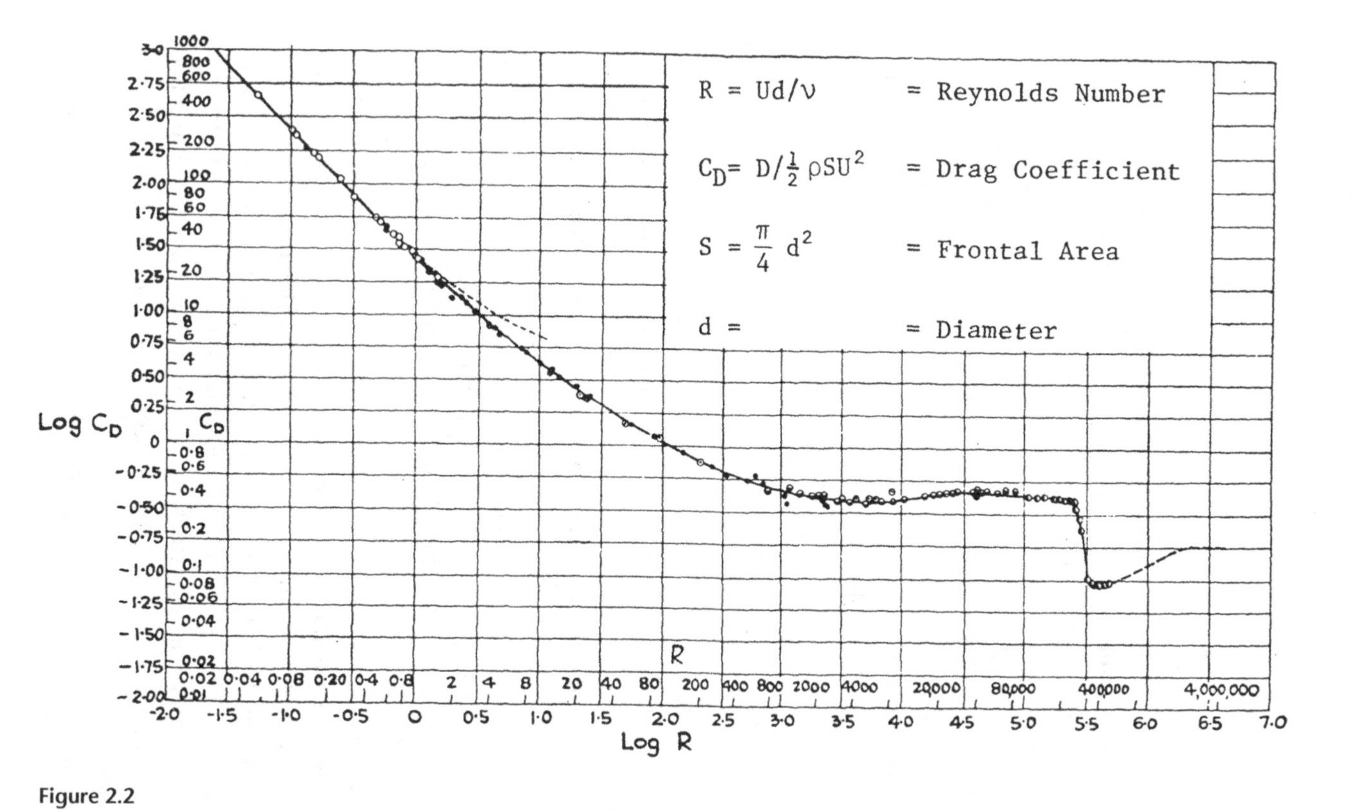

Figure 2.2
The drag coefficient of a sphere (reproduced from Goldstein 1938). Note the transition from laminar to turbulent flow at a Reynolds number of approximately 3×10^5. Dashed line for $R < 10$ is a low-Reynolds number theory due to Oseen.

pressure compared to the free-stream value at infinity. Downstream, within the separated wake, the fluid is relatively stagnant but has a reduced pressure dictated by the requirement of continuity with the free-stream pressure outside the separated region. Thus, as a result of separation, there is a substantial pressure difference between the forebody and afterbody; from dimensional considerations it must be on the order of magnitude of ρU^2, and the resulting drag coefficient is of order of magnitude unity. This situation ensues, in figure 2.2, over the range of Reynolds numbers between 10^3 and 3×10^5. In this regime the viscous flow in the boundary layer on the forebody is *laminar.*

At a critical Reynolds number of about 3×10^5, the boundary-layer flow becomes *turbulent,* and the resulting increase of momentum convection postpones separation. Thus, the separated region behind the sphere is diminished, and the drag is reduced in a dramatic fashion that coincides with transition of the upstream boundary-layer flow from the laminar to turbulent regime. The precise point where transition to turbulence occurs, with a resulting decrease in the drag of a bluff body, depends not only on the body shape but also on the ambient turbulence of the flow and on the roughness of the body. The indented surface of a golf ball is intended to stimulate early transition and reduced drag.

2.5 Viscous Drag on a Flat Plate

As an example of a highly streamlined body shape, we consider a rectangular flat plate of length l, breadth B, and negligible thickness, moving with velocity U in the longitudinal direction parallel to its length dimension. The drag coefficient will depend not only on the Reynolds number but also on the aspect ratio B/l of the plate. In general, we must anticipate a result of the form

$$\frac{D}{\frac{1}{2}\rho S U^2} = C_D(R, B/l). \tag{13}$$

Here S is the surface area of the plate; we have chosen this in preference, say, to B^2 or l^2 because we expect the viscous drag to be roughly proportional to the surface area over which viscous shear stresses act. This expectation is confirmed by the fact that, for Reynolds numbers on the order of 10^5 to 10^{10}, the drag coefficient is relatively insensitive to the ratio B/l. This is not a foreseeable consequence of dimensional analysis; instead, it must be explained from the characteristic thickness of the boundary layer.

Experimentally determined *frictional-drag* coefficients C_F are shown in figure 2.3 for various flat plates, along with the semiempirical equation determined by Schoenherr:

$$0.242 / \sqrt{C_F} = \log_{10}(RC_F). \tag{14}$$

Here the coefficient has been changed from C_D to C_F because we must distinguish subsequently between total drag and frictional drag. Once again, the validity of the dimensional analysis is confirmed by the collapse of data from diverse experiments both in water and in air.

There is a noticeable scatter of data in the transition range of Reynolds numbers between 10^5 and 2×10^6. In that range, the flow changes from a smooth laminar regime to the turbulent regime that persists throughout the range of higher Reynolds numbers. In this transition range an important mechanism that triggers turbulence is the smoothness of the body surface. Thus, the drag of rough plates shifts to the turbulent value at lower Reynolds numbers; for very smooth plates, laminar flow can be maintained longer. Similar results occur if the ambient flow is not uniform.

The principal consequence of turbulence is to increase the momentum defect of the boundary layer and the resulting frictional drag on the flat plate. This is precisely opposite to the effect noted for a sphere where the increase in frictional drag is insignificant compared with the decrease in pressure drag. That the magnitude of the frictional drag coefficient shown in figure 2.3 is two orders of magnitude less than that of the drag coefficient for a sphere emphasizes the practical importance of streamlined body shape.

2.6 Viscous Drag on General Bodies

The drag on more general bodies can be determined from tests of geometrically similar models, or *geosims,* provided that the Reynolds number for model and full-scale bodies is the same. However, practical difficulties may prevent this scaling procedure from being carried out. Since it is difficult to find nonvolatile liquids appreciably less viscous than water, the ratio of model velocity to full-scale velocity must be inversely proportional to the ratio of the lengths. As an example, let us consider a ship or submarine of

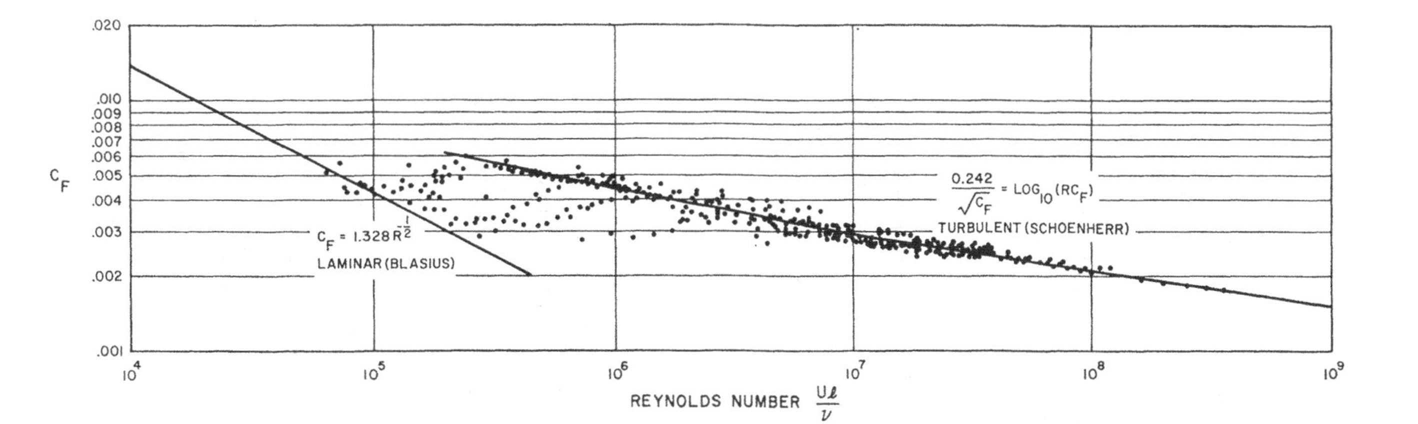

Figure 2.3

Schoenherr's flat-plate frictional drag coefficient, compared with various experimental results and the Blasius boundary-layer theory. Note that transition occurs between $R = 10^5$ and 2×10^6, depending on the plate roughness and ambient turbulence. For references to the experimental data see Todd (1967), figure 3.

100 m length, moving at 10 m/s or 20 knots; if a 10 m model of the same vessel is to be tested in water at the same Reynolds number, it must move with a velocity of 100 m/s or 200 knots.

This difficulty is usually overcome by separating the total viscous drag into two components, frictional drag and pressure drag. These can be defined as the longitudinal components of the forces acting on the body due respectively to tangential shear stresses and normal pressure stresses. We assume that the tangential stress is the only component affected by the Reynolds number and, moreover, that the resulting frictional-drag force is equal to that of a flat plate of equal area and Reynolds number. The remaining force component, due to normal pressure stresses acting on the body surface, will depend on the form of the body; for this reason it is often called the *form drag* or *pressure drag*. In the case of a streamlined body the form or pressure drag can be assumed independent of the Reynolds number over the range where the boundary layer is thin compared to the body dimensions, typically for $R > 10^5$. On the other hand, for a bluff body the possible dependence of the point of separation on R must be recognized.

With these assumptions, the total drag coefficient can be written in the form

$$C_D(R) = C_F(R) + C_P, \tag{15}$$

where C_F is the flat-plate frictional-drag coefficient defined in the last section, and plotted for a wide range of Reynolds numbers in figure 2.3. The pressure-drag coefficient C_p can be determined, for the body in question, by performing a model test at a convenient Reynolds number and by using data such as that in figure 2.3 to predict the corresponding value of C_F.

For streamlined three-dimensional bodies with maximum thickness less than about one fifth of the length, the frictional drag is dominant and the drag coefficient can be predicted from the flat-plate drag coefficient C_F. The pressure drag is more important for two-dimensional cylinders or struts, because of the increased obstruction of the flow past the body in two dimensions. Figure 2.4 shows the total drag coefficient for a circular cylinder, two streamlined struts with thickness-length ratios of 1/2 and 1/4, and the limiting case of zero thickness given by twice the flat-plate frictional drag coefficient. This figure illustrates the conflicting roles of transition to

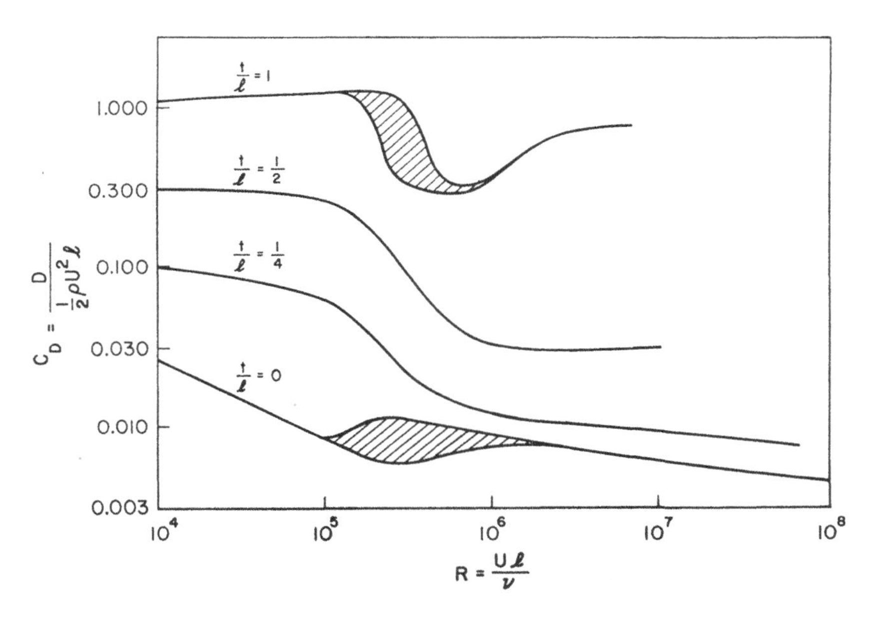

Figure 2.4

Drag coefficient as a function of Reynolds number for two-dimensional cylinders of maximum thickness-to-length ratio t/l. For the flat plate $(t/l = 0)$ the curve shown is twice the skin-friction coefficient of figure 2.3, and the shaded areas of this and the circular cylinder indicate the general range of uncertainty depending on ambient turbulence and roughness of the surface. For the intermediate cylinders, the precise values of the drag coefficients will depend on the shape, especially near transition.

turbulence, which decreases the drag of bluff bodies and increases the drag of fine bodies.

If the body has sharp edges not aligned with the flow, separation will generally occur at these edges irrespective of the Reynolds number or the state of the upstream boundary layer. For example, a circular disc moving normal to its plane will experience separation at the periphery; the drag coefficient for this body is about 1.1 for all Reynolds numbers greater than 10^3.

In summary, the drag coefficient of a body moving in a viscous fluid will depend in general on the Reynolds number; but for a bluff body at large Reynolds numbers the drag coefficient is insensitive to Reynolds number except insofar as this affects the point of separation.

2.7 Hydrofoil Lift and Drag

Hydrofoils, rudders, propeller blades, yacht sails, and keels are diverse examples of *lifting surfaces*. Generally these are thin streamlined bodies, intended to develop a hydrodynamic *lift* force L. The term *hydrofoil* will be used in a generalized sense to encompass all possible applications of lifting surfaces, as opposed to the strict definition of the supporting devices for a hydrofoil boat.

The performance of hydrofoils in water is analogous to that of airplane wings (and sails) in air, the principal distinction between these being the density of the fluid medium. The term *lift* derives from the aeronautical context and is used for the corresponding force component on a lifting surface regardless of its orientation in space. In particular, the lift force on a rudder, yacht sail, or keel is generally horizontal.

Inevitably, hydrofoils experience a drag force D, and the total force $\mathbf{F}$ is a vector quantity, with components (D, L). In a reference frame moving with the hydrofoil, the drag component is in the same direction as the free stream, as shown in figure 2.5. By definition, the lift force is perpendicular to the free stream, regardless of the angle of attack, α, between the orientation of the hydrofoil and the free-stream direction.

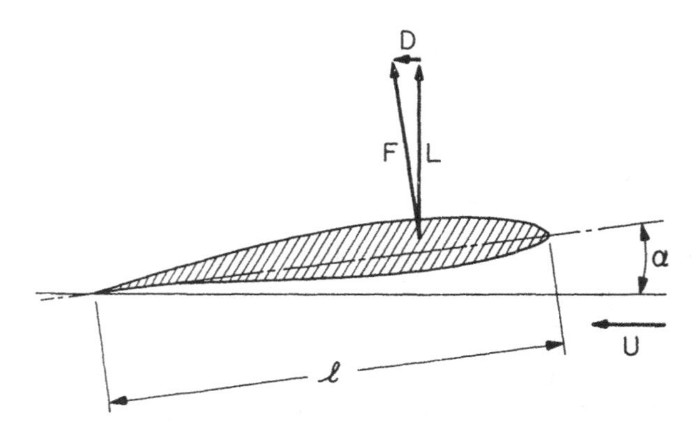

Figure 2.5
Flow past a hydrofoil section. The angle of attack α is generally defined with respect to the "nose-tail line," between the center of the minimum radius of curvature of the leading edge and the sharp trailing edge. L and D denote the lift and drag components of the total force F, and are defined respectively to be perpendicular and parallel to the free-stream velocity vector.

In this section the dimensional analysis of lifting surfaces is outlined in a form that can be used with model tests to predict the lift and drag forces acting on full-scale hydrofoils. Geometrical similarity is assumed, as in the earlier sections of this chapter.

For the two-dimensional case shown in figure 2.5, the length scale of the hydrofoil or its model can be described by the *chord length* l. In three dimensions, the lateral extent of the hydrofoil is its *span* s. For geometrically similar hydrofoils, either l or s can be used as the characteristic length for dimensional analysis of the lift and drag force. However, as in the case of the viscous drag on a flat plate, the results will be more general if we anticipate that the force acting on a lifting surface is approximately proportional to the *planform* area S. For a planar hydrofoil this can be defined as the projected area of the foil in the direction of the lift force, with $\alpha = 0$.

We shall assume that the hydrofoil operates in a steady-state manner with constant velocity and angle of attack, in an unbounded fluid, and that the ambient pressure is sufficiently high to preclude cavitation. Thus the relevant physical parameters affecting the hydrodynamic forces are the planform area S, velocity U, angle of attack α, fluid density ρ and the kinematic viscosity υ). The lift and drag forces can be expressed in the nondimensional forms

$$\frac{L}{\frac{1}{2}\rho U^2 S} = C_L(R,\alpha), \tag{16}$$

$$\frac{D}{\frac{1}{2}\rho U^2 S} = C_D(R,\alpha). \tag{17}$$

As in the case of the flat plate, viscous effects depend primarily on the length scale of the body parallel to the free-stream direction. For this reason the chord length l is used to define the Reynolds number $R = Ul/\upsilon$.

Values of the lift and drag coefficients can be obtained experimentally by placing a scale model in a wind tunnel or water tunnel. The angle of attack can be varied by a suitable arrangement and the force components measured with dynamometers. Here again, however, there is a practical difficulty in preserving the full-scale Reynolds number. Thus the drag is generally treated as for a streamlined nonlifting body, accounting for the Reynolds number dependence simply in terms of the frictional drag of a flat plate at zero angle of attack. The remaining pressure drag is assumed to depend only on the angle of attack, and therefore

$$C_D(R,\alpha) \simeq C_F(R) + C_P(\alpha). \tag{18}$$

Usually the lift coefficient is assumed to depend only on the angle of attack, and to be independent of the Reynolds number, in the form

$$C_L(R,\alpha) \simeq C_L(\alpha). \tag{19}$$

In practice, many hydrofoils are of large span, relative to their chord. Under these circumstances the flow at each section along the span is approximately two-dimensional, and the total three-dimensional force coefficients can be estimated by integration of the *sectional* lift and drag coefficients along the span. The differential projected area dS is proportional to the product of the local chord length and the differential element of the distance along the span. Thus the two-dimensional sectional lift and drag coefficients nondimensionalized in terms of the local chord lengths are consistent with the definition of the total force coefficients based on the planform area S.

Typical experimental results for the sectional lift and drag coefficients are reproduced in figures 2.6 and 2.7, for the foil geometry shown in figure 2.5. The validity of the approximations (18–19), over a limited range of Reynolds numbers, can be judged from figures 2.6 and 2.7. For small angles of attack the lift coefficient shown in figure 2.6 is insensitive to the Reynolds number, and C_L increases in a linear manner with the angle of attack. As the angle of attack increases, however, the streamlined effect of the foil diminishes, and ultimately separation, or *stall,* occurs with a dramatic reduction in the lift coefficient. The point at which stall takes place is sensitive to the Reynolds number, as well as to the ambient turbulence of the fluid and to the roughness of the foil surface; this situation is analogous to the drag coefficient for a sphere or other bluff body.

Figure 2.7 shows the drag coefficient. For small positive or negative angles of attack where the lift force is small, the drag coefficient is insensitive to the Reynolds number and takes a value comparable to the flat plate frictional drag coefficient. As stall is approached, the drag goes up appreciably and in a manner sensitive to the Reynolds number. Since the magnitude of the drag coefficient is small by comparison to the lift, these effects are of less importance.

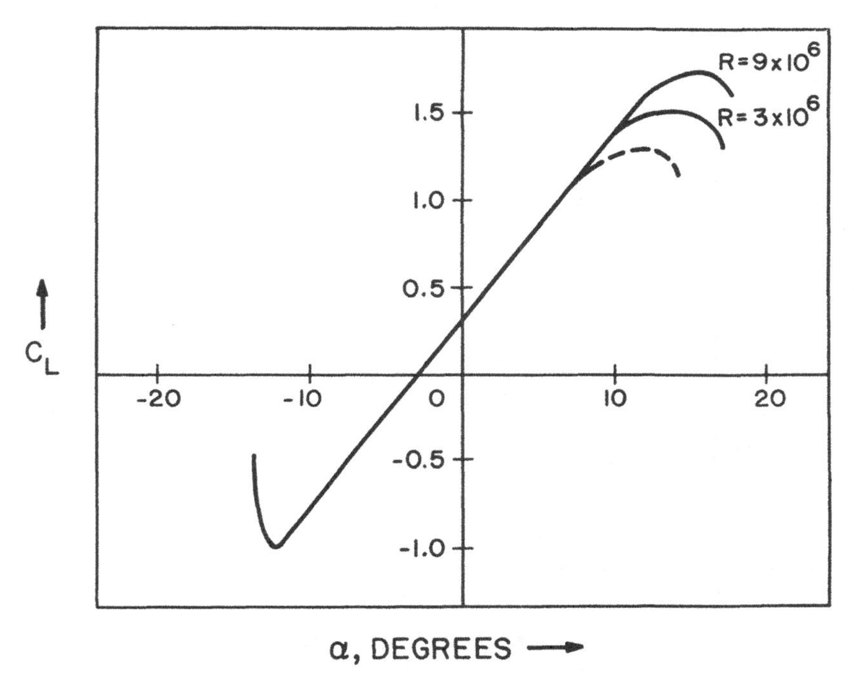

Figure 2.6
Lift coefficient (16) for a two-dimensional hydrofoil. The results here are for the NACA 63–412 section which is shown in figure 2.5. The dashed curve (— — —) shows the behavior at stall for a foil with artificial roughness near the leading edge, for $R = 6 \times 10^6$. (Adapted from Abbott and von Doenhoff 1959)

2.8 Screw Propeller

A screw propeller consists of several hydrofoil-type lifting surfaces, arranged in a helicoidal fashion to produce lift and thrust when rotated about the axis. The dimensional analysis of propellers is similar to that of the simpler hydrofoil, except for a change in the conventional set of physical parameters. We now consider the required shaft torque Q and resulting axial thrust force T associated with a propeller of diameter d, which rotates with constant angular velocity n while simultaneously moving in the forward (axial) direction with velocity U, as shown in figure 2.8. Note that n is the number of shaft revolutions per unit time; thus, the peripheral velocity of the blade at a radius r is $2\pi nr$. A velocity diagram of the blade element at r is as shown

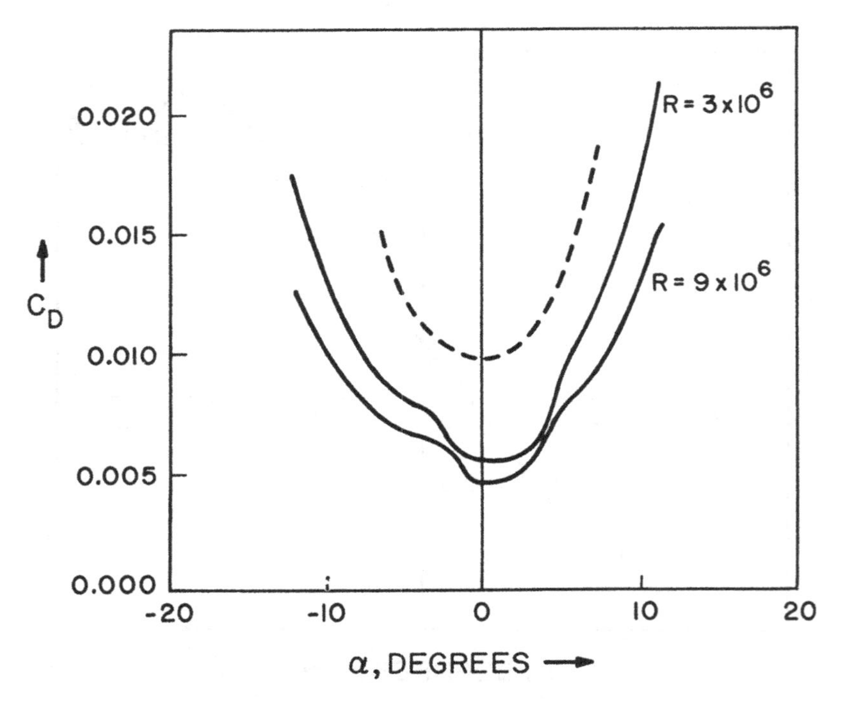

Figure 2.7
Drag coefficient (17) for the NACA 63–412 section as described in figure 2.6. (Adapted from Abbott and von Doenhoff 1959)

in figure 2.9, which may be compared with figure 2.5. The angle of attack is measured now by the nondimensional *advance ratio* $J = U/nd$.

The results of section 2.7 suggest that Reynolds-number effects will be negligible, provided the propeller is not operated in a regime of excessive angle of attack. Similarly, we assume no cavitation. It follows that the thrust and torque can be nondimensionalized so as to depend only on the advance ratio, in the form

$$\frac{T}{\rho n^2 d^4} = K_T(J), \tag{20}$$

$$\frac{Q}{\rho n^2 d^5} = K_Q(J). \tag{21}$$

The propeller efficiency η_p is the ratio of the work done by the propeller in developing a thrust force UT, divided by the work required to overcome the shaft torque $2\pi nQ$. Thus, it follows that

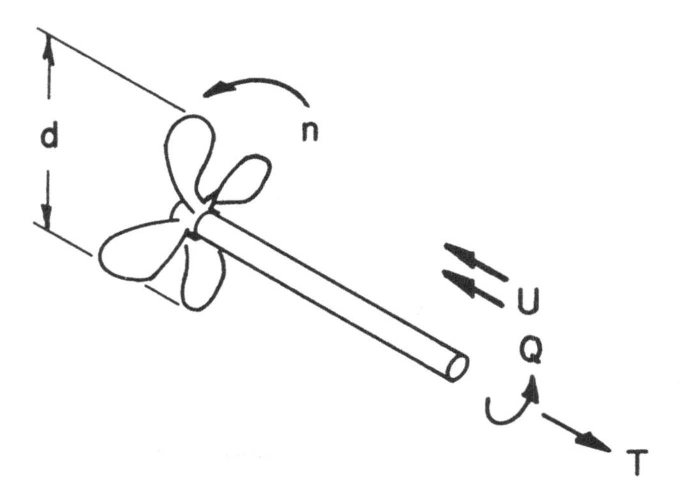

Figure 2.8
Perspective sketch of a propeller and its shaft.

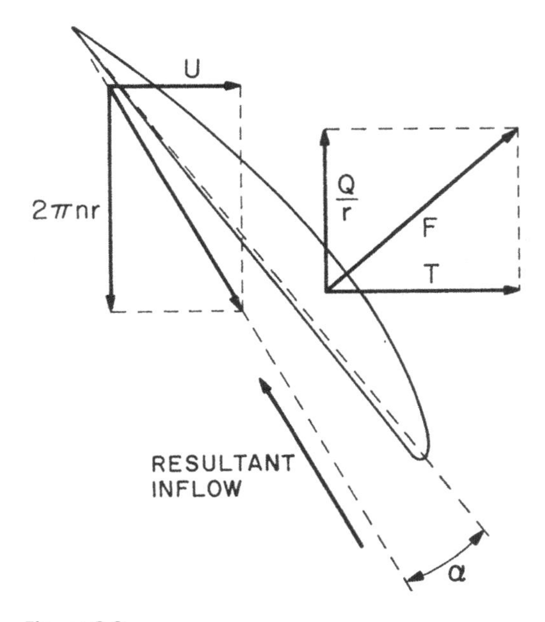

Figure 2.9
Two-dimensional view of a propeller blade section, moving to the right with velocity U and in the peripheral direction with velocity $2\pi nr$. The local angle of attack α is the difference between the inflow angle $\tan^{-1}(U/2\pi nr)$ and the pitch angle of the blade. The resultant force F contains an axial component (thrust), and a peripheral component (torque/radius).

$$\eta_P = \frac{UT}{2\pi nQ} = \frac{J}{2\pi}\frac{K_T}{K_Q}. \tag{22}$$

Figure 2.10 shows values of the thrust and torque coefficients and the efficiency for a series of propellers of varying *pitch,* which is the axial distance of advance in one complete revolution along the helix of the blades. The thrust and torque coefficients are decreasing functions of the advance ratio J, and the thrust coefficient vanishes at a value of J such that the propeller blades are at zero angle of attack with respect to the incoming fluid. At this same condition there is still a small residual torque, due to frictional drag on the blades and hub. At a slightly higher value of J, the propeller "windmills," requiring no torque but with a small drag force. For propeller blades of pure helicoidal shape, the thrust would vanish in the absence of viscous friction at the point where the advance ratio is precisely equal to the pitch-diameter ratio. However, the blades have some camber and, therefore, develop a small amount of lift and thrust at zero geometrical angle of attack.

Decreasing the advance ratio J increases the angle of attack, as shown in figure 2.9. Thus, in figure 2.10, the thrust and torque coefficients increase monotonically with decreasing values of J. At $J = 0$ the propeller operates in a regime of large thrust and torque, corresponding to the bollard-pull test of a tug boat exerting a force on a fixed object. The efficiency of this regime is zero, as defined by (22), and the results of figure 2.10 show that the optimum efficiency of a propeller is achieved at a relatively small angle of attack, with maximum efficiencies on the order of 0.6 to 0.8.

If cavitation occurs, the thrust coefficient will decrease due to the corresponding loss of lift of each blade section. Under these circumstances the torque is also reduced, but not to the same extent, and thus the propeller efficiency decreases. Supercavitating propellers, designed to operate efficiently in the cavitating regime, have been developed for use in high-speed vessels. As a rule, these propellers are somewhat less efficient than a conventional subcavitating propeller but more efficient than a conventional propeller that is cavitating. Supercavitating propellers are discussed by Venning and Haberman (1962) and Todd (1967). Charts for the characteristics of conventional propellers, operating in both regimes are given by van Lammeren, Troost, and Konig (1948), and extended by van Lammeren, van Manen, and Oosterveld (1969).

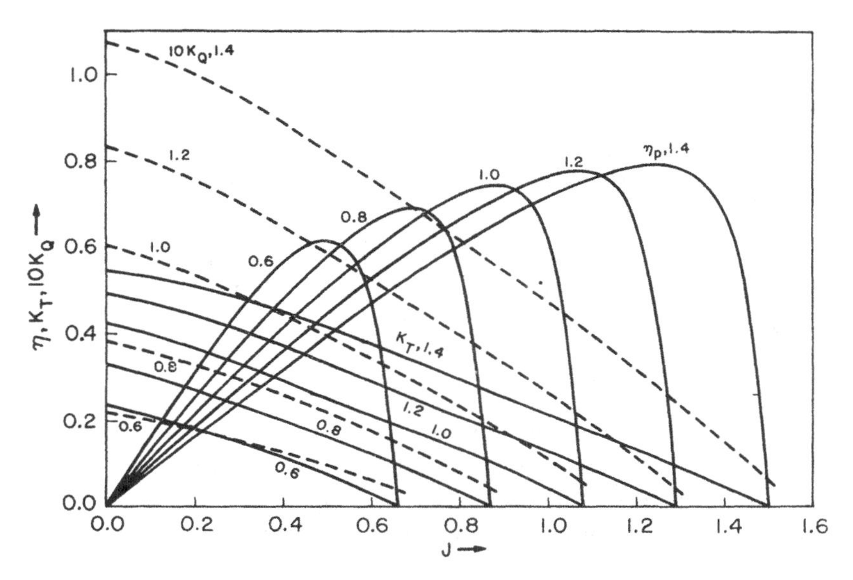

Figure 2.10
Thrust, torque, and efficiency coefficients for a series of three-bladed propellers with pitch/diameter ratios 0.6, 0.8, 1.0, 1.2, and 1.4. (Adapted from van Lammeren, Troost, and Konig 1948)

2.9 Drag on a Ship Hull

Next we consider a ship hull moving with constant velocity U on the free surface. As in the previous examples, geometrical similarity is assumed with the geometry of the hull characterized by its length l. Since the steady motion of a ship on the free surface generates a system of waves dependent on the gravitational acceleration g, this parameter must be included in the dimensional analysis along with the drag D, length l, velocity U, density ρ, and kinematic viscosity υ. These six dimensional quantities can be reduced to three nondimensional ratios; the usual form for the drag coefficient is given by

$$\frac{D}{\frac{1}{2}\rho S U^2} = C_D(R, F). \tag{23}$$

Here S is the *wetted-surface* area of the hull, and the Froude number $F = U/(gl)^{1/2}$ represents the effect of gravity.

Aside from the practical difficulty of testing small models at the full-scale Reynolds number, it is impossible now to scale simultaneously both the Reynolds and Froude numbers, as noted in chapter 1. Therefore, one cannot determine the drag of a ship hull from a suitable experiment with a small-scale model, unless additional assumptions are made. We assume that the drag coefficient (23) can be expressed as the sum of a frictional-drag coefficient, depending on the Reynolds number, plus a *residual-drag* coefficient which now must depend on the Froude number. Thus, we adopt Froude's hypothesis in the form

$$C_D(R,F) \simeq C_F(R) + C_R(F). \tag{24}$$

Here C_F is the flat-plate frictional-drag coefficient, and C_R is the residual-drag coefficient. Without such an assumption, one can express C_D as the sum of a flat-plate frictional-drag coefficient C_F, which is a function only of the Reynolds number, plus a residual-drag coefficient C_R defined as

$$C_R(R,F) = C_D(R,F) - C_F(R), \tag{25}$$

but it is an approximation to assume that the resulting C_R is independent of the Reynolds number.

Before testing Froude's hypothesis experimentally, one might consider the physical motivations for this decomposition of the total drag. In section 2.6 a similar assumption was made; the additional component was a constant independent of Reynolds number and associated with the form drag or normal pressure force. Here the pressure force is augmented significantly by the drag associated with the formation of waves on the free surface. Work must be done by the ship hull on the surrounding fluid to generate the waves, and there is an associated drag component known as the *wave resistance*. Since the characteristics of the waves are governed by gravity, the wave drag will depend on the Froude number; hence, the residual drag is not a constant but a function of the additional nondimensional parameter F.

Part of the residual drag may be due to viscous form drag, which should not depend on the Froude number; but since this part is a constant, it can be added to the residual drag without violating the assumption that the latter is independent of the Reynolds number. Moreover, as we shall see, on all but the slowest of ships the dominant portion of the residual drag is associated with wave resistance rather than viscous pressure forces.

An experimental validation of Froude's hypothesis can be made by plotting C_D vs. R, for different values of the hull length. Figure 2.11 shows results of experimental measurements from geosim models ranging in length from 1.2 m to 9.1 m, together with the results from fullscale drag measurements of the *Lucy Ashton,* a ship of 58 m length. For each model length experimental points are shown for several values of the Froude number, which is identified here in terms of the full-scale ship's speed in knots. According to Froude's hypothesis, all experimental points for the same Froude number (represented in the graph by the same symbol) should be situated equal distances above the flatplate frictional-resistance curve, labeled here as the Schoenherr line, and Froude's hypothesis can be evaluated according to the degree to which this is true. Rather than attempt to evaluate the discrete points, we can refer instead to the smoothed cross-curves labeled according to the various ships' speeds. These represent values of C_D vs. R for constant values of F. These curves are not strictly parallel to the Schoenherr line, but it is clear from figure 2.11 that Froude's method does succeed to a considerable extent in correlating the resistance of geosim ship hulls of widely differing lengths.

The low-Froude-number asymptotic tendency in figure 2.11 is indicated by coalescence of the results for full-scale speeds of 4 to 5 knots. At these speeds the wave resistance is negligible, and the principal contribution to the residual drag is viscous form drag. Thus the differences between the flat-plate "Schoenherr" curve in figure 2.11 and the curve immediately above it can be attributed to viscous form drag, whereas the difference between the latter curve and the appropriate curve for each larger Froude number is attributable to wave resistance. The most noticeable departure from the Froude hypothesis in figure 2.11 is that the form drag is not constant but increases with decreasing Reynolds number. This tendency has led to the suggestion that the form drag should be assumed proportional to the frictional drag or, alternatively, that the frictional drag curve should be more steeply sloped than the Schoenherr line. Two such empirical alternatives, the "Hughes line" and the "I.T.T.C. line," are shown in figure 2.12. The I.T.T.C. (International Towing Tank Conference) line has become the most widely accepted extrapolation to use with Froude's procedure.

For purposes of extrapolating model results to obtain full-scale resistance coefficients, adding a constant to the extrapolator will not affect the

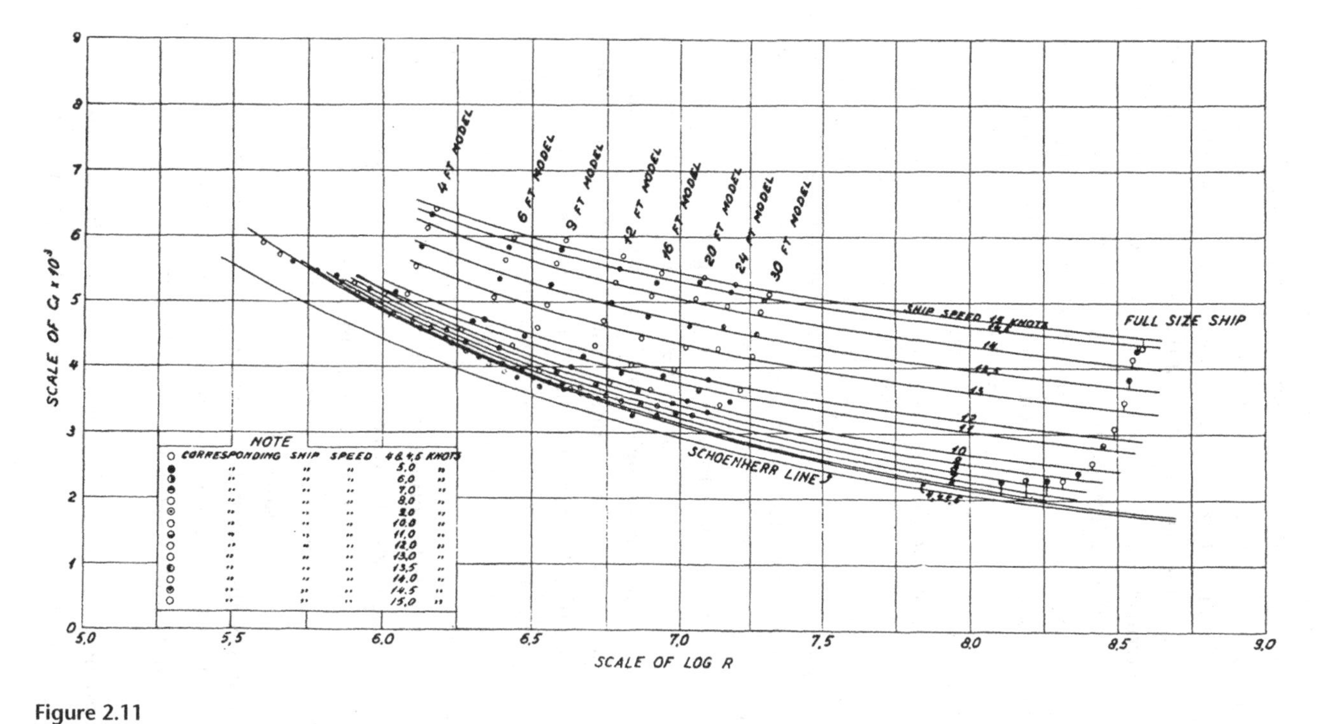

Figure 2.11

Total drag coefficients of the *Lucy Ashton* and several geosim models of the same vessel (from Troost and Zakay 1954). The faired curves represent constant values of the Froude number and, if Froude's hypothesis were strictly valid, these would be parallel with spacing independent of the Reynolds number. Note that, even for this small full-scale vessel (58 m long), there is a large gap between the largest model results and the full-scale results.

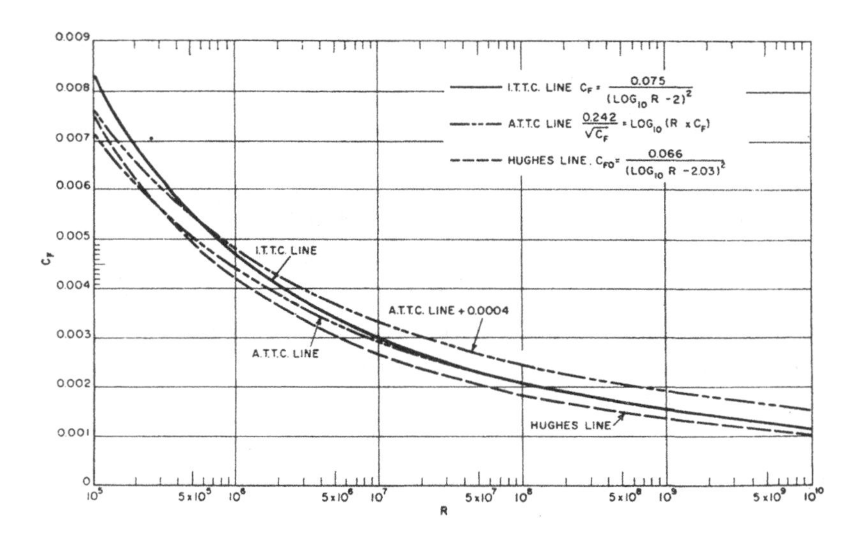

Figure 2.12
Frictional drag coefficients or *extrapolators*. The ATTC and ITTC lines are those recommended by the American and International Towing Tank Conferences; the former is identical to the Schoenherr line. (From Todd 1967; reproduced by permission of the Society of Naval Architects and Marine Engineers)

ultimate prediction of C_D This fact can be emphasized by writing Froude's hypothesis in the form

$$(C_D)_{\text{ship}} = (C_D)_{\text{model}} - (C_F)_{\text{model}} + (C_F)_{\text{ship}}. \tag{26}$$

Therefore, the difference in level between the I.T.T.C. and Hughes line in figure 2.12 is not significant to the resulting predictions based on the alternative employment of one curve or the other.

Further study of the geosim results in figure 2.11 shows that while the viscous form-drag coefficient, as measured at low Froude numbers, increases with decreasing Reynolds number, the drag coefficient curves for higher Froude numbers display the opposite tendency. Thus, the residual-drag coefficient actually decreases while the form drag is increasing. This is a fortunate circumstance, since the corresponding errors in Froude's hypothesis will cancel. However, it implies that both the wave drag and form drag depend separately on Reynolds number. Fortunately these effects are small, particularly for the larger models shown in figure 2.11, and it is apparent from this discussion why large towing tanks are built to accommodate models of 7 m to 10 m length.

Finally, the use of a *roughness allowance*, which is generally taken to be an additive contribution of +0.0004 to the full-scale value of C_F, is attributed to the significant surface roughness of full-scale hulls, due to rivet heads, welding seams, marine fouling, and so forth. Such roughness effects are significant, and accounting for them in this way is not irrational if guided by empirical results from full-scale trials. On the other hand, it is disconcerting to find different values of the roughness coefficient used as a catchall to account for various observed scale effects, based upon the performance of ships; it is especially troublesome when the resulting value is negative. As a recognition of this inconsistency, it is now common to refer to the *correlation* allowance, instead of to the *roughness* allowance.

2.10 Propeller-Hull Interactions

Having discussed propellers and ship hulls separately in sections 2.8 and 2.9, we should also comment briefly on their interactions. The most obvious of these is that the propeller operates in the wake of the hull; there the velocity is generally retarded to a degree depending on the fullness of the hull and the position of the propeller. A more subtle interaction results when the suction effect of the propeller modifies the flow past the hull and the resulting drag on the hull. Similar interactions may occur in other contexts; in particular, when a second vessel operates in the wake of a body, the characteristics of the flow in the wake are a principal determinant of the second vessel's performance.

The conventional experiment to study propeller-hull interactions is a *self-propelled* test, which uses a model fitted with and propelled by a scale-model propeller. The first difficulty with such a test is that with Froude scaling, the drag coefficients of the model and full-scale vessel differ by the Reynolds-number dependent difference in the respective frictional-drag coefficients. Thus $(C_D)_{\text{model}} > (C_D)_{\text{ship}}$, and the extra thrust required to propel the model must be furnished from the model propeller or from an external force. To preserve the correct thrust and torque coefficients of the model propeller, this extra differential thrust force is generally provided by the towing carriage. However, this cannot account for all of the Reynolds-number dependent differences between the model and full-scale ship; in particular, the boundary layer of the model will be thicker, and thus the wake retardation of the model will be larger.

It is assumed that the model drag D is known from a towing test without the propeller and, similarly, that the *open-water* characteristics of the propeller have been determined from a separate experiment or from charts, such as figure 2.10. For the self-propelled test the propeller thrust T is prescribed from the balance of propulsion and drag forces, and the necessary torque Q and shaft rotation rate n can be measured from the model.

The retardation of the wake behind the model is measured by the *wake fraction* w. Thus, if U is the forward velocity of the hull through the undisturbed water and U_A is the effective velocity of advance experienced by the propeller in the presence of the hull, Then

$$w = 1 - U_A / U. \tag{27}$$

Typically, $0 < w < 0.4$; the larger value applies for very full hulls. The wake fraction can be measured directly, with the propeller removed, or it can be deduced indirectly from the thrust-identity condition, which yields the *Taylor wake* w_T. The Taylor wake is determined by substituting the thrust coefficient K_T, as determined from the self-propelled test, in the open-water propeller chart, to determine the advance ratio J and the corresponding advance velocity

$$U_A = ndJ. \tag{28}$$

With this value of U_A, the Taylor wake w_T is defined by (27).

Since the wake behind the hull is not a parallel inflow with constant uniform velocity, the correction of the propeller performance based simply on a wake fraction is only approximate. While the thrust coefficients K_T are defined to be equal, there may be a small difference in the torque coefficients K_Q, as shown in figure 2.13. This difference is accounted for by the *relative rotative efficiency*

$$\eta_R = \frac{(K_Q)_{OW}}{(K_Q)_{SP}}, \tag{29}$$

where the subscripts OW and SP denote the open-water and self-propelled conditions. In general, $1.0 < \eta_R < 1.1$.

In the converse effect of the propeller on the hull, the suction of the propeller generally reduces the pressure at the stern and hence increases the drag force

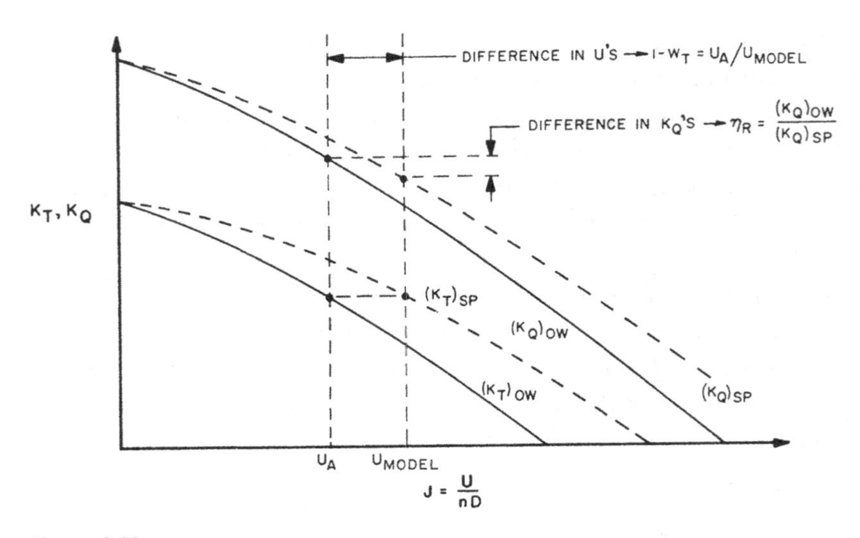

Figure 2.13
Relationship between self-propelled (SP) and open-water (OW) propeller tests, using the thrust-identity condition to relate these two experiments.

$$D = (1-t)T. \tag{30}$$

The coefficient t is the *thrust-deduction coefficient,* which is typically less than 0.2. In exceptional circumstances the thrust-deduction coefficient may be negative. For example, the suction effect of the propeller may delay the point of separation and reduce the resulting form drag. Another exception is a supercavitating propeller, where the thickness of the cavities results in a source-like flow and a positive pressure ahead of the propeller.

These interactions can be integrated to give an overall *propulsive efficiency.* The power input is the product of the propeller torque and rotation rate, while the useful work done is the product of the drag of the hull and its velocity. The propulsive efficiency is thus given by

$$\eta = \frac{DU}{2\pi nQ} = \frac{(1-t)UK_T}{2\pi nd(K_Q)_{SP}} = \frac{(1-t)}{(1-w_T)}\eta_P\eta_R, \tag{31}$$

where η_p is the open-water propeller efficiency, as defined by (22). The ratio $(1 - t)/(1 - w_T)$ is known as the *hull efficiency* η_H; as a rule it takes a value between 1.0 and 1.2. Thus the propulsive efficiency is generally greater than the open-water propeller efficiency.

2.11 Unsteady Force on an Accelerating Body

If the relative motion between a body and the surrounding fluid is unsteady, the hydrodynamic force $F(t)$ exerted on the body will vary with time. To simplify this problem as much as possible, we consider first the special case of a body accelerated impulsively from a state of rest, at time $t = 0$, with a constant velocity U thereafter. The fluid is assumed to be unbounded, as in sections 2.4 to 2.6 where the corresponding steady-state problem was discussed.

The time t may be nondimensionalized in the form Ut/l, the number of body lengths traveled in a time t. We nondimensionalize the force as before and anticipate that this will act predominantly in the direction opposite to the body motion; then

$$\frac{-F}{\frac{1}{2}\rho U^2 l^2} = C_F(R, Ut/l). \tag{32}$$

Here R denotes the Reynolds number Ul/ν, and F is the force acting in the same direction as the body motion.

From the standpoint of dimensional analysis, equation (32) cannot be simplified further, and experiments must be conducted with the two parameters R and Ut/l varied independently to determine the nondimensional force coefficient C_F. Such experimental information is sparse and depends on the details of the initial acceleration. Thus, we focus our attention on the two limiting cases of large and small values of time. Results from these cases will add considerable physical information and enable us to predict the unsteady force for more general body motions with realistic acceleration histories.

For large values of the nondimensional time Ut/l, the force coefficient (32) should approach the steady-state drag coefficient. Thus

$$C_F(R, Ut/l) \simeq C_D(R), \qquad Ut/l \gg 1. \tag{33}$$

The complementary limit of small time is less obvious but of comparable importance. During the initial stage of rapid acceleration of the body and surrounding fluid, inertial effects will dominate the viscous stresses in the fluid, and (32) can be approximated by the inviscid limit

$$C_F(R, Ut/l) \simeq C_F(\infty, Ut/l), \qquad Ut/l \ll 1. \tag{34}$$

Considerable insight into this unsteady flow problem can be gained from the classical photographs made by Prandtl and his coworkers (figure 2.14). These show the flow past a circular cylinder accelerated impulsively to a constant velocity. The first of these photographs, taken shortly after the motion commences, confirms that the flow past the cylinder is a symmetrical attached flow; it is essentially identical to the flow to be analyzed in chapter 4 on the assumption that the fluid is *ideal* with vanishing viscosity. For subsequent times shown in figure 2.14, the effects of viscosity

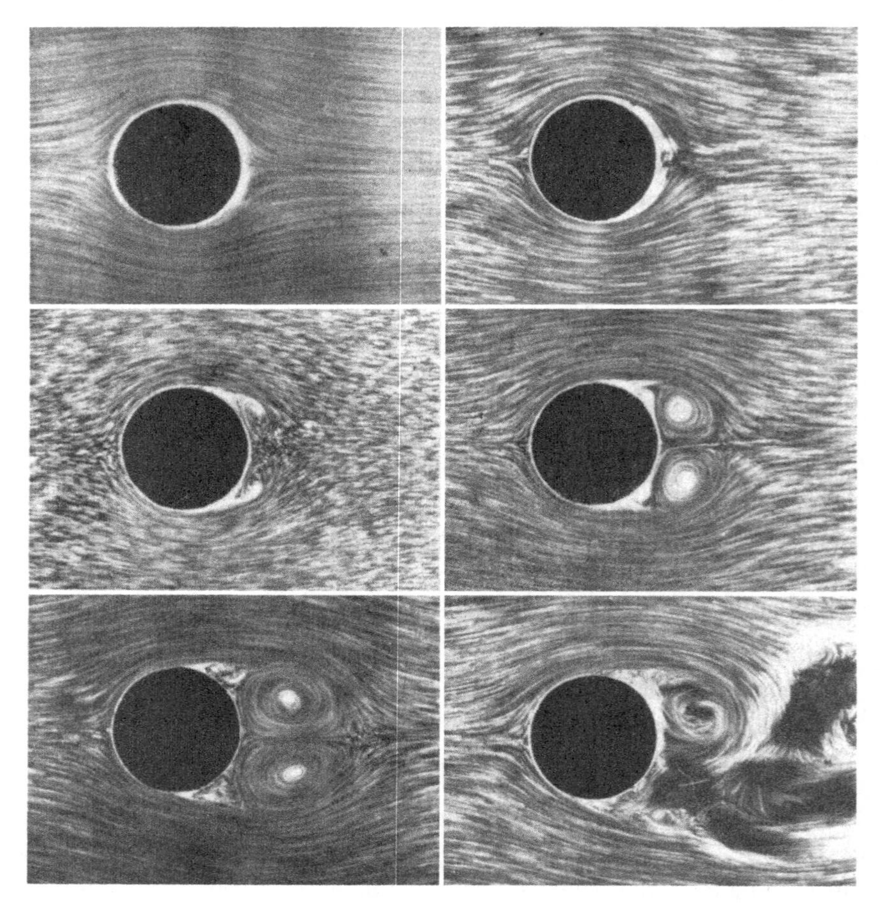

Figure 2.14
Initial stages of the flow past a circular cylinder which is accelerated impulsively from a state of rest to constant velocity. (From Prandtl 1927)

increase; the boundary layer along the cylinder surface grows and, more importantly, the wake separates downstream. Initially the separated wake consists of two symmetric vortices, but with increasing time these vortices grow in strength. Ultimately this arrangement becomes unstable; the vortices are shed alternately from the body and convected downstream in a staggered configuration. The oscillatory nature of this flow for large time contradicts the assumption of a steady state, but for practical purposes the drag force approaches an essentially constant limiting value, corresponding to the steady-state drag coefficient (33).

The force coefficient (34), which applies for small time t, is the result of accelerating the body in an ideal fluid. As will be confirmed in chapter 4, this force is simply proportional to the acceleration of the body, in manner analogous to Newton's equation $F = ma$. The coefficient of proportionality is the *added mass* or the effective mass of the fluid that surrounds the body and must be accelerated with it. We will denote the added mass by m_{11}. Here the letter refers to the analogy with the body mass m in Newton's equation, and the subscripts denote the directions of the force and body motion. During the initial stage where viscous forces are negligible, it follows that

$$F = -m_{11}\dot{U}, \tag{35}$$

where $\dot{U}$ denotes the acceleration, and the force coefficient (34) takes the form

$$C_F(\infty, Ut/l) = \frac{m_{11}\dot{U}}{\frac{1}{2}\rho U^2 l^2}. \tag{36}$$

The relative magnitudes of the viscous-drag force and the added-mass force are measured by forming the ratio of the coefficients (33) and (36). Since the added mass is proportional to ρl^3, it follows that

$$\frac{C_F(R,\infty)}{C_F(\infty, Ut/l)} = \frac{\frac{1}{2}\rho U^2 l^2 C_D(R)}{m_{11}\dot{U}} \propto U^2 / \dot{U}l. \tag{37}$$

For impulsive acceleration from a state of rest to a constant velocity U, the parameter $U^2 / \dot{U}l$ is small during the period of acceleration and large subsequently. Thus added-mass forces will dominate initially, and viscous forces subsequently, with a gradual transition between these regimes as shown in figure 2.15.

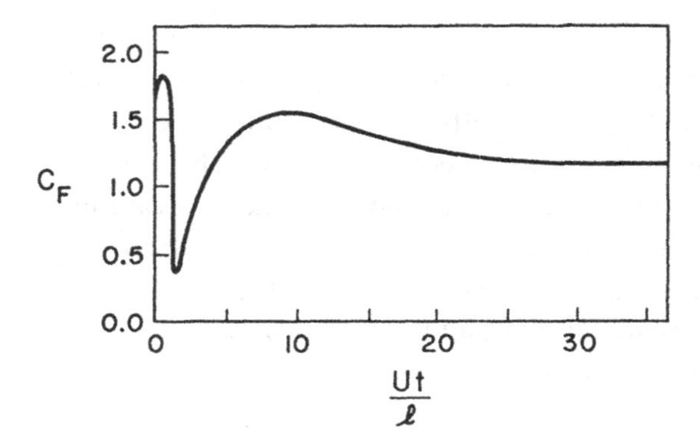

Figure 2.15
Force coefficient for a circular cylinder accelerated from an initial state of rest to a constant velocity *U*, based on experiments of Sarpkaya (1966).

The results in figure 2.15 are for a circular cylinder, synthesized from experiments of Sarpkaya (1966) in which the acceleration is practically constant for a short period of time and the velocity is constant thereafter. The Reynolds numbers for this experiment are $\boldsymbol{R} \simeq 10^5$. The details of the initial force peak depend on the magnitude and duration of the acceleration.

2.12 Vortex Shedding

When a bluff cylindrical body such as a circular cylinder is moved with constant velocity normal to its axis, one might expect the resulting flow to be steady and laterally symmetrical. In fact, neither of these assumptions is correct, because of the oscillatory shedding of vortices into the wake shown in the last photograph of figure 2.14.

The ultimate configuration of vortices downstream in the wake is antisymmetrical as predicted from a stability analysis by von Kármán, and the staggered array of shed vortices is known as a *Kármán street.* The vortices are shed into the wake at a frequency $f = \omega/2\pi$, and as a result of their antisymmetrical arrangement a lateral *lift* force will act on the body with the same frequency. The magnitude of the lift force and the frequency can be nondimensionalized in the forms

$$L_{\max} / \tfrac{1}{2}\rho U^2 l = C_L(R), \tag{38}$$

$$fl / U = S(R), \tag{39}$$

where L_{max} denotes the maximum value of the lift force, per unit length along the cylinder, and S is the *Strouhal number.* These parameters have been measured experimentally, especially for circular cylinders where C_L is typically equal to 0.5 and the Strouhal number is about 0.22 for Reynolds numbers in the laminar regime. As noted by Roshko (1961), the Strouhal number for circular cylinders appears to be proportional to the inverse of the drag coefficient. For Reynolds numbers between 10^2 and 10^7, but excluding the transitional regime $\boldsymbol{R} \simeq 10^6$, the Strouhal number is approximated closely by $S = 0.23/C_D$.

From an engineering standpoint, the oscillatory lift force may be of more practical importance than the drag, especially if a hydroelastic resonance occurs between the Strouhal frequency and a structural mode of vibration of the body. Strumming oscillations of cables are a consequence of this phenomena, as well as lateral vibrations induced upon pilings and other fixed structures in a current. The magnitude of the oscillatory lift force can be reduced by cable fairing or simply by installing a small *splitter plate* on the after-side of the cylinder. For long slender cylinders with axis normal to the flow direction, the correlation of the vortices along the cylinder axis is particularly important, and small devices that destroy this correlation are effective in reducing the total lift force. On the other hand, lateral vibrations of the body in response to the lift force will serve to correlate the phase of the vortex shedding, thereby increasing the magnitude of the lift. Extensive bibliographies of this subject are given by Mair and Maull (1971) and Berger and Wille (1972). More recent work is described by Griffin and Ramberg (1976).

2.13 Wave Force on a Stationary Body

Ocean waves are of particular interest to ocean engineers and naval architects because of the interactions between these waves and structures on or beneath the free surface. Our first example of such a problem is the unsteady force acting upon a fixed structure in the presence of waves. For simplicity, we treat only the x-component of this force, but a similar analysis applies for the total force vector.

Assuming geometrical similarity, we can describe the structure by a length scale l, which may be the length or the diameter of the body. If we neglect surface tension effects and assume a plane progressive wave system similar to that described in section 2.3, then the magnitude of the unsteady wave force can depend only on the density ρ, gravity g, viscosity υ, depth h, body length l, and time t, as well as the wave amplitude A, wavelength λ, and the angle of incidence β of the waves relative to the body axis. Thus, a total of ten dimensional quantities must be related in the form

$$F = f(\rho, g, \nu, A, \lambda, \beta, h, l, t). \tag{40}$$

Alternatively, seven nondimensional parameters can be defined to replace (40); one choice is the force coefficient

$$\frac{F}{\rho g l^3} = C_F(A/\lambda, h/\lambda, l/\lambda, \beta, R, \omega t). \tag{41}$$

Here we find it convenient to use the wave frequency ω, which is not an additional independent parameter but one related to g, A, and λ by (6).

In this problem the Reynolds number R should be the ratio Ul/υ where U is a typical velocity scale of the fluid relative to the body. Since the oscillatory displacement of the fluid particles is proportional to the wave amplitude A, the magnitude of the fluid velocity is ωA, and an appropriate Reynolds number can be defined by

$$R = \omega A l / \nu. \tag{42}$$

In principle, there is no difficulty in conducting a model test with the appropriate values of A/λ, h/λ, l/λ, and β. If the resulting wave force is measured over one or more cycles, all relevant values of ωt will be included. However, the Reynolds number cannot be scaled properly, since it follows from (6) and (42) that $R \propto l^{3/2}$. This situation is essentially analogous to the steady drag force acting on a ship hull.

To overcome this dilemma, we need a simplifying assumption analogous to Froude's hypothesis for the drag on a ship hull. For this purpose we estimate the relative magnitudes of the viscous and inertial forces, as in (37). Here, with U replaced by ωA and the fluid acceleration $\dot{U}$ by $\omega^2 A$, it follows that the ratio of viscous forces to inertial forces is proportional to

$$U^2 / \dot{U} l = A / l. \tag{43}$$

First, we restrict our attention to the case of a large structure or a vessel such as a ship hull where the ratio in (43) is small and viscous effects are negligible. If we assume in addition, that the wave amplitude A is small compared to the wavelength λ and depth h, then the force coefficient (41) will be linearly proportional to A, with nonlinear contributions proportional to A^2 neglected. The linearized fluid motion is sinusoidal in time, with frequency ω, and (41) can be expressed in the form

$$C_F = C_{FO} \cos(\omega t + \varepsilon), \tag{44}$$

where ε is a phase angle, C_{FO} a force coefficient, and both depend on h/λ, l/λ, and β but not on time. Alternatively, with $U(t)$ the oscillatory velocity of fluid relative to the body at some prescribed position such as the body centroid, (44) can be replaced by the equivalent expression

$$C_F = C_M \dot{U} + C_d U. \tag{45}$$

C_M and C_d are known respectively as the *apparent mass* and *apparent damping coefficients*. These depend on l/λ, or on the wave frequency ω, and can be determined experimentally by measuring the amplitude and phase of the wave force acting on the body.

Since (45) derives from (44), $U(t)$ must be harmonic in time despite the appearance of (45). However, a spectrum of waves with arbitrary time dependence can be generated by superposition, and since (44–45) are linear, the resulting wave forces can be obtained by a similar process of superposition using these expressions.

The inertia and damping coefficients can be predicted theoretically. A particularly simple limit occurs for a submerged body that is small compared with the wavelength, $l/\lambda \ll 1$. Thus, from a theorem to be derived in section 4.17, the force acting on the body is proportional to the local acceleration of the fluid relative to the body. Here, as opposed to the case of an accelerating body in a fixed fluid, the constant of proportionality includes an additional "buoyancy" force proportional to the displaced volume of fluid $\forall$ and associated with the pressure gradient of the fluid in the absence of the body. For $l/\lambda \ll 1$ it follows that

$$C_M \simeq (m_{11} + \rho\forall) / \rho g l^3, \tag{46}$$

$$C_d \simeq 0. \tag{47}$$

Thus, in the special case where $A \ll l \ll \lambda$, the wave force is given by the approximation

$$F \simeq (m_{11} + \rho\forall)\dot{U}. \tag{48}$$

When the body is small compared to the wave amplitude, however, the situation is fundamentally different. From (43) and (37) viscous drag forces will be dominant, and an obvious representation of the wave force is given by

$$F = \tfrac{1}{2}\rho l^2 U|U|C_D(R). \tag{49}$$

In this expression the square of the velocity U has been replaced by the product $U|U|$ to ensure that the viscous drag force (49) acts in the same direction as the fluid velocity. The utility of the approximation (49) is diminished because the Reynolds number is itself a function of the instantaneous velocity. However, (49) is relatively useful for bluff bodies, where the viscous drag coefficient is not sensitive to the Reynolds number.

The intermediate case, where A/l is a quantity of order one, is one of the most important and least understood problems of this field. It is important because many structures, such as the risers of offshore platforms and other cylindrical pilings, have diameters of about the same magnitude as the typical wave amplitude. In this regime viscous and inertial effects are of comparable magnitude, and one must return to the exact force coefficient (41). However, the inherent interactions between viscous and inertial effects have not prevented engineers from simplifying expressions for the wave force. The most common approximation is *Morison's formula,* which assumes that the total wave force is the sum of the inertial force (48) and the viscous force (49),

$$F = (m_{11} + \rho\forall)\dot{U} + \tfrac{1}{2}\rho l^2 C_D U|U|. \tag{50}$$

Since the validity of each term in (50) is restricted to a regime where A/l is respectively small or large, the justification for Morison's formula is strictly pragmatic and must rest with experimental confirmation.

Considerable experimental effort has been devoted to validating Morison's formula (50). For submerged bodies an approximation similar to (50) appears valid for engineering purposes, provided the coefficients are determined experimentally and for appropriate values of the Reynolds number and the parameter A/l. For bodies such as vertical surface-piercing pilings, no satisfactory experimental confirmation of (50) exists; the reasons for this are not understood adequately. Surveys of this subject have been made by Hogben (1974) and Milgram (1976).

2.14 Body Motions in Waves

If ocean waves are incident upon a freely floating body, the unsteady wave force discussed in section 2.13 causes the body to oscillate. The body motion is often more important than the components of the hydrodynamic force, although a study of the forces is essential to the theoretical prediction of the body motions.

In this section we focus our attention on the vertical *heave* oscillations of an unrestrained body in waves. The body may be floating on the surface, or it may be submerged beneath the surface; but in both cases, the unrestrained body is neutrally buoyant, and the body mass m is equal to the displaced mass of water $\rho\forall$.

If coupling effects are neglected, the heave motion $y(t)$ results from equilibrium between the vertical hydrodynamic force and the product of the body mass multiplied by the vertical acceleration $\ddot{y}$. Thus, the body mass m must be added to the nine physical parameters upon which the force (40) depends. In nondimensional form, it follows that

$$y / A = f(A / \lambda, h / \lambda, l / \lambda, \beta, R, \omega t, m / pl^3). \quad (51)$$

Since $m = \rho\forall$, the last parameter in (51) is equal to $\forall / l^3$. For a prescribed body shape this parameter is independent of the length scale; thus the last parameter in (51) may be deleted in our study of unrestrained bodies.

The dependence of (51) on the Reynolds number is once again a source of fundamental problems for model testing, as well as for making theoretical predictions of body motions. For large bodies, where A/l is small, it follows from (43) that viscous effects are negligible. Moreover, for small bodies where A/l is of order one, or large compared to one, the situation is considerably simpler than that discussed in conjunction with the wave force on a fixed structure. Here, a small unrestrained body will bob up and down like a cork upon the sea, with little or no relative motion between itself and the surrounding fluid. Thus, there will be no significant viscous drag for small unrestrained bodies in waves, and viscosity may be neglected in general.

Subject to these assumptions, (51) reduces to the simpler form

$$y / A = f(A / \lambda, h / \lambda, l / \lambda, \beta, \omega t). \quad (52)$$

If the wave amplitude is small compared to the wavelength and depth, nonlinear effects can be neglected. The linearized approximation to (52) then takes the form

$$y / A = f(0, h / \lambda, l / \lambda, \beta, \omega t) = f_0(h / \lambda, l / \lambda, \beta)\cos(\omega t + \varepsilon), \tag{53}$$

where for a prescribed body shape, the amplitude f_0 and phase angle ε depend only on h/λ, l/λ, and β. In deep water there is no dependence on the depth ratio h/λ.

For axisymmetric bodies with vertical axis, the heave response is independent of the angle of incidence β. As a particular example, figure 2.16 shows the heave-amplitude ratio for a slender spar buoy of draft T in deep water. We note that $f_0 \to 1$ in the limit of $T/\lambda \ll 1$, in accordance with our presumption that a very small body will move with the same velocity as the surrounding fluid. In the opposite limit, $T/\lambda \gg 1$, the body is large compared to the wavelength, and $f_0 \to 0$. At an intermediate wavelength of about six times the draft, a sharp resonance occurs which is analogous to the motion of a weakly-damped mechanical oscillator. This severe resonant motion for the spar buoy is a consequence of vertical slenderness. We shall see in the next section as well as in chapter 6 that most floating bodies experience relatively moderate resonant motions by comparison with the spar buoy shown in figure 2.16.

Two important exceptions occur where viscous forces significantly affect the motions of unrestrained bodies in waves. First, if the body shape is such that the inertial forces are small, then frictional forces due to viscous shear will be important, as in the case of the steady drag on a flat plate or streamlined body. Second, cross-flow drag will be significant for a long slender body if the length is such that the motions do not coincide with the local wave velocity.

An obvious example of the first exception is the angular motion of a body of revolution about its axis. Thus, the yaw moment on an axisymmetric buoy is due entirely to viscous effects. The rolling motion of ships and submarines will be affected similarly if the body sections are nearly circular. Viscous shear stresses are significant also in the heaving motions of a slender spar buoy if the draft is very large compared to the diameter. In the latter examples viscous damping is particularly important at resonance, where the inertial and hydrostatic forces cancel each other.

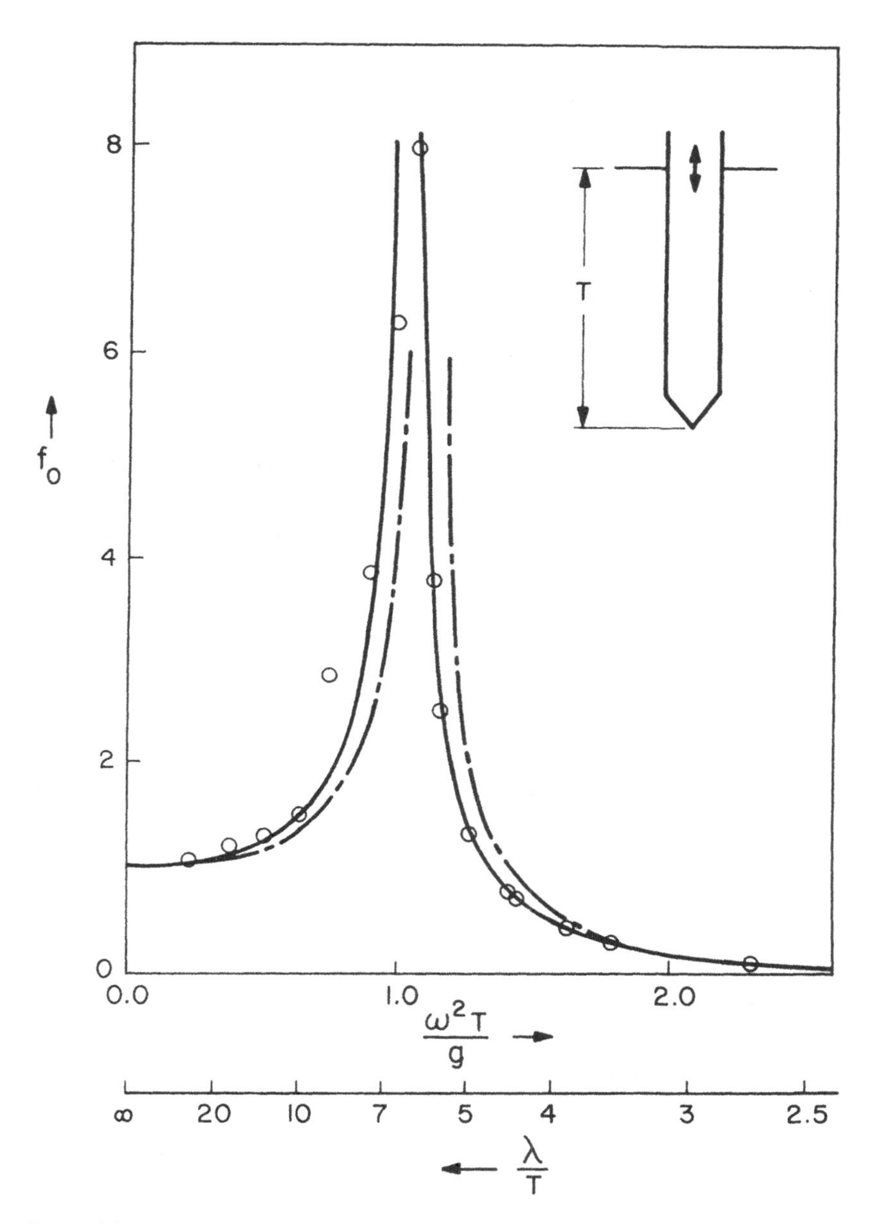

Figure 2.16
Heave response of a slender spar buoy in regular waves, from Adee and Bai (1970). The ordinate is the ratio of heave amplitude to wave amplitude, and the buoy is a circular cylinder, with a conical bottom, as shown to scale in the sketch. The dashed line is a theoretical prediction that neglects the hydrodynamic forces due to the motions of the body. The solid line includes a correction for the added mass. The circles denote experimental measurements.

The second exception occurs if the body is slender, with its length comparable to the wavelength and its transverse dimensions comparable to the wave amplitude. If the body is rigid, it will move relative to the waves, unlike a small body that is free to move with the local wave field. This situation is analogous to the case of a fixed body. Thus, the cross-flow drag force can be estimated from (49), with U the relative lateral velocity component between the body section and the wave field. This situation occurs for the horizontal force acting on a slender spar buoy of large draft and for the vertical force on the hull of a semisubmerged stable platform. Cross-flow drag is discussed in the latter context by Söding and Häusler (1976).

If the body is restrained by a mooring, the dynamics of this restraint will affect the body motions in waves. In the simplest case the mooring may be regarded as a linear elastic restraint and lumped with the hydrostatic restoring force acting on the body. For small bodies this restraint can induce relative motion between the body and waves, and then viscous drag forces must be included once again. If the mooring cable is very long, viscous forces on the cable are significant as well. The complexity of the latter problem is aggravated by the difficulty of modeling the cable in a wave tank of limited depth. A comprehensive account of moored buoys is given by Berteaux (1976).

2.15 Ship Motions in Waves

As our final example, we consider the case of a ship hull moving with constant forward velocity U through a system of waves. In contrast with the situation of section 2.14, this example adds one physical parameter U and a corresponding nondimensional Froude number $U/(gl)^{1/2}$. If the ship is unrestrained, if viscous effects are negligible, and if the wave amplitude is sufficiently small for linearization to be valid, then the heave-amplitude ratio is a generalization of (53) in the form

$$y / A = f_0(h / \lambda, l / \lambda, \beta, U / (gl)^{1/2}) \cos(\omega t + \varepsilon). \qquad (54)$$

Similar nondimensional expressions hold for the other five degrees of freedom.

Typical experimental results are shown in figure 2.17 for roll (angular motion about the longitudinal axis) and pitch (about the lateral axis) of an aircraft carrier in deep water. Here the Froude number is 0.23, corresponding

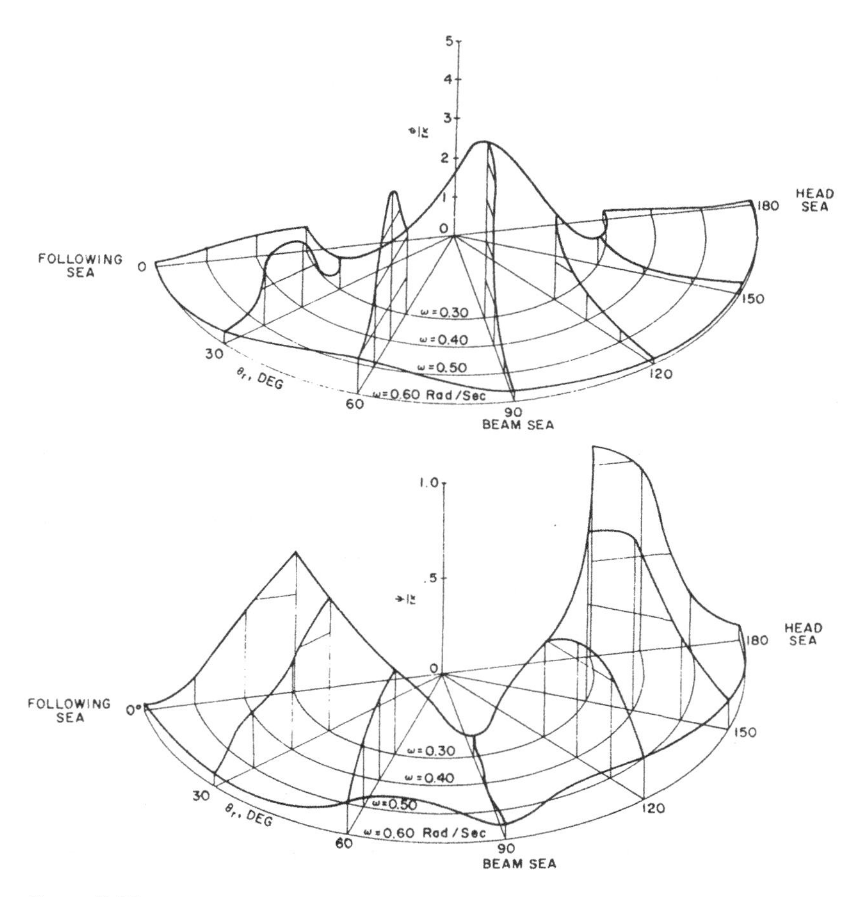

Figure 2.17
Roll and pitch response of a 319 m ship at 25 knots. The motions are nondimensionalized in terms of the maximum wave slope $2\pi A/\lambda$. (From Wachnik and Zarnick 1965; reproduced by permission of the Society of Naval Architects and Marine Engineers.)

to a full-scale speed of 25 knots. The roll amplitude is a maximum in beam seas and, except for small experimental errors, is zero in head and following seas. For a given angle of incidence, the roll response is similar to the buoy response shown in figure 2.16, although in some cases the resonance is outside the range of frequencies shown. Similar comments apply to pitch, except that the maxima now occur in head and following seas with zero response in beam seas.

Comprehensive discussions of this subject are given by Korvin-Kroukovsky (1961), Vossers (1962), and Lewis (1967). More recent contributions are included in Bishop and Price (1975).

Problems

1. In shallow water ($h/\lambda \ll 1$), waves of small amplitude propagate with a phase velocity independent of the wavelength. From a dimensional analysis show that under these circumstances $V_P \propto (gh)^{1/2}$ and thus that $T \propto \lambda(gh)^{-1/2}$. Deduce the qualitative change in wavelength that will occur as waves of constant period move toward a beach of gradually decreasing depth.

2. Find the maximum density and weight of a spherical instrument package of diameter 0.5 m whose terminal velocity does not exceed 0.3 m/s when it is dropped in salt water at a constant temperature of 5°C. What are the terminal velocities if this density is in error by ± 1%?

3. The drag of an oceanographic research submarine 9 m long is to be estimated from wind tunnel tests with a scale model of length 1.5 m and wetted surface area 1.2 m^2. The measured drag force in the wind tunnel is 7.9 N at a wind speed of 30 m/s and air temperature of 25°C. If the full-scale submarine is deeply submerged in salt water at 15°C, at what speed can its drag be most accurately predicted? What is the best estimate of its drag at a speed of 2 m/s? Can this estimate be refined by raising or lowering the temperature in the wind tunnel?

4. A hydrofoil vessel weighing 100,000 N is supported by two identical hydrofoils, with NACA 63–412 sections, set at a 3-degree angle of attack. If the lift coefficients of these hydrofoils are as shown in figure 2.6 and the chord length is 0.5 m, what must their span be to support the vessel at a velocity of 20 knots?

5. A propeller is operating initially at $J = 1.0$. For the blade section at 70 percent of the maximum radius, find the value of the advance ratio where the angle of attack is reduced by 5 degrees.

6. A motorboat with a drag of 8,000 N at the design speed of 6 m/s is to be powered by an engine that can deliver 100 kw at a shaft speed of 1800 revolutions per minute. The maximum propeller diameter that can be

used is 0.6 m. (1) What is the optimum choice of propeller from the series shown in figure 2.10 for operation at 1800 rpm? (Extrapolate if necessary.) (2) What is the optimum if a reduction gear is used, and what is the resulting increase in propeller efficiency? (3) What percentage of the available engine power is used in each case, assuming no mechanical losses in the shaft or reduction gear?

7. A small motor vessel has the following drag at corresponding velocities: velocity, m/s: 2.0 3.0 4.0 5.0 6.0; drag force, N: 1200 2700 4800 9500 19,500.

An optimum propeller is to be chosen for this vessel with three blades, diameter 0.6 m, and a fixed rotation rate of 10 revolutions per second. Plot the required thrust coefficient as a function of the advance ratio J on the same scale as figure 2.10. By observing the equilibrium points where this thrust coefficient is equal to that of different pitch propellers, determine the propeller pitch for maximum velocity. Discuss this choice from the standpoint of propeller efficiency.

8. The total drag of a ship 200 m long, moving at 20 knots in salt water (15°C), is to be determined from towing-tank tests of a 2 m model in fresh water (15°C). The ship's wetted surface is 6000 m^2, and its displacement is 190 MN. (1) Find the weight of the model. (2) What is the wetted-surface area of the model? (3) What speed should be used in the model test? (4) If the model drag at this speed is 1.6 N, what is the predicted full-scale drag based on the ITTC friction line? (5) What model speed would preserve the full-scale Reynolds number?

9. For a hydrofoil submerged 1.0 m beneath the free surface, what is the minimum velocity at which cavitation can be anticipated, if the cavitation number for inception of cavitation is $\sigma = 0.1$. For a 1/10 scale model of the same hydrofoil, operating at the same Froude number, find the atmospheric pressure in a "vacuum towing tank" necessary to produce the same cavitation number. Assume the temperature range 0–30°C, with the worst possible temperature full scale and the best possible temperature in the towing tank.

10. A model of a sailing yacht is towed with a small angle of attack, or leeway angle, about the vertical axis. The drag force and side force are measured separately. If the yacht hull operating in this manner can be

regarded as a vertical hydrofoil in the presence of the free surface, what are the appropriate scaling laws to apply to each force component? Distinguish between parameters that should be scaled in principle and those that would be most important in a model test. In fresh water (15°C), towing-tank tests of a model 2 m long with wetted-surface area 1.5 m^2 give a measured drag force of 10 N and a side force of 28 N at a speed of 1.4 m/s. What are the corresponding full-scale results for a yacht 10 m long in salt water (15°C)?

11. Construct a simple mathematical model for the propulsive equilibrium of a canoe, assuming the canoe velocity U_c and paddle velocity U_p are constants. If these velocities are both defined relative to the surrounding water, show that the propulsive efficiency is

$$\eta = \frac{U_c}{U_c + U_p} = \left\{ 1 + \left[\frac{(SC_D)_c}{(SC_D)_p} \right]^{1/2} \right\}^{-1}$$

where $(SC_D)_{c,p}$ is the product of the area and the drag coefficient for the canoe or paddle respectively. How does the efficiency depend on the size of the paddle?

12. A submerged spherical body of volume 0.5 m^3 and density equal to half the density of water is moored in the deep ocean with a taut vertical cable 1000 m long. Neglecting all hydrodynamic forces, but including the upward buoyancy force on the body, find the tension in the cable and the natural period of this inverted pendulum. Comment on how the unsteady force in section 2.11 would affect this period. Estimate the horizontal excursion of the buoy and the angle of the mooring cable in a steady current of 0.5 knots.

13. Resonant strumming of a taut cable with circular cross-section is observed at a flow velocity of 10 knots perpendicular to the cable. The diameter of the cable is 2 cm. What is the natural frequency of the cable? Is this situation likely to become more serious at higher speeds?

14. For an oscillatory flow $U(t) = A\omega \sin \omega t$, normal to a circular cylinder of diameter d, find the ratio A/d such that the viscous and inertial forces in Morison's equation (50) are equal. Assume that the added mass m_{11} is equal to the displaced mass and that the drag coefficient based on the diameter is equal to 1.0.

15. A floating spar buoy of draft 2 m contains an accelerometer to detect its heave motion relative to an inertial reference frame, while floating freely in waves. Based on the results shown in figure 2.16, predict the range of wavelengths for which this buoy can measure the wave height to an accuracy of 20 percent without making corrections for hydrodynamic effects.

References

Abbott, I. H., and A. E. von Doenhoff. 1959. *Theory of wing sections*. New York: Dover.

Adee, B. H., and Bai, K. J. 1970. *Experimental studies of the behavior of spar type stable platforms in waves*. College of Engineering, University of California, Berkeley report NA-70-4.

Berger, E., and R. Wille. 1972. Periodic flow phenomena. *Annual Review of Fluid Mechanics* 4:313–340.

Berteaux, H. O. 1976. *Buoy engineering*. New York: Wiley-Interscience.

Birkhoff, G. 1955. *Hydrodynamics—a study in logic, fact, and similitude*. New York: Dover.

Bishop, R. E. D., and W. G. Price eds. 1975. *International symposium on the dynamics of marine vehicles and structures in waves*. London: Institution of Mechanical Engineers.

Goldstein, S., ed. 1938. *Modern developments in fluid dynamics*. 2 Vols. Oxford: Oxford University Press. Reprinted 1965, New York: Dover.

Griffin, O. M., and S. E. Ramberg. 1976. Vortex shedding from a vibrating cylinder, *J. Fluid Mech.* 75: 267–271 + 5 plates.

Hogben, N. 1974. *Fluid loading of offshore structures, a state of art appraisal: Wave loads*. Royal Institution of Naval Architects Marine Technology Monograph.

Korvin-Kroukovsky, B. V. 1961. *Theory of seakeeping*. New York: Society of Naval Architects and Marine Engineers.

Lewis, E. V. 1967. The motion of ships in waves. In *Principles of naval architecture*, ed. J. P. Comstock, 607–717. New York: Society of Naval Architects and Marine Engineers.

Mair, W. A., and D. J. Maull. 1971. Bluff bodies and vortex shedding—A report on Euromech 17. *Journal of Fluid Mechanics* 45:209–224.

Milgram, J. H. 1976. Waves and wave forces. In *Proceedings of the conference on the behaviour of offshore structures* (BOSS '76). pp. 11–38. Trondheim: The Norwegian Institute of Technology.

Prandtl, L. 1927. The generation of vortices in fluids of small viscosity. *Journal of the Royal Aeronautical Society* 31:720–741.

Roshko, A. 1961. Experiments on the flow past a circular cylinder at very high Reynolds number. *Journal of Fluid Mechanics* 10:345–356.

Sarpkaya, T. 1966. Separated flow about lifting bodies and impulsive flow about cylinders. *J. American Institution of Aeronautics and Astronautics* 4:414–420.

Sedov, L. I. 1959. *Similarity and dimensional methods in mechanics*. New York: Academic Press.

Söding, H., and F. U. Häusler. 1976. Minimization of vertical motions of floating structures. In *Proceedings of the conference on the behaviour of offshore structures* (BOSS '76). pp. 307–319. Trondheim: The Norwegian Institute of Technology.

Todd, F. H. 1967. Resistance and propulsion. In *Principles of naval architecture,* ed. J. P. Comstock, 228–462. New York: Society of Naval Architects and Marine Engineers.

Troost, L., and A. Zakay. 1954. A new evaluation of the Lucy Ashton ship model and full scale tests. *International Shipbuilding Progress* 1:5–9.

van Lammeren, W. P. A., L. Troost, and J. G. Konig. 1948. *Resistance, propulsion and steering of ships*. Haarlem, Netherlands: H. Starn.

van Lammeren, W. P. A., J. D. van Manen, and M. W. C. Oosterveld. 1969. The Wageningen B-screw series. *Society of Naval Architects and Marine Engineers Transactions* 77:269–317.

Venning, E., and W. L. Haberman. 1962. Supercavitating propeller performance. *Society of Naval Architects and Marine Engineers Transactions* 70:354–417.

Vossers, G. 1962. *Behaviour of ships in waves*. Haarlem, Netherlands: H. Stam.

Wachnik, Z. G., and E. E. Zarnick. 1965. Ship motion prediction in realistic short-crested seas. *Society of Naval Architects and Marine Engineers Transactions* 73:100–134.

3 The Motion of a Viscous Fluid

To develop analytical representations for the flow of a viscous fluid, it is necessary first to describe the properties of the flow, especially the velocity field, and the relevant forces that act on the fluid particles. Subsequently, the physical laws expressing conservation of mass and momentum must be invoked, in terms of this description, to derive the governing equations of the flow. In addition, the role of viscosity must be described by a suitable hypothesis that relates the stresses between adjacent fluid particles to the kinematic description of their relative motion. In this manner we shall derive the equations of motion for a viscous fluid, as a system of coupled nonlinear partial differential equations. These words may sound frightening to those not well trained in advanced calculus, and still greater apprehension will be felt by the opposite group who recognize that such a system of equations generally does not possess simple solutions. Of course, most fluid flows are not simple, so we must expect the governing equations to be somewhat complicated.

Because of this complexity, we will restrict our discussion to very simple flows, particularly those involving infinitely long, flat, or cylindrical boundary surfaces. These flows are of limited *direct* applicability, but they will give us a qualitative "feel" for the role of viscosity and an important quantitative tool in the form of boundary-layer theory. The importance of viscosity for many problems of engineering interest is confined to a thin layer adjacent to boundary surfaces. Furthermore, if the radii of curvature of these surfaces are large compared to the boundary-layer thickness, the flow will appear locally plane in the boundary layer.

There are exceptions to this convenient state of affairs; for example, in low-Reynolds-number flows, the role of viscosity is not confined to a thin boundary layer. This situation would hold for bodies of microscopic size, as in fluid suspensions or the swimming of microorganisms, but these are outside the usual scope of ocean engineering and naval architecture. A more significant exception here is *separation,* for flows past bodies not suitably streamlined. Separated flows are practically impossible to analyze. When these exist, the boundary-layer theory is useful only for predicting the flow upstream of the separation point and ultimately for predicting the occurrence of separation.

3.1 Description of the Flow

Adopting the Eulerian approach, we define a velocity vector $\mathbf{V}(x, y, z, t)$ to be equal to the velocity of the fluid particle at the point $\mathbf{x} = (x, y, z)$ in a Cartesian, rectangular coordinate system at time t. In this coordinate system the three components of $\mathbf{V}$ will be denoted by u, v, and w. Other physical parameters that will be required are the fluid density ρ, which ultimately will be assumed constant; the external force field $\mathbf{F}$ acting on the fluid particles, notably the gravitational force $\rho\mathbf{g}$; and the surface stresses, or forces per unit area, which act upon adjacent surfaces of the fluid.

Each component of the surface stress must be defined not only by the direction in which it acts but also by the orientation of the surface upon which it is acting; hence, a total of $3 \times 3 = 9$ stress components must be defined on a cubical surface aligned with the coordinate system, as shown in figure 3.1. Thus, for example, the stress τ_{yx} acts in the y-direction upon a surface of constant x.

This characterization of the stress components is actually too restrictive, since it does not account for the more general situation where the orientation of the surface differs from one of the three principal planes of a Cartesian coordinate system. By a familiar argument that plays a similar role in structural mechanics, however, the cube in figure 3.1 can be replaced by an infinitesimal tetrahedron, with three orthogonal faces normal to the Cartesian coordinates and one oblique face of arbitrary orientation. The tetrahedron is assumed to be sufficiently small that the stresses are effectively constant along each face, and since the volume will be negligible compared with the surface area, the surface forces will dominate the body forces. Thus, the forces exerted on the four faces of the tetrahedron by the surface stress components must balance. If $\mathbf{n} = (n_x, n_y, n_z)$ denotes the unit normal vector on the oblique face, each of its three components equals the

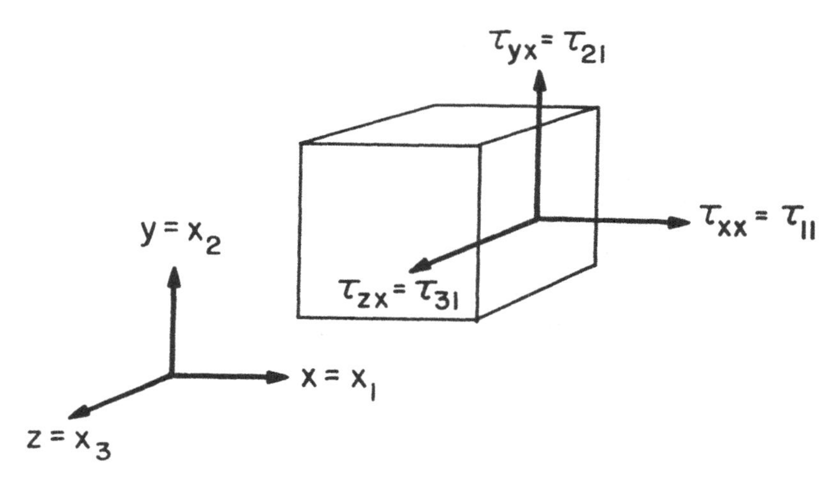

Figure 3.1
Definition of Cartesian coordinates and stress tensor.

ratio of the area of the corresponding face to the area of the oblique face. Thus, for the surface forces to be balanced, the component of the stress on the oblique face, say in the x-direction, must be equal to $\tau_{xx}n_x + \tau_{xy}n_y + \tau_{xz}n_z$. In other words, regardless of the angular orientation of a surface element, the stress acting on it in a given direction may be decomposed into three orthogonal components, as if it were a vector. Thus, the stress components can be described in general by a second-order *tensor.*

A similar argument regarding the balance of moments can be used to show that the stress tensor is symmetric. Thus, if the cube shown in figure 3.1 is sufficiently small so that surface moments dominate body moments and changes in the magnitude of the stress components may be neglected across the cube, then a positive value of τ_{yx} on the face shown will cause a counterclockwise moment about the centroid. The moment on the opposite face will have the same sign, since the normal vector has the opposite sense and hence a positive τ_{yx} acts downward. The only source of a balancing clockwise moment is the stress τ_{xy} acting in the x-direction on the top and bottom faces. From this it follows that $\tau_{yx} = \tau_{xy}$; by similar reasoning, $\tau_{xz} = \tau_{zx}$, $\tau_{yz} = \tau_{zy}$.

To avoid subsequent unwieldy algebra, we shall denote the Cartesian coordinates by subscripts (1, 2, 3) with the convention that $x = x_1$, $y = x_2$, and $z = x_3$. Similarly, for the velocity components, $u = u_1$, $v = u_2$, and $w = u_3$.

Both notations will be used. The complete stress tensor may then be written in any of the three following forms:

$$\tau_{ij} = \begin{Bmatrix} \tau_{11} & \tau_{12} & \tau_{13} \\ \tau_{21} & \tau_{22} & \tau_{23} \\ \tau_{31} & \tau_{32} & \tau_{33} \end{Bmatrix} = \begin{Bmatrix} \tau_{xx} & \tau_{xy} & \tau_{xz} \\ \tau_{yx} & \tau_{yy} & \tau_{yz} \\ \tau_{zx} & \tau_{zy} & \tau_{zz} \end{Bmatrix}. \tag{1}$$

Since the stress tensor is symmetric, $\tau_{ij} = \tau_{ji}$.

Adopting a similar notation for the unit vector, $\mathbf{n} = (n_1, n_2, n_3)$, the stress component in the ith direction, on a surface element with unit normal $\mathbf{n}$, is given by the sum

$$\sum_{j=1}^{3} \tau_{ij} n_j \equiv \tau_{ij} n_j, \qquad i = 1,2,3. \tag{2}$$

It is customary to delete the summation sign, as has been done in the last form of this equation, with the *summation convention*: any term of an equation that contains the same index twice should be summed over that index. It is also customary to assume that an equation such as this is valid for all possible values of a free index, in this case i. Both these conventions are employed hereafter, unless otherwise stated.

3.2 Conservation of Mass and Momentum

The conservation laws of physics can be related to a fluid provided we focus our attention on a group of fluid particles or a *material volume* of fluid so that we always examine the same group of particles. Thus we define a volume of fluid $\mathcal{V}(t)$ subject to the above restriction. If the fluid density is denoted by ρ, the total mass of fluid in this volume is given by the integral $\iiint \rho \, d\mathcal{V}$. Conservation of mass requires that this integral be constant, or

$$\frac{d}{dt} \iiint_{\mathcal{V}} \rho \, d\mathcal{V} = 0. \tag{3}$$

Similarly, the momentum density of a fluid particle is equal to the vector $\rho \mathbf{V}$, with components ρu_i. Conservation of momentum requires that the sum of all forces acting on the fluid volume be equal to the rate of change of its momentum, with respect to a Newtonian frame of reference, or

$$\frac{d}{dt} \iiint_{\mathcal{V}} \rho u_i \, d\mathcal{V} = \iint_{S} \tau_{ij} n_j \, dS + \iiint_{\mathcal{V}} F_i \, d\mathcal{V}. \tag{4}$$

Here the surface integral is the *i*th component of the surface forces acting on S, and the last volume integral is the sum of the body forces, such as that due to gravity. The conventions stated at the end of section 3.1 apply.

Equation (4) can be put in a more convenient form, involving only volume integrals, by using the divergence theorem

$$\iiint_{\mathcal{V}} \nabla \cdot \mathbf{Q} d\mathcal{V} = \iint_{S} \mathbf{Q} \cdot \mathbf{n}\, dS. \tag{5a}$$

This can be rewritten in the indicial notation as

$$\iiint_{\mathcal{V}} \frac{\partial Q_i}{\partial x_i} d\mathcal{V} = \iint_{S} Q_i n_i\, dS. \tag{5b}$$

Here Q is any vector that is continuous and differentiable in the volume $\mathcal{V}$, and the unit normal $\mathbf{n}$ is the exterior normal vector pointing out of $\mathcal{V}$ on the surface S. Using (5b) to transform the surface integral in (4),

$$\frac{d}{dt}\iiint_{\mathcal{V}} \rho u_i\, d\mathcal{V} = \iiint_{\mathcal{V}} \left[\frac{\partial \tau_{ij}}{\partial x_j} + F_i\right] d\mathcal{V}. \tag{6}$$

Equations (3) and (6) express the conservation laws of mass and momentum for the fluid, in terms of an arbitrary prescribed material volume $\mathcal{V}(t)$. The need to consider this volume integral and especially its time derivative is inconvenient, however. To overcome this problem, we first consider the evaluation of the time derivative, bearing in mind that the volume of integration is itself a function of time.

3.3 The Transport Theorem

Let us consider a general volume integral of the form

$$I(t) = \iiint_{\mathcal{V}(t)} f(\mathbf{x},t)\, d\mathcal{V}. \tag{7}$$

Here f is an arbitrary differentiable scalar function of position $\mathbf{x}$ and time t to be integrated over a prescribed volume $\mathcal{V}(t)$, which may also vary with time. Therefore the boundary surface S of this volume will change with time, and its normal velocity is denoted by U_n.

In the usual manner of elementary calculus we consider the difference

$$\Delta I = I(t+\Delta t) - I(t) = \iiint\limits_{\mathcal{V}(t+\Delta t)} f(\mathbf{x},t+\Delta t)d\mathcal{V} - \iiint\limits_{\mathcal{V}(t)} f(\mathbf{x},t)d\mathcal{V}. \tag{8}$$

Neglecting second-order differences proportional to $(\Delta t)^2$, we have

$$f(\mathbf{x},t+\Delta t) = f(\mathbf{x},t) + \Delta t \frac{\partial f(\mathbf{x},t)}{\partial t}.$$

A similar decomposition can be made for the volume $\mathcal{V}(t)$. Thus $\mathcal{V}(t+\Delta t)$ differs from $\mathcal{V}(t)$ by a thin volume $\Delta\mathcal{V}$ contained between the adjacent surfaces $S(t+\Delta t)$ and $S(t)$ and proportional to Δt. From equation (8) it follows that

$$\begin{aligned}\Delta I &= \iiint\limits_{\mathcal{V}+\Delta\mathcal{V}} \left(f + \Delta t \frac{\partial f}{\partial t}\right) d\mathcal{V} - \iiint\limits_{\mathcal{V}} f\, d\mathcal{V} \\ &= \Delta t \iiint\limits_{\mathcal{V}} \frac{\partial f}{\partial t} d\mathcal{V} + \iiint\limits_{\Delta\mathcal{V}} f\, d\mathcal{V} + O[(\Delta t)^2],\end{aligned} \tag{9}$$

where the last term denotes a second-order error proportional to $(\Delta t)^2$.

To evaluate the integral over the volume $\Delta\mathcal{V}$, we note that this thin region has a thickness equal to the distance between $S(t)$ and $S(t+\Delta t)$. This thickness is the normal component of the distance traveled by $S(t)$, in the time Δt, which is equal to the product $U_n\Delta t$. Thus the contribution from the last integral is of first order, or proportional to Δt; to the same degree of accuracy the integrand f may be assumed constant across the thin region in the direction normal to S. Integrating in this direction only, we then have

$$\Delta I = \Delta t \iiint\limits_{\mathcal{V}} \frac{\partial f}{\partial t} d\mathcal{V} + \iint\limits_{S} (U_n \Delta t) f\, dS + O[(\Delta t)^2]. \tag{10}$$

Finally, we obtain the desired result by dividing both sides by Δt and taking the limit as this tends to zero; thus

$$\frac{dI}{dt} = \iiint\limits_{\mathcal{V}} \frac{\partial f}{\partial t} d\mathcal{V} + \iint\limits_{S} f U_n\, dS. \tag{11}$$

Equation (11) is known as the *transport theorem*. The surface integral in this equation represents the transport of the quantity of f out of the volume $\mathcal{V}$, as a result of the movement of the boundary. In the special case where S is fixed and $U_n = 0$, equation (11) reduces to the simpler form where differentiation under the integral sign is justified.

In another special case of particular interest, $\mathcal{V}$ is a material volume, always composed of the same fluid particles; hence the surface S moves with the same normal velocity as the fluid and $U_n = \mathbf{V} \cdot \mathbf{n} = u_i n_i$. In this case it follows from (11) and the divergence theorem (5b) that

$$\begin{aligned}\frac{d}{dt}\iiint_{\mathcal{V}(t)} f d\mathcal{V} &= \iiint \frac{\partial f}{\partial t} d\mathcal{V} + \iint_S f u_i n_i dS \\ &= \iiint_{\mathcal{V}(t)} \left\{\frac{\partial f}{\partial t} + \frac{\partial}{\partial x_i}(f u_i)\right\} d\mathcal{V}.\end{aligned} \tag{12}$$

3.4 The Continuity Equation

Returning now to equation (3), expressing conservation of mass, we have immediately from equation (12) that

$$\frac{d}{dt}\iiint_{\mathcal{V}} \rho d\mathcal{V} = \iiint_{\mathcal{V}} \left[\frac{\partial \rho}{\partial t} + \frac{\partial}{\partial x_i}(\rho u_i)\right] d\mathcal{V} = 0. \tag{13}$$

Since the last integral is evaluated at a fixed instant of time, the distinction that $\mathcal{V}$ is a material volume is unnecessary at this stage. Moreover, this volume can be composed of an arbitrary group of fluid particles; hence, the integrand itself is equal to zero throughout the fluid. Thus, the volume integration in equation (13) can be replaced by a partial differential equation expressing conservation of mass in the form

$$\frac{\partial \rho}{\partial t} + \frac{\partial}{\partial x_i}(\rho u_i) = 0. \tag{14}$$

Our intention from now on is to assume that the fluid is incompressible and the density constant. Thus (14) can be simplified to give the continuity equation

$$\frac{\partial u_i}{\partial x_i} = 0, \tag{15a}$$

or, in vector form,

$$\nabla \cdot \mathbf{V} = 0. \tag{15b}$$

3.5 Euler's Equations

If the transport theorem (12) is applied to the conservation of momentum (6), it follows that,

$$\iiint_{\mathcal{V}} \left[\frac{\partial}{\partial t}(\rho u_i) + \frac{\partial}{\partial x_j}(\rho u_i u_j) \right] d\mathcal{V} = \iiint_{\mathcal{V}} \left(\frac{\partial \tau_{ij}}{\partial x_j} + F_i \right) d\mathcal{V}. \tag{16}$$

Once again the volume in question is arbitrary; hence equation (16) must hold for the integrands alone, in the form

$$\frac{\partial}{\partial t}(\rho u_i) + \frac{\partial}{\partial x_j}(\rho u_i u_j) = \frac{\partial \tau_{ij}}{\partial x_j} + F_i. \tag{17}$$

Finally, if the derivatives of products on the left side of this equation are expanded by the chain rule, and conservation of mass is invoked from equation (15), we obtain Euler's equations in the form

$$\frac{\partial u_i}{\partial t} + u_j \frac{\partial u_i}{\partial x_j} = \frac{1}{\rho} \frac{\partial \tau_{ij}}{\partial x_j} + \frac{1}{\rho} F_i. \tag{18}$$

The left-hand side of Euler's equations can be interpreted as the acceleration of a material particle of fluid, since the *substantial derivative*

$$\frac{D}{Dt} \equiv \frac{\partial}{\partial t} + \mathbf{V} \cdot \nabla = \frac{\partial}{\partial t} + u_j \frac{\partial}{\partial x_j} \tag{19}$$

expresses the time rate-of-change in a coordinate system moving with the fluid particle.[1]

3.6 Stress Relations in a Newtonian Fluid

Finally, we must relate the stress tensor τ_{ij} to the kinematic properties of the fluid. The task here is analogous to specifying the stress-strain relations in solid mechanics.

If the fluid is at rest, and more generally if there are no shear stresses, a normal *pressure* stress will exist within the fluid. Equilibrium of the forces acting across a small tetrahedron requires that this pressure be isotropic. Thus, in the absence of viscous shear the stress tensor is

$$\tau_{ij} = -p\delta_{ij}, \tag{20}$$

where δ_{ij} is the Kroenecker delta function, equal to 1 if $i = j$ and 0 if $i \neq j$. By hypothesis, there are no viscous forces if the fluid moves as a rigid mass without deformation, or with the velocity field

$$\mathbf{V} = \mathbf{A} + \mathbf{B} \times \mathbf{r}. \tag{21}$$

Here **A** and **B** are constant vectors equal to the translation and rotation velocities, and **r** is the position vector from the origin of rotation.

Viscous stresses will occur when the fluid velocity differs from (21), with relative motion between adjacent fluid particles. The simplest example is a uniform shear flow, shown in figure 3.2. Here, and for more general velocity fields, the fundamental assumption of a *Newtonian* fluid is that the stress tensor is a linear function of the nine gradients $\partial u_k/\partial x_l$. This ensures vanishing of the viscous stress components for uniform translation of the fluid. The rotational term in (21) will be stress-free provided the gradients occur only in the form of sums $\partial u_k/\partial x_l + \partial u_l/\partial x_k$. For an isotropic fluid, the values of the stress components must be independent of the choice of coordinates, and for flow in one plane there can be no shear stress in the direction

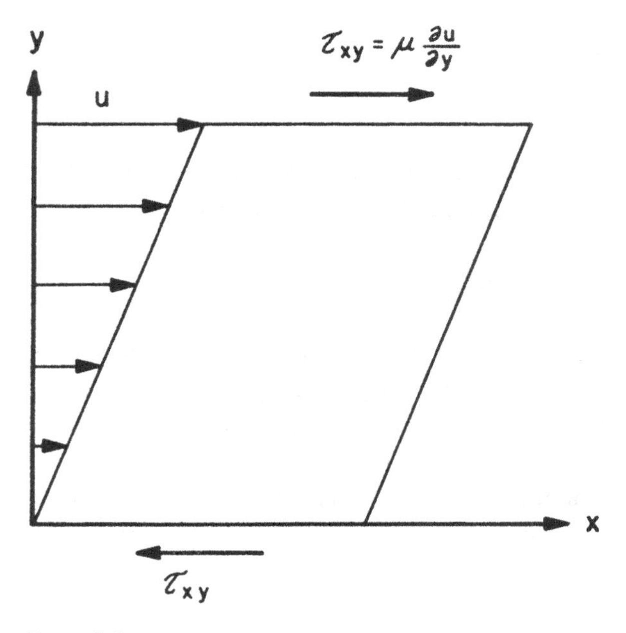

Figure 3.2
Stress and strain for simple shear flow of a Newtonian fluid.

normal to this plane. The most general linear function of the velocity gradients, consistent with these conditions and the requirement that τ_{ij} is symmetric, is of the form

$$\tau_{ij} = \mu(\partial u_i / \partial x_j + \partial u_j / \partial x_i), \quad \text{for } i \neq j. \tag{22}$$

The coefficient μ is the *viscous shear coefficient,* or simply the *coefficient of viscosity.*

In order for τ_{ij} to be a tensor, the only possible addition to (22) for $i = j$ is a second constant times the divergence $\partial u_i/\partial x_i$, which vanishes for an incompressible fluid. Thus, the total stress tensor for an incompressible fluid is given by

$$\tau_{ij} = -p\delta_{ij} + \mu\left(\frac{\partial u_i}{\partial x_j} + \frac{\partial u_j}{\partial x_i}\right), \tag{23a}$$

or, in Cartesian components,

$$\{\tau_{ij}\} = \begin{Bmatrix} -p & 0 & 0 \\ 0 & -p & 0 \\ 0 & 0 & -p \end{Bmatrix} + \mu \begin{Bmatrix} 2\dfrac{\partial u}{\partial x} & \dfrac{\partial u}{\partial y} + \dfrac{\partial v}{\partial x} & \dfrac{\partial u}{\partial z} + \dfrac{\partial w}{\partial x} \\ \dfrac{\partial v}{\partial x} + \dfrac{\partial u}{\partial y} & 2\dfrac{\partial v}{\partial y} & \dfrac{\partial v}{\partial z} + \dfrac{\partial w}{\partial y} \\ \dfrac{\partial w}{\partial x} + \dfrac{\partial u}{\partial z} & \dfrac{\partial w}{\partial y} + \dfrac{\partial v}{\partial z} & 2\dfrac{\partial w}{\partial z} \end{Bmatrix}. \tag{23b}$$

The first matrix in (23) is the normal pressure stress. The second matrix is the viscous stress tensor, proportional to the viscosity coefficient μ. The diagonal elements of the viscous stress are associated with elongations of fluid elements, and the off-diagonal elements are due to shearing deformations of the form shown in figure 3.2.

Most common fluids, including water and air, are found to conform to the Newtonian stress relations (23) for all practical purposes. An exception occurs if dilute solutions of polymers are formed which disrupt the isotropic nature of the fluid. The resulting stress relations are said to be "non-Newtonian." In certain cases the frictional drag on a body can be reduced by this means, as described in the surveys by Lumley (1969) and Bark, Hinch, and Landahl (1975).

3.7 The Navier-Stokes Equations

It is now a simple matter to obtain the Navier-Stokes equations, which express conservation of momentum for a Newtonian fluid, by substituting the stress-strain relations (23) in Euler's equations (18). The required derivatives of the stress tensor are

$$\begin{aligned}\frac{\partial \tau_{ij}}{\partial x_j} &= -\frac{\partial p}{\partial x_i} + \mu \frac{\partial}{\partial x_j}\left(\frac{\partial u_i}{\partial x_j} + \frac{\partial u_j}{\partial x_i}\right) \\ &= -\frac{\partial p}{\partial x_i} + \mu \frac{\partial^2 u_i}{\partial x_j \partial x_j},\end{aligned} \tag{24}$$

since, from the continuity equation (15a), $\dfrac{\partial^2 u_j}{\partial x_j \partial x_i} = \dfrac{\partial}{\partial x_i}\dfrac{\partial u_j}{\partial x_j} = 0$.

Thus, we obtain the desired Navier-Stokes equations:

$$\frac{\partial u_i}{\partial t} + u_j \frac{\partial u_i}{\partial x_j} = -\frac{1}{\rho}\frac{\partial p}{\partial x_i} + \nu \frac{\partial^2 u_i}{\partial x_j \partial x_j} + \frac{1}{\rho} F_i,$$

where $\nu = \mu/\rho$ is the kinematic viscosity coefficient. Written in vector form,

$$\frac{\partial \mathbf{V}}{\partial t} + (\mathbf{V} \cdot \nabla)\mathbf{V} = -\frac{1}{\rho}\nabla p + \nu \nabla^2 \mathbf{V} + \frac{1}{\rho}\mathbf{F}, \tag{25b}$$

or, finally, in Cartesian coordinates:

$$\begin{aligned}\frac{\partial u}{\partial t} + u\frac{\partial u}{\partial x} + v\frac{\partial u}{\partial y} + w\frac{\partial u}{\partial z} &= -\frac{1}{\rho}\frac{\partial p}{\partial x} + \nu \nabla^2 u + \frac{1}{\rho}F_x \\ \frac{\partial v}{\partial t} + u\frac{\partial v}{\partial x} + v\frac{\partial v}{\partial y} + w\frac{\partial v}{\partial z} &= -\frac{1}{\rho}\frac{\partial p}{\partial y} + \nu \nabla^2 v + \frac{1}{\rho}F_y \\ \frac{\partial w}{\partial t} + u\frac{\partial w}{\partial x} + v\frac{\partial w}{\partial y} + w\frac{\partial w}{\partial z} &= -\frac{1}{\rho}\frac{\partial p}{\partial z} + \nu \nabla^2 w + \frac{1}{\rho}F_z.\end{aligned} \tag{25c}$$

This system of three partial differential equations, together with the continuity equation (15), governs the motion of a viscous fluid subject only to the assumptions of constant density and a Newtonian stress-strain relationship. The latter is justified for all practical purposes in the case of most fluids, including water and air.

The difficulty comes in attempting to solve the Navier-Stokes equations; they form a coupled system of nonlinear partial differential equations and have been solved analytically only for some very simple geometrical configurations, principally those in which the nonlinear convective acceleration terms $(\mathbf{V} \cdot \nabla) \mathbf{V}$ can be assumed to vanish. We shall examine some examples of such flows subsequently.

3.8 Boundary Conditions

Before attempting to solve the Navier-Stokes equations, it is necessary to impose appropriate physical conditions on the boundaries of the fluid domain. In fact, it is precisely these conditions that distinguish different flow problems.

The simplest boundary condition is that which must be imposed on a solid boundary, such as the boundary between the fluid and a rigid body. In this case, the fluid velocity equals the velocity of the body, since in a viscous fluid the existence of a shear stress requires both the normal and tangential velocity components of the fluid and the boundary to be equal.

An alternative physical boundary is a free surface, such as will exist between the surface of the ocean and the atmosphere or between the fluid and vapor domains of a cavitating flow. In these instances the fluid density is so large, by comparison with the neighboring gas, that tangential stresses are negligible; the only stress at such a surface is the normal pressure, which is generally known and specified (atmospheric pressure or vapor pressure for the two cases cited). This is the *dynamic* boundary condition at a free surface. In addition, it is necessary to impose the *kinematic* boundary condition that the normal velocity of the fluid and of the free surface are equal.

3.9 Body Forces and Gravity

In our derivation of the Euler and Navier-Stokes equations, an external *body* force per unit volume, $\mathbf{F}$, has been included without specification or restriction. The most common and important body force in the context of marine hydrodynamics is the force due to gravity $\rho\mathbf{g}$, where $\mathbf{g}$ is the gravitational acceleration vector. For all practical purposes $\mathbf{g}$ is a constant, acting downward with magnitude 9.8 m/s^2 or 32.2 ft/s^2. If this is the only body force acting upon the fluid, or more generally if the body force is *conservative,* it can be represented as the gradient of a scalar function, in the form $\mathbf{F} = \rho\nabla\Omega(\mathbf{x})$ where, for gravity, $\Omega = \mathbf{g} \cdot \mathbf{x}$. From the Navier-Stokes equations (25), this force can be absorbed with the pressure p, by means of the substitution $\tilde{p} = p - \rho\Omega$. With this modification the Navier-Stokes equations no longer include the body-force terms explicitly. Hence, as far as conservation of momentum is concerned, the only effect of a gravitational force or any conservative force field is to change the pressure by an additive amount $-\rho\Omega$.

Generally we regard p as the *total* pressure, $p - \rho\Omega$ as the *hydrodynamic* pressure, and the difference as the *hydrostatic* pressure. If the fluid is in a state of rest, the total and hydrostatic pressures are equal.

In view of this additive decomposition, the body force has no effect on the fluid motion, unless it affects the problem via a boundary condition. Except for this possibility, the effect of gravity can be ignored throughout, with the understanding that this gives rise to an additive hydrostatic pressure.

Gravity will play an important role when we consider wave motions on the free surface. Here the dynamic boundary condition states that the total pressure is equal to a specified value, and the hydrostatic component must be included. When the boundary conditions are entirely kinematic, however, with prescribed values of the normal velocity, there can be no interaction between gravitational forces acting on the fluid and the dynamics of the fluid motion. With this justification, we will hereafter ignore body forces in general and the gravitational force in particular until chapter 6, where we consider the problem of waves on a free surface.

3.10 The Flow between Two Parallel Walls (Plane Couette Flow)

The simplest solutions of the Navier-Stokes equations are those pertaining to the steady flow of a viscous fluid between two parallel walls separated by a distance h, under the assumption that the horizontal dimensions of the walls are very large compared to h. It is reasonable then to assume that the flow is independent of the coordinates parallel to the walls and depends only on the transverse coordinate. Let us suppose that the two walls occupy the planes $y = 0$ and $y = h$, and that the motion is in the x-direction, driven either by a pressure gradient or by the imposed uniform motion of one wall relative to the other or by a combination of these two effects.

Let us suppose that the boundary $y = 0$ is at rest, while the boundary $y = h$ moves in the positive x-direction with velocity U. (Such a situation can be realized in general by means of a coordinate transformation that brings the lower boundary to rest.) Since the motion is independent of both the x- and z-coordinates, it follows from the continuity equation that $\partial v/\partial y = 0$; thus, since v must equal zero at the two boundaries, this velocity must be identically zero throughout the flow region $0 < y < h$. The velocity component w is considered to vanish since there is no driving mechanism for a flow in the z-direction.

It follows that the fluid velocity vector is parallel to the x-direction, that is, u is the only nonzero component of the velocity vector and the flow is everywhere parallel to the x-axis. In addition, since the flow is independent of the x-coordinate, the convective acceleration terms of the Navier-Stokes equations are all equal to zero, and these equations reduce to the relatively simple system

$$\frac{1}{\rho}\frac{\partial p}{\partial x} = \nu\frac{\partial^2 u}{\partial y^2}, \qquad \frac{\partial p}{\partial y} = 0, \qquad \frac{\partial p}{\partial z} = 0, \tag{26}$$

subject to the boundary conditions $u(0) = 0$ and $u(h) = U$. From (26) the pressure is independent of y and z, and $\partial p/\partial x = dp/dx$ is constant since u is independent of x. The solution, obtained by integration, is

$$u = \frac{1}{2\mu}\frac{dp}{dx}y(y-h) + \frac{U}{h}y. \tag{27}$$

The pressure gradient gives rise to a parabolic profile, and the relative velocity between the two walls generates a linear shear flow. In general, a combination of these two flows will exist, as shown in figure 3.3.

3.11 The Flow through a Pipe (Poiseuille Flow)

A similar solution can be obtained for the flow in a tube of circular cross-section; for laminar flows, this solution is physically important for the determination of pressure drops in pipes. Strictly speaking, the solution should be obtained in circular cylindrical coordinates. To avoid the details of transforming the Navier-Stokes equations into such a coordinate system, we shall find the solution from the Cartesian form of the Navier-Stokes equations, with the x-axis coincident with the axis of the tube; we assume that $u = u(r)$ depends only on the radial coordinate r, and hence on the combination $y^2 + z^2$, while $v = w = 0$. The reduction of the Navier-Stokes and

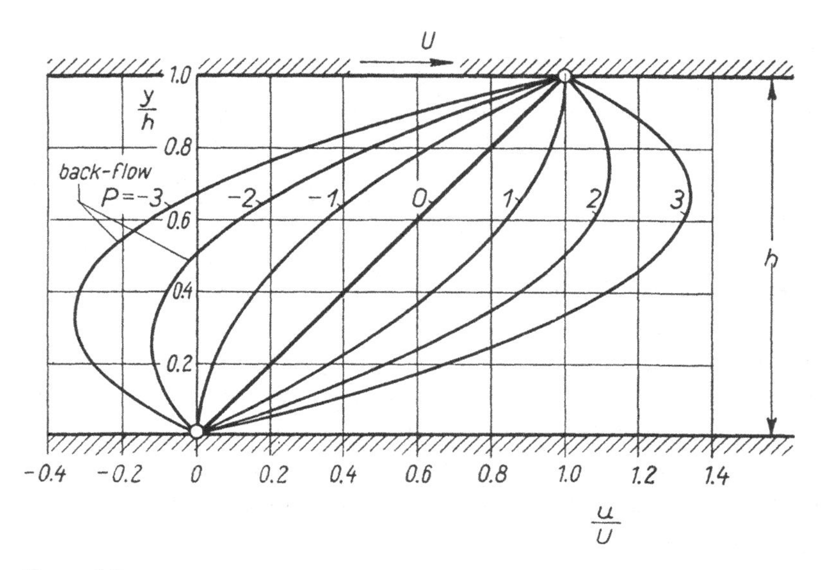

Figure 3.3
Couette flow between two parallel walls. The curves are labeled with values of the nondimensional pressure gradient $P = -(h^2/2\mu U)(dp/dx)$. Note the occurrence of a secondary backflow, analogous to separation, when this parameter is less than –1. (From Schlichting 1968)

continuity equations is then identical to that carried out for plane Couette flow. Hence the axial velocity satisfies the equation

$$\frac{1}{\rho}\frac{dp}{dx} = \nu\left(\frac{\partial^2 u}{\partial y^2} + \frac{\partial^2 u}{\partial z^2}\right), \tag{28}$$

subject to the boundary condition $u(r_0) = 0$.

The appropriate solution of (28) is given by

$$u(r) = -\frac{1}{4\mu}\frac{dp}{dx}(r_0^2 - y^2 - z^2) = -\frac{1}{4\mu}\frac{dp}{dx}(r_0^2 - r^2). \tag{29}$$

Once again, the pressure gradient dp/dx drives a flow parabolic in profile across the tube. It is simple to compute the volume rate-of-flow or flux Q of fluid through the pipe, by integrating across the circular cross-section from $y = 0$ to $y = r_0$,

$$Q = 2\pi\int_0^{r_0} u(r)r\,dr = \frac{\pi r_0^4}{8\mu}\left(-\frac{dp}{dx}\right). \tag{30}$$

Reynolds and others have determined experimentally that such a flow will exist in practice only if the Reynolds number based upon the diameter of the tube is less than a critical value.[2] Otherwise, the flow becomes unsteady or *turbulent* and is characterized by rapid velocity fluctuations in both space and time. The comparatively smooth flow existing at Reynolds numbers less than the critical value is a *laminar* flow.

3.12 External Flow Past One Flat Plate

The solutions obtained above illustrate steady laminar *internal* flow where the fluid is restricted in width. In our subsequent consideration of *external* flow, for example past a streamlined body in two or three dimensions, the situation is significantly different. Not only may the form of the body be relatively complicated, but also the fluid domain is unbounded in the direction normal to the body surface.

A technique for dealing with this type of problem was provided by Prandtl's introduction of the boundary-layer approximation. For sufficiently large values of the Reynolds number, it is reasonable to suppose that within the bulk of the fluid, viscous forces will be negligible by comparison with inertial forces, and the corresponding flow may be considered inviscid. This is a significant simplification, for as we shall see in the following chapters, various techniques can be employed to analyze inviscid flows. Nevertheless, a consequence of viscous shear stresses is that however large the Reynolds number may be, the fluid velocity on any rigid boundary of the fluid must still be equal to the velocity of the boundary. Thus, there must exist significant viscous shear in a thin boundary layer at the surface of any body that moves relative to the bulk of the fluid. To simplify the Navier-Stokes equations, however, we can exploit the thinness of this region.

This approach provides a scheme for calculating viscous effects, at least for unseparated flows at high Reynolds number; equally important, it gives a rationale for neglecting viscous stresses outside the boundary layer. Moreover, if the body is sufficiently regular in shape, its radii of curvature will be much larger than the boundary-layer thickness, and the local flow within the boundary layer will be effectively plane. Thus, if one imagines looking at the boundary-layer flow, say by enlarging it with a magnifying glass or microscope, the details of the flow within this thin region will become visible, but the overall shape of the body is lost to view and the boundary of the body will appear practically flat within the region of view.

This approach leads us to expect that the most important and relevant viscous flow problem is the external flow past a single flat plate. This might be taken to coincide with the plane $y = 0$, as for the lower wall of the plane Couette flow, but with the fluid occupying the entire upper half-space $y > 0$. We might hope to obtain the desired result from the solution (27) of the corresponding problem for two plates, situated at $y = 0$ and $y = h$, simply by letting the separation distance h tend to infinity while focusing attention on the flow near the lower plate at $y = 0$. A cursory examination of equation (27) shows that this will not work, however. The solution for Couette flow is linear if driven by the movement of one plate and parabolic if driven by a pressure gradient. In either event the presence of the second boundary affects the flow near the first, and viscous effects persist throughout the entire fluid.

This analogy suggests that for steady laminar flow past a single infinitely long flat plate, viscous effects will persist infinitely far away from the plate in the normal direction. In principle this is correct and should not be surprising. Viscous effects are diffusive, and in this problem, diffusion has an infinite amount of time to take effect. This is not very useful in our quest for a boundary-layer theory where viscous effects are confined to a thin region, and in some sense we must remove the opportunity for diffusion to occur without limit in time and space.

Two possible modifications can be imposed to make this problem more realistic physically and to limit the time for viscous diffusion to occur. First, we might consider an unsteady problem where the time is limited by the duration of the motion. Alternatiyely, for steady-state flows, the length of the plate should be finite to limit the elapsed time during which a single particle of fluid is traveling past the plate. The latter problem is the most important and relevant one for us to consider, but it is also the more mathematically complicated. Thus, we shall first solve the unsteady problem for a plate of infinite length, which is not without some practical interest in any event.[3]

3.13 Unsteady Motion of a Flat Plate

As we have assumed above, the plate coincides with the plane $y = 0$ and the fluid occupies the half-space $y > 0$. If the plate moves in its own plane parallel to the x-axis with prescribed velocity $U(t)$, then the flow is again independent of the x- and z-coordinates and the z-component of the fluid velocity is zero. From continuity and the boundary condition on the plate, it follows that $v = 0$ and the velocity field is a parallel flow in the x-direction which depends only on the coordinate y and on time t. Thus from the Navier-Stokes equations, in the absence of any external pressure gradients, the unknown velocity $u(y, t)$ is governed by the partial differential equation

$$\frac{\partial u}{\partial t} = v\frac{\partial^2 u}{\partial y^2}. \tag{31}$$

This equation is called the diffusion equation, or heat equation, since it governs the diffusion of heat in a conducting medium.

Appropriate boundary conditions are the no-slip condition on the plate,

$$u(0,t) = U(t) \quad \text{on } y = 0 \tag{32}$$

and, in addition, the condition that the fluid motion tends to zero far away from the plate,

$$u(y,t) \to 0 \quad \text{as } y \to \infty. \tag{33}$$

The last condition is somewhat pragmatic; it may be replaced by the initial condition that the fluid is at rest everywhere, at some initial time before the plate is set in motion. But (33) enables us to consider as the simplest possible example sinusoidal time dependence, which in fact violates the initial condition. We shall return to this point subsequently.

For the simplest unsteady motion, assume that the motion of the plate is sinusoidal, say $U(t) = U_0 \cos \omega t$ where ω is the radian frequency. The resulting flow is thus sinusoidal, and it is convenient to write the unknown velocity $u(y, t)$ in the complex form

$$u(y,t) = \mathrm{Re}[f(y)e^{i\omega t}], \tag{34}$$

where the real part of f is the velocity component in phase with $\cos \omega t$, and the imaginary part is in phase with $\sin \omega t$. Substituting (34) in the governing equation (31) gives an ordinary differential equation for f in the form

$$i\omega f = \nu \frac{d^2 f}{dy^2}. \tag{35}$$

The solution is well known and can be written as

$$f = Ae^{ky} + Be^{-ky}, \tag{36}$$

where the constants A and B are to be determined subsequently and

$$k = (i\omega / \nu)^{1/2} = (1+i)(\omega / 2\nu)^{1/2}. \tag{37}$$

(In view of the form of (36), one can take the positive or negative square root, as this would only reverse the roles of the two constants A and B.)

Since the parameter k is complex with positive real part, the boundary condition (33) at infinity can only be satisfied if $A = 0$ in (36). The boundary condition (32) on the plate determines the value of the remaining constant B, and the final solution of the problem is

$$\begin{aligned} u(y,t) &= \mathrm{Re}\{U_0 \exp(-ky + i\omega t)\} \\ &= U_0 \exp[-(\omega / 2\nu)^{1/2} y] \cos[(\omega / 2\nu)^{1/2} y - \omega t]. \end{aligned} \tag{38}$$

The motion corresponding to (38) is a damped transverse wave; it oscillates in the x-direction while propagating in the y-direction away from the plate. The wavelength is $\lambda = 2\pi(2\nu/\omega)^{1/2}$, and this will be extremely short for all but very low frequencies. For example, if $\nu = 10^{-6} \mathrm{m}^2/\mathrm{s}$ (for water), and $\omega = 2\pi$ (1-second period), the wavelength is 0.35 cm. Moreover, the exponential factor in (38) will attenuate the wave amplitude in this distance by a factor $e^{-2\pi} = 0.002$. These numbers suggest that if the motion is not extremely slow, the boundary layer will be very thin.

Since the governing equation (31) and boundary conditions (32–33) are linear, more general time-dependent solutions can be constructed by superposition. Moreover, if the frequency ω is permitted to be complex, the motion of the plate will increase or decrease exponentially with time, according as $\mathrm{Im}(\omega)$ is negative or positive, respectively. The only restriction here is that for the condition at $y = \infty$ to be satisfied, Re(k) must be positive as assumed above. Hence if ω is complex, it must be negative imaginary corresponding in (34) to an exponentially *growing* velocity of the plate. Indeed, this is the only reasonable alternative from the standpoint that the motion is initially at rest at $t = -\infty$. Laplace transforms can then be used to solve more general initial-value problems, including the particular case of a step-function plate velocity, which is initially at rest but impulsively accelerated to a constant velocity U.

There is a simpler way to solve the latter problem, because no characteristic time scale is associated with the acceleration of the plate. Thus from dimensional analysis, the unknown velocity $u(y, t)$ must depend on only one nondimensional *similarity parameter,* which will be taken as

$$\eta = \frac{y}{2(\nu t)^{1/2}}, \tag{39}$$

with the time t measured from the instant when the plate is accelerated. Using the chain rule to evaluate the derivatives in (31) gives the ordinary differential equation

$$\frac{d^2 f}{d\eta^2} + 2\eta \frac{df}{d\eta} = 0 \tag{40}$$

for the nondimensional velocity $f = u/U$. The boundary conditions for the function $f(\eta)$ are that $f(0) = 1$ and $f(\infty) = 0$. The solution of the differential equation (40) subject to these boundary conditions is

$$f = \frac{u}{U} = 1 - \operatorname{erf}(\eta) = 1 - \frac{2}{\sqrt{\pi}} \int_0^{\eta} e^{-\eta^2} d\eta. \tag{41}$$

The function $\operatorname{erf}(\eta)$ is known as the error function; its values can be determined from various tables (Abramowitz and Stegun 1964). From such a source it is easy to plot the nondimensional velocity profile shown in figure 3.4.

Let us now try to measure the *boundary-layer thickness—that* is, the width of the region in which the fluid velocity is affected by the velocity of the plate. Clearly this region extends to infinity, but the attenuation of the velocity with increasing distance from the plate is quite rapid; for example, the nondimensional velocity at a value of $y/2(\nu t)^{1/2} = \eta = 2.0$ is approximately $u/U = 0.005$, a very difficult velocity to measure experimentally. A common engineering criterion for the boundary-layer thickness is the distance from the plate where the velocity is reduced to 0.01 of its value on the boundary. This thickness is found from the above tables to be approximately

$$\delta = 3.64(\nu t)^{1/2}. \tag{42}$$

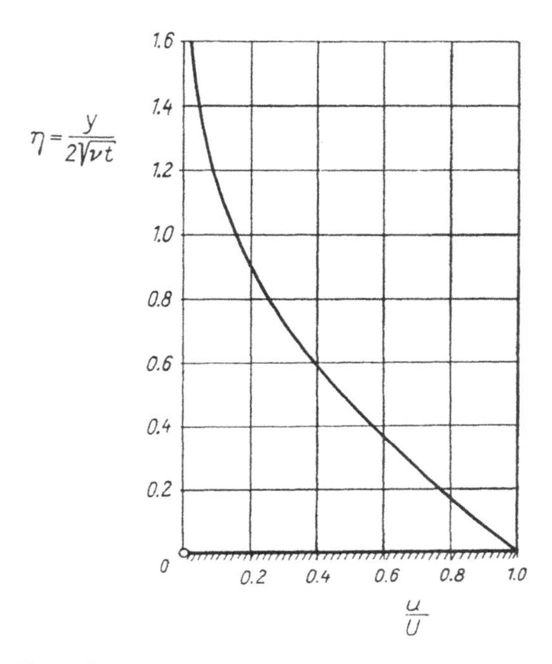

Figure 3.4
Velocity distribution near an impulsively accelerated wall. (From Schlichting 1968)

Again, the boundary-layer thickness is relatively small except over extremely large time scales where, ultimately, the effects of diffusion are more widespread. For example, the boundary-layer thickness predicted from (42) for water is 0.4 cm after one second and 0.24 m after one hour!

3.14 Laminar Boundary Layers: Steady Flow Past a Flat Plate

Let us turn now to the complementary problem of steady-state flow past a flat plate of finite length l. Once again the motion is assumed two dimensional, independent of the z-coordinate, and the plate is taken to coincide with the portion $0 < x < l$ of the plane $y = 0$. With $w = 0$, the continuity equation and (steady-state) Navier-Stokes equations take the form

$$\frac{\partial u}{\partial x} + \frac{\partial v}{\partial y} = 0, \tag{43}$$

$$u\frac{\partial u}{\partial x} + v\frac{\partial u}{\partial y} = -\frac{1}{\rho}\frac{\partial p}{\partial x} + \nu\left(\frac{\partial^2 u}{\partial x^2} + \frac{\partial^2 u}{\partial y^2}\right), \tag{44}$$

$$u\frac{\partial v}{\partial x}+v\frac{\partial v}{\partial y}=-\frac{1}{\rho}\frac{\partial p}{\partial y}+\nu\left(\frac{\partial^2 v}{\partial x^2}+\frac{\partial^2 v}{\partial y^2}\right). \tag{45}$$

The boundary conditions require that $u = v = 0$ on the portion of the plane $y = 0$ occupied by the plate, say $0 < x < l$, and that the velocity at large distances from the plate must be equal to the undisturbed flow $u = U$, $v = 0$.

The above equations are exact, regardless of the magnitudes of the dimensional parameters U, l, and ν or the Reynolds number Ul/ν. Our interest is focused now on the approximate solution of these equations when the Reynolds number is large, anticipating that the boundary layer will be very thin under this circumstance. We begin by making a qualitative estimate of the boundary-layer thickness, based on the unsteady solution of the preceding section. If the characteristic time during which viscous effects act on a fluid particle is $t = l/U$, the corresponding boundary-layer thickness obtained from (40) will be $\delta = 3.64\ (\nu l/U)^{1/2}$, or in nondimensional form, $\delta/l = 3.64\ R^{-1/2}$. A more careful analysis will reveal that the form of this equation is correct, but the constant is slightly less than 5.0. Nevertheless, the above estimate is sufficient for our present purposes because it suggests the order of magnitude of the boundary-layer thickness.

Since $u \simeq U$ immediately outside the boundary layer, it follows that the horizontal velocity component u must increase from 0 to U over the very small transverse distance δ. Therefore, the derivative $\partial u/\partial y$ will be large, of order (U/δ), compared with changes parallel to the plate, of order (U/l). Thus $\partial u/\partial y \gg \partial u/\partial x$. From the continuity equation it follows that $\partial u/\partial y \gg \partial v/\partial y$, or that the change in normal velocity v is relatively small in passing across the boundary layer. Since $v = 0$ on the plate it follows that this component of the velocity must be small throughout the boundary layer, or $u \gg v$.

As a result of these inequalities, the Navier-Stokes equations can be simplified to the forms

$$u\frac{\partial u}{\partial x}+v\frac{\partial u}{\partial y}=-\frac{1}{\rho}\frac{\partial p}{\partial x}+\nu\frac{\partial^2 u}{\partial y^2}, \tag{46}$$

$$0=\frac{1}{\rho}\frac{\partial p}{\partial y}. \tag{47}$$

Here the concept that u changes most significantly in the transverse direction has been used to discard the second derivative with respect to the horizontal coordinate, or $\partial^2 u/\partial x^2 \ll \partial^2 u/\partial y^2$.

From (47) it follows that the pressure p is not changed significantly in passing through the boundary layer, as one might anticipate on physical grounds. As a result, the longitudinal pressure gradient $\partial p/\partial x$ can be set equal to its value outside the boundary layer. For a flat plate the inviscid flow outside the boundary layer is uniform, with velocity U = constant. Thus the pressure gradient $\partial p/\partial x = 0$, and the laminar boundary-layer equations for a flat plate are the two relatively simple equations

$$u\frac{\partial u}{\partial x} + v\frac{\partial u}{\partial y} - \nu\frac{\partial^2 u}{\partial y^2} = 0, \tag{48}$$

$$\frac{\partial u}{\partial x} + \frac{\partial v}{\partial y} = 0. \tag{49}$$

It remains to nondimensionalize these equations. Taking advantage of the disparate length scales, we stretch the transverse coordinate in proportion to δ; hence, appropriate independent variables are

$$x' = x / l, \qquad y' = yR^{1/2} / l.$$

Then, guided by the continuity equation, we nondimensionalize the velocity components in the form

$$u' = u / U, \qquad v' = vR^{1/2} / U.$$

With these substitutions, equations (48) and (49) are replaced by

$$u'\frac{\partial u'}{\partial x'} + v'\frac{\partial u'}{\partial y'} - \frac{\partial^2 u'}{\partial y'^2} = 0, \tag{50}$$

$$\frac{\partial u'}{\partial x'} + \frac{\partial v'}{\partial y'} = 0, \tag{51}$$

with the boundary conditions

$$u' = v' = 0 \quad \text{on } y' = 0,$$

$$u' \to 1 \qquad \text{for } y' \to \infty.$$

Since (50–51) do not involve the viscosity coefficient or Reynolds number explicitly, the solution of these equations will be a *similarity solution* valid for all Reynolds numbers, in the regime where the laminar boundary-layer equations (48–49) are valid.

To reduce these partial differential equations to ordinary differential equations, let us assume that the length l is large compared to the distance downstream x and that the flow is unaffected locally by the actual length l.[4] The solution of these equations then must be independent of l, so that the dependence on the space coordinates x' and y' can only involve the ratio $y'/\sqrt{x'} = y\,(U/\nu x)^{1/2}$. Similarly, to remove the dependence on l in the definition of v', it follows that $v'\sqrt{x'}$ must depend only on the ratio $y'/\sqrt{x'}$. Thus, we seek a solution in the form

$$\frac{u}{U} = F(\eta), \qquad v(x/U\nu)^{1/2} = G(\eta),$$

where $\eta = y(U/\nu x)^{1/2}$. With these substitutions the resulting mathematical problem is

$$-\frac{1}{2}\eta F\frac{dF}{d\eta} + G\frac{dF}{d\eta} - \frac{d^2F}{d\eta^2} = 0, \tag{52}$$

$$-\frac{1}{2}\eta\frac{dF}{d\eta} + \frac{dG}{d\eta} = 0, \tag{53}$$

$$F(0) = 0, \quad F(\infty) = 1, \qquad G(0) = 0.$$

These are coupled, ordinary differential equations for the two unknowns F and G. The continuity equation (53) can be used to eliminate one unknown, and in this manner we seek a stream function $f(\eta)$ such that

$$F = df/d\eta, \qquad G = \frac{1}{2}\left(\eta\frac{df}{d\eta} - f\right).$$

Then the continuity equation is satisfied automatically, and the resulting problem is as follows:

$$\begin{gathered} f\frac{d^2f}{d\eta^2} + 2\frac{d^3f}{d\eta^3} = 0, \\ f = \frac{df}{d\eta} = 0 \quad \text{on} \quad \eta = 0, \qquad \frac{df}{d\eta} = 1 \quad \text{at} \quad \eta = \infty. \end{gathered} \tag{54}$$

Further progress requires that this nonlinear differential equation be integrated numerically. The resulting solution is identified with Blasius, who computed the boundary-layer solution for a flat plate in 1908. The resulting plot of the velocity distribution in the boundary layer is shown in figure 3.5.

This figure is qualitatively similar to figure 3.4, for the velocity distribution on an accelerating plate. Blasius' solution predicts the value $u/U = 0.99$ when $\eta = 4.9$, and the boundary-layer thickness as defined before is approximately $\delta = 4.9(\nu x/U)^{1/2}$. An alternative definition of the boundary-layer thickness, which can be defined unambiguously, is the *displacement thickness*

$$\delta^* = \int_0^\infty (1 - u/U)dy. \tag{55}$$

Calculations reveal that

$$\delta^* = 1.72(\nu x/U)^{1/2}. \tag{56}$$

This value is shown as a dotted line in figure 3.5 and corresponds graphically to the top edge of the rectangle, which has the same enclosed area as the total area under the curve.

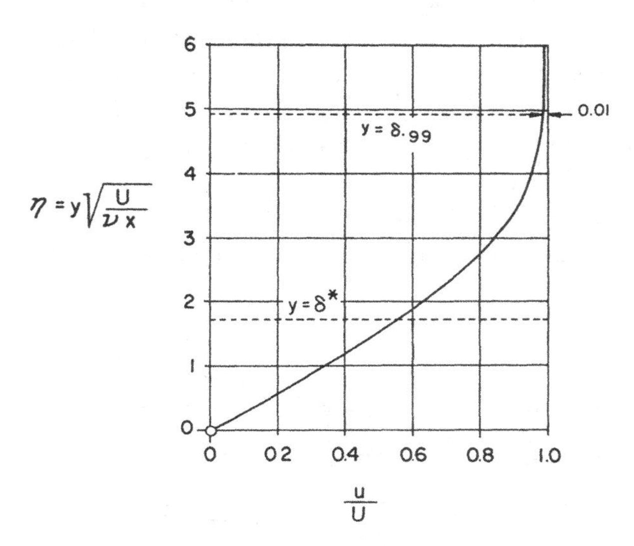

Figure 3.5
Velocity distribution in a Blasius laminar boundary layer. The total area under the curve is equal to the area under the dashed line, corresponding to the flux defect and the displacement thickness δ^*.

Physically, the displacement thickness is a useful concept since it defines a region of width equal to the retardation of fluid flux in the boundary layer, divided by the stream velocity U. Thus δ^* represents the effective amount by which the body is "thickened" because of the boundary layer, and in this sense adding δ^* to the body thickness will correct the body shape to account for viscous effects. Corresponding to this loss in flux will also be a loss in momentum, which defines the *momentum thickness*

$$\theta = \int_0^\infty \frac{u}{U}(1 - u / U)dy. \tag{57}$$

Numerical evaluation results in a value for the flat plate of

$$\theta = 0.664(\nu x / U)^{1/2}. \tag{58}$$

This description of the velocity distribution and boundary-layer thicknesses is independent of the x-coordinate and Reynolds number, by virtue of the nondimensionalization of the y-coordinate as shown in figure 3.5. Figure 3.6 shows these results in a dimensional form but with a greatly magnified lateral scale. Here the development of the boundary layer with distance downstream is more apparent. Upstream of the plate the velocity profile is uniform, whereas downstream of the leading edge the velocity profile of the boundary layer grows in the lateral direction proportional to $x^{1/2}$. The outer limit of the boundary layer is defined roughly by the 0.99 thickness parameter, whereas the displacement thickness represents the outward displacement of the inviscid streamlines. In this figure, the lateral scale is distorted by an amount that depends on the Reynolds number. To

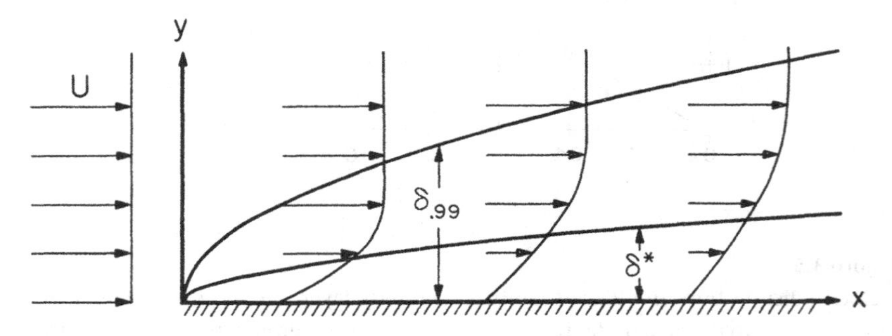

Figure 3.6
Development of laminar boundary layer along a flat plate. The lateral scale is magnified as explained in the text.

get some idea of the magnification, note that if the longitudinal length scale is as represented in this figure full-scale, and if water were flowing past with a velocity of 1 m/s, or 2 knots, the magnification factor for the lateral dimension would be about 1:40. In other words, the displacement thickness at the right-hand side of this figure would be about one quarter of a millimeter.

Finally, we can consider the drag force exerted on the plate by viscous shear stresses. The appropriate shear stress component is

$$\tau_{xy} = \mu(\partial u / \partial y)_{y=0} = \mu U(U / \nu x)^{1/2}(d^2 f / d\eta^2)_{\eta=0}, \tag{59}$$

or, after a calculation,

$$\tau_{xy} = 0.332\rho U^2 R_x^{-1/2}, \tag{60}$$

where $R_x = Ux/\nu$ is the *local* Reynolds number based on distance from the leading edge. The total drag, acting on one side of the plate, is

$$D = .332\mu bU(U / \nu)^{1/2}\int_0^l \frac{dx}{x^{1/2}} = .664b(\mu\rho l U^3)^{1/2}, \tag{61}$$

where b is the width of the plate. The resulting frictional drag coefficient is

$$C_F = \frac{D}{\frac{1}{2}\rho U^2(bl)} = 1.328(\nu / Ul)^{1/2} = 1.328R^{-1/2}. \tag{62}$$

This curve is shown in figure 2.3 along with experimental values that support the theory in the laminar regime. Experimental confirmation for the velocity distribution (figure 3.5) can be found in Schlichting (1968), for Reynolds numbers ranging from 10^5 to 7×10^5. For Reynolds numbers greater than 2×10^5 but less than about 3×10^6, the flow may be either laminar or turbulent, depending on the degree of roughness of the plate and also on the ambient turbulence level in the incident flow. When the flow becomes turbulent, the Blasius solution is no longer valid and must be replaced by one based upon the analysis of turbulent boundary layers.

3.15 Laminar Boundary Layers: Steady Two-Dimensional Flow

The analysis of the laminar boundary layer on a flat plate can be extended to more general steady flows past two-dimensional bodies, such as the flow past a streamlined strut of large aspect-ratio. If the radius of curvature of the body surface is large compared to the boundary-layer thickness, the same analysis carries over independent of the gradual change in direction due to the curvature of the body surface. The boundary-layer equation (46), together with the continuity equation (43), remains valid in this case provided that x and y are taken to be locally tangential and normal coordinates at each point of the body. However, the pressure gradient $\partial p/\partial x$ is now significant, since the inviscid flow will be affected by the body and will possess a nontrivial inviscid pressure distribution. Thus, one must first determine the inviscid flow past the body, compute the corresponding pressure gradient on the body surface, and then attempt to solve the boundary-layer equation (46) with the pressure gradient included. If more accuracy is desired, this procedure can be carried further by recalculating the inviscid flow, including the additional effects of the displacement thickness due to the first-order boundary layer, and then finding a second-order boundary-layer solution.

In view of this synthesis, we can discuss the qualitative features of the two-dimensional boundary-layer theory for general two-dimensional flow by discussing the flow past a flat plate with an arbitrary external pressure gradient imposed. For the usual situation of the flow past a streamlined body, the relevant pressure typically will be a maximum at the leading and trailing edges, where there are stagnation points, and will fall to a minimum value along the sides of the body where the velocity is a maximum. Thus, a typical pressure gradient will include a negative region over the forward portion of the body and a positive region over the afterbody.

The most significant feature of the pressure gradient is its effect upon the boundary layer and, especially, the circumstances under which it can lead to separation. If separation occurs at some point on the body surface, the streamlines will break away from the body at this point, enclosing a separated wake downstream of relatively high vorticity and low pressure. The separation point, if it exists, will be a stagnation point at which the flow from ahead and behind meets and abruptly changes direction. Ahead of this point the tangential velocity will be positive, and behind it the velocity will be negative, as shown in figure 3.7. It is clear from this figure that separation will occur when the shear stress is zero, or

$$(\partial u / \partial y)_{y=0} = 0. \qquad (63)$$

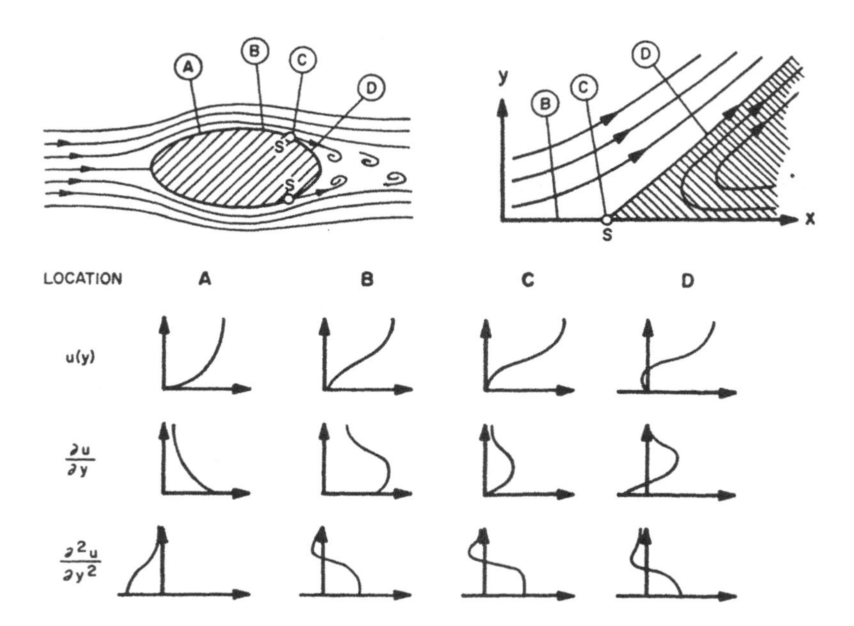

Figure 3.7
Flow past a body with separation. The second figure shows the local flow near the separation point, in terms of local coordinates that are tangential (x) and normal (y) to the body at the separation point. The graphs show the different velocity profiles, in the boundary layer, for a favorable pressure gradient on the forebody (A), an unfavorable pressure gradient on the afterbody (B), at the separation point (C), and downstream of the separation point (D). Note the qualitative similarity for each of these velocity profiles with the Couette flows shown in figure 3.3.

It is also apparent that a *necessary* condition is that

$(\partial^2 u / \partial y^2)_{y=0} > 0.$

This requirement can be related to the pressure gradient from the boundary-layer equation (46). For this purpose we set $y = 0$; thus $u = v = 0$, and we obtain from (46) the relation

$$\mu \frac{\partial^2 u}{\partial y^2} = \frac{\partial p}{\partial x} \quad \text{on} \quad y = 0. \tag{64}$$

This qualitative discussion is obviously not sufficient to determine the drag due to shear stresses acting on the body surface. Before describing an approximate computational method, we first note that from conservation of momentum in the boundary layer, the shear stress on the body surface can be related to the displacement thickness, momentum thickness, and the pressure gradient of the velocity field outside the boundary layer, in a simple manner due to von Kármán. For this purpose we integrate equation (46) in the normal direction, across the boundary layer:

$$\int_0^\delta \left(u\frac{\partial u}{\partial x} + v\frac{\partial u}{\partial y} \right) dy = -\frac{1}{\rho}\int_0^\delta \frac{\partial p}{\partial x} dy + \nu\int_0^\delta \frac{\partial^2 u}{\partial y^2} dy. \tag{65}$$

Here the upper limit of integration δ is somewhat arbitrary; ultimately we shall require only that it be sufficiently large that the velocity attains its asymptotic value $u = U$. From Bernoulli's equation (4.21) for the pressure outside the boundary layer,

$$p = -\frac{1}{2}\rho U^2 + \text{constant}. \tag{66}$$

Thus the pressure gradient (constant across the boundary layer) is

$$\frac{\partial p}{\partial x} = -\rho U \frac{dU}{dx}. \tag{67}$$

Now, regrouping the terms in (65),

$$\begin{aligned} \int_0^\delta \left(u\frac{\partial u}{\partial x} + v\frac{\partial u}{\partial y} - U\frac{dU}{dx} \right) dy &= \nu\int_0^\delta \frac{\partial^2 u}{\partial y^2} dy \\ &= \nu\left(\frac{\partial u}{\partial y}\right)_{y=\delta} - \nu\left(\frac{\partial u}{\partial y}\right)_{y=0} \\ &= -\tau_{xy} / \rho, \end{aligned} \tag{68}$$

since $\partial u/\partial y \to 0$ outside the boundary layer. Next we integrate the continuity equation in a similar manner to obtain

$$\int_0^y \left(\frac{\partial u}{\partial x} + \frac{\partial v}{\partial y} \right) dy = 0 \tag{69}$$

or, since $v = 0$ on the boundary $y = 0$,

$$v(x,y) = -\int_0^y \frac{\partial u}{\partial x} dy. \tag{70}$$

Substituting this expression for v in (68), and integrating by parts, we obtain

$$\begin{aligned}\frac{-\tau_{xy}}{\rho} &= \int_0^\delta \left[u\frac{\partial u}{\partial x} - \frac{\partial u}{\partial y}\left(\int_0^y \frac{\partial u}{\partial x}dy\right) - U\frac{dU}{dx}\right]dy \\ &= \int_0^\delta \left[u\frac{\partial u}{\partial x} + u\frac{\partial u}{\partial x} - U\frac{dU}{dx}\right]dy - U\int_0^\delta \frac{\partial u}{\partial x}dy \\ &= \int_0^\delta \left[2u\frac{\partial u}{\partial x} - U\frac{\partial u}{\partial x} - U\frac{dU}{dx}\right]dy \\ &= -\int_0^\delta \frac{\partial}{\partial x}[u(U-u)]dy - \int_0^\delta \frac{dU}{dx}(U-u)dy \\ &= -\frac{d}{dx}\int_0^\delta u(U-u)dy - \frac{dU}{dx}\int_0^\delta (U-u)dy.\end{aligned} \tag{71}$$

Recalling that the upper limit of integration δ is a point outside of the boundary layer, and using the definitions of the displacement and momentum thickness from (55) and (57), we obtain the following result:

$$\frac{\tau_{xy}}{\rho} = \frac{d}{dx}(U^2\theta) + U\delta^*\frac{dU}{dx}. \tag{72}$$

For the flat plate with zero pressure gradient, the last term of this equation is zero, and (72) can be integrated with respect to x to give a relation between the drag coefficient and the momentum thickness. This relation is consistent with equations (58) and (61) and expresses the physical relationship that exists between the hydrodynamic force acting on the plate and the loss of momentum in the fluid.

Finally, we shall describe a scheme, due to Pohlhausen, for computing the characteristics of a two-dimensional laminar boundary layer with a pressure gradient. We suppose that the velocity distribution in the boundary layer can be approximated by a fourth-degree polynomial, in the form

$$(u/U) = a(y/\delta) + (y/\delta)^2 + c(y/\delta)^3 + d(y/\delta)^4. \tag{73}$$

Here the condition that $u = 0$ on the boundary $y = 0$ has been satisfied by eliminating the constant term in the polynomial. The remaining four boundary conditions that can be imposed on physical grounds are that on the inner boundary $y = 0$, the boundary layer equation (46) holds, or

$$\nu\frac{\partial^2 u}{\partial y^2} = \frac{1}{\rho}\frac{\partial p}{\partial x} = -U\frac{dU}{dx}, \quad \text{on } y = 0,$$

and that, on the outer boundary,

$$u = U, \qquad \frac{\partial u}{\partial y} = 0, \qquad \frac{\partial^2 u}{\partial y^2} = 0, \quad \text{on } y = \delta.$$

(The first derivative $\partial u/\partial y$ must vanish if u is to approach U asymptotically, and the second derivative then vanishes as a consequence of equation (46).) These four conditions determine the four constants a, b, c, and d in equation (73) in terms of the thickness parameter δ. If we introduce the nondimensional parameter

$$\Lambda = \frac{\delta^2}{\nu}\frac{dU}{dx}, \tag{74}$$

the four coefficients in (73) are determined as

$$a = 2 + \Lambda/6, \quad b = -\Lambda/2, \quad c = -2 + \Lambda/2, \quad d = 1 - \Lambda/6.$$

The velocity distribution within the boundary layer is approximated then by a one-parameter family of curves that depends only on the *shape factor* Λ. A plot of this family appears in figure 3.8. Certain values of the shape factor Λ have special significance. The value $\Lambda = 0$ corresponds to a boundary layer with zero pressure gradient, i.e., the case of the flat plate treated in the preceding section. Positive values of Λ correspond to favorable (decreasing) pressure gradients as described in figure 3.8, and negative values of Λ correspond to unfavorable pressure gradients. The criterion (63) for inception of separation corresponds to the value $\Lambda = -12$; larger negative values suggest the reverse flow region further downstream. Positive values of Λ greater than 12 are excluded, since these imply a velocity within the boundary layer larger than that outside.

For different values of the shape factor Λ, values of the displacement and momentum thickness and of the shear stress can be computed with the following results:

$$\delta^* = \int_0^\delta \left(1 - \frac{u}{U}\right) dy = \delta\left(\frac{3}{10} - \frac{\Lambda}{120}\right), \tag{75}$$

$$\theta = \int_0^\delta \frac{u}{U}\left(1 - \frac{u}{U}\right) dy = \delta\left(\frac{37}{315} - \frac{\Lambda}{945} - \frac{\Lambda^2}{9{,}072}\right), \tag{76}$$

$$\tau_{xy} = \mu\left(\frac{\partial u}{\partial y}\right)_{y=0} = \frac{\mu U}{\delta}\left(2 + \frac{\Lambda}{6}\right). \tag{77}$$

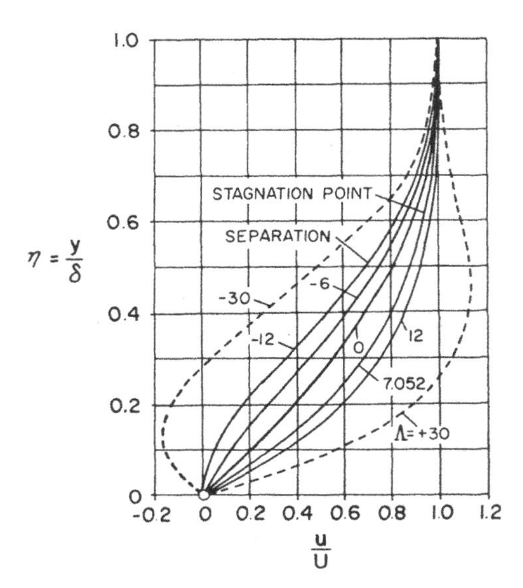

Figure 3.8
Pohlhausen velocity profiles, for various values of the shape factor Λ.

If these values are substituted in the momentum equation (72), we obtain a differential equation involving both $\Lambda(x)$ and $\delta(x)$, but the latter can be eliminated by using the definition (74). The final result of these substitutions is the following differential equation for Λ:

$$\frac{d\Lambda}{dx} = \frac{dU/dx}{U} g(\Lambda) + \frac{d^2U/dx^2}{dU/dx} h(\Lambda), \tag{78}$$

where

$$g(\Lambda) = \frac{7257.6 - 1336.32\Lambda + 37.92\Lambda^2 + 0.8\Lambda^3}{213.12 - 5.76\Lambda - \Lambda^2},$$

$$h(\Lambda) = \frac{213.12\Lambda - 1.92\Lambda^2 - 0.2\Lambda^3}{213.12 - 5.76\Lambda - \Lambda^2}.$$

Further progress requires that this differential equation be integrated numerically with respect to x, for any given inviscid velocity function $U(x)$, starting upstream at the stagnation point where the boundary layer commences on the body surface. At the stagnation point, we must have $U = 0$, while $dU/dx > 0$; to avoid a singularity at this point from the first term on the right side of equation (78), $g(\Lambda)$ must be zero. Of the three possible roots, the appropriate one is $\Lambda = 7.052$, and this criterion determines the initial value of the shape factor Λ for numerical integration of the differential equation (78). The corresponding initial velocity distribution is indicated in figure 3.8.

As a simple estimate of the accuracy of this method, the results may be compared with Blasius' solution for the flat plate, corresponding to the curve $\Lambda = 0$ in the above method. The displacement and momentum thicknesses from Pohlhausen's approximation are found to be

$$\delta^* = 1.75(\nu x / U)^{1/2}, \tag{79}$$

$$\theta = 0.685(\nu x / U)^{1/2}. \tag{80}$$

Thus, the error incurred in this case by assuming the polynomial profile (73) is, by comparison with equations (56) and (58), approximately 2 to 3 percent.

3.16 Laminar Boundary Layers: Closing Remarks

Three-dimensional laminar boundary layers are complicated not only by the changes in the external pressure gradient but also by cross-flow components in the third dimension within the boundary layer, and simple qualitative conclusions cannot be made. An exceptional case is the axisymmetric flow past a body of revolution, where a transformation due to Mangler reduces the continuity equation to its two-dimensional form and preserves the form of the boundary-layer equation. Unsteadiness is another complication, but it is of less practical concern for most problems of marine hydrodynamics. Discussions of these and other aspects of laminar boundary layers can be found in the references by Landau and Lifshitz (1959), Batchelor (1967), Schlichting (1968), Yih (1969), and White (1974).

3.17 Turbulent Flow: General Aspects

Turbulence is characterized by the superposition of a highly irregular and oscillatory velocity pattern upon an otherwise "smooth" flow. Thus, flows that would otherwise be steady in time and slowly varying in space become unsteady and rapidly fluctuating. If the Navier-Stokes equations are to remain valid under these conditions, the temporal and spatial variations of the velocity vector must be included. On the other hand, the principal characteristics of practical importance are quantities such as the mean velocity profile or the mean drag acting on a body, and this suggests that one should average the velocity, stresses, etc., with respect to space or time. With this in mind, we can separate the velocity vector into mean and fluctuating parts as follows:

$$u_i = \bar{u}_i + u_i'. \tag{81}$$

Here, a bar denotes the mean or average value, and u_i' the fluctuating component.

This average may be defined with respect to time, space, or an ensemble of occurrences. If the intent of separating turbulent and mean flow components is to be realized, the average with respect to time must be taken over a temporal period that is large compared to the characteristic period of turbulent oscillations but small compared to the time scale of the basic flow. Similarly, spatial averages must be carried out over length scales small compared to the length scale of the basic or mean flow. If the basic flow is steady-state, a time-average is the simplest concept to adopt.

If we substitute the velocity components into the Navier-Stokes and continuity equations and then average these equations, we can progress toward a decomposition of the overall problem. Before doing so, we recall some familiar averaging properties of oscillatory systems. First, the average of the velocity u_i is, by definition, $\bar{u}_i$. Thus, from (81),

$$\overline{u_i'} = 0, \tag{82}$$

with a similar result for derivatives of the velocity,

$$\overline{\frac{\partial u_i}{\partial x_j}} = \frac{\partial \bar{u}_i}{\partial x_j}; \qquad \overline{\frac{\partial u_i'}{\partial x_j}} = 0. \tag{83}$$

On the other hand, for a product of two velocities, or derivatives of the velocities, it is necessary to include the average of the products of the fluctuating components, such as

$$\overline{u_i u_j} = \overline{(\bar{u}_i + u_i')(\bar{u}_j + u_j')} = \bar{u}_i \bar{u}_j + \overline{u_i' u_j'}. \tag{84}$$

Now let us examine the consequences of substituting (81) in the continuity equation (15a),

$$\frac{\partial \bar{u}_i}{\partial x_i} + \frac{\partial u_i'}{\partial x_i} = 0. \tag{85}$$

If averages are taken, this equation applies independently to the mean velocity,

$$\frac{\partial \bar{u}_i}{\partial x_i} = 0, \tag{86}$$

and hence, by subtraction of (86) from (85), also to the fluctuating components:

$$\frac{\partial u_i'}{\partial x_i} = 0. \tag{87}$$

If we follow the same procedure with the Navier-Stokes equations (25), there is a contribution to the averaged equations from the nonlinear terms, involving the fluctuating components. Thus

$$\frac{\partial \bar{u}_i}{\partial t} + \bar{u}_j \frac{\partial \bar{u}_i}{\partial x_j} + \overline{u_j' \frac{\partial u_i'}{\partial x_j}} = -\frac{1}{\rho}\frac{\partial \bar{p}}{\partial x_i} + \nu \nabla^2 \bar{u}_i. \tag{88}$$

Using the continuity equation, we can transform the additional contribution represented by the third term in (88) as follows:

$$\begin{aligned} \overline{u_j' \frac{\partial u_i'}{\partial x_j'}} &= \frac{\partial}{\partial x_j} \overline{u_i' u_j'} - \overline{u_i' \frac{\partial u_j'}{\partial x_j}} \\ &= \frac{\partial}{\partial x_j} \overline{u_i' u_j'}. \end{aligned} \tag{89}$$

If this is transferred to the right side of (88), we obtain

$$\frac{\partial \bar{u}_i}{\partial t} + \bar{u}_j \frac{\partial \bar{u}_i}{\partial x_j} = -\frac{1}{\rho}\frac{\partial \bar{p}}{\partial x_i} + \nu \nabla^2 \bar{u}_i - \frac{\partial}{\partial x_j} \overline{u_i u_j}. \tag{90}$$

If (90) is regarded as an equation for the mean velocity $\bar{u}_i$, the only effect of the velocity fluctuations is to contribute an additional stress system. By comparison with Euler's equations (18), this *Reynolds stress* is equal to $-\rho \overline{u_i' u_j'}$. In other words, the total stress tensor can be written in the form

$$\bar{\tau}_{ij} = -\bar{p}\delta_{ij} + \mu\left(\frac{\partial \bar{u}_i}{\partial x_j} + \frac{\partial \bar{u}_j}{\partial x_i}\right) - \rho \overline{u_i' u_j'}. \tag{91}$$

Physically, the Reynolds stress can be identified with the transport of momentum associated with the fluctuating components of the velocity. If this stress were known, equation (91) could be solved for the mean velocity $\bar{u}_i$ in a manner analogous to the solution of the Navier-Stokes equations for laminar flow. However, there is no rational method of determining the Reynolds stress system, for example by imposing additional boundary conditions, and so it is necessary to resort to empirical methods based on experimental observations.

In general, Reynolds stresses are more important than the ordinary viscous stresses. This is apparent from (91) since the Reynolds number is inversely proportional to the viscosity coefficient μ. Thus, the Reynolds stress system will be dominant at high Reynolds numbers, except in the *viscous sublayer* very close to the boundaries, where the mean velocity gradient is large and, simultaneously, the fluctuating components are forced to vanish by the boundary condition $u_i' = 0$ on the wall.

The increased stress level due to the Reynolds stress suggests that the frictional drag on a body will be greater in the turbulent regime than is the case for laminar flow, and this is confirmed by experimental data, as shown in figure 2.3. An alternative physical explanation is suggested by the convection of momentum, represented by the Reynolds stress. Since this momentum convection will cause the velocity profile to become more uniform, as shown in figure 3.9, it must result ultimately in a larger velocity gradient and shear stress on the body surface.

3.18 Turbulent Boundary Layer on a Flat Plate

Now let us consider the characteristics of the mean velocity distribution for the flow in a turbulent boundary layer. To keep this very complex problem as simple as possible, we shall assume a flat plate with zero pressure gradient. Also, we shall be concerned only with the mean velocity $\bar{u}$, and the bar will be deleted with the understanding that the average value is implied.

The tangential velocity component u is assumed to be of the form

$$u = f(x, y, \rho, \mu, U), \tag{92}$$

where x and y are the usual space coordinates with the plate occupying the positive x-axis, ρ and μ are the density and viscosity of the fluid, and U is the velocity of the free stream outside the boundary layer. The subsequent

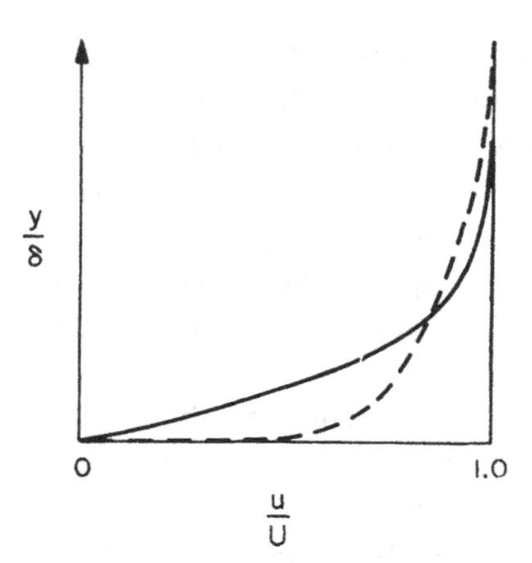

Figure 3.9
Comparison of laminar (———) and mean turbulent (-----) velocity profiles, for the boundary layer on a flat plate. Since the boundary layer thickness is substantially greater in the turbulent case, the difference in scales should be noted in this comparison.

development of nondimensional variables is facilitated if the coordinate x is replaced by the boundary-layer thickness δ, under the assumption that $d\delta/dx > 0$, so that to each value of x corresponds a unique value of δ and vice versa. The exact definition of the boundary-layer thickness is not important; it will be regarded somewhat loosely as the value of y where $u = U$, even though this equality is only asymptotic. Similarly, the velocity U will be replaced by the *friction velocity*,

$$u_\tau \equiv [\tau_{xy}(x,0)/\rho]^{1/2} \equiv (\tau_0/\rho)^{1/2}. \tag{93}$$

Again, to justify this· substitution of variables, we rely on a physical argument: for each value of U and fixed x, there is a unique value of the shear stress on the wall, and vice versa.[5] With these substitutions, the functional form of equation (92) is replaced by

$$u = f(\delta, y, \rho, \mu, u_\tau). \tag{94}$$

After nondimensionalization, it follows that

$$\frac{u}{u_\tau} = f(u_\tau y/\nu, y/\delta). \tag{95}$$

Since Reynolds stresses dominate viscous stresses except in the viscous sublayer, viscous stresses identified with the viscosity coefficient ν and hence the parameter $u_\tau y/\nu$ in (95), are significant only in the immediate vicinity of the boundary $y = 0$, where the turbulent fluctuations and associated Reynolds stress approach zero. In this viscous sublayer, $y/\delta \ll 1$ and (96) may be replaced by the *inner law* or *law of the wall*

$$\frac{u}{u_\tau} \simeq f\left(\frac{u_\tau y}{\nu}, 0\right) \equiv f_1\left(\frac{u_\tau y}{\nu}\right). \tag{96}$$

In other words, the nondimensional velocity very close to the wall depends only on the Reynolds number based on the friction velocity u_τ and the distance from the boundary y. This conclusion is not surprising since, in the viscous sublayer, the turbulent outer portion of the boundary layer is remote, and the boundary-layer thickness δ is relatively unimportant compared to the viscosity coefficient, shear stress, and distance from the wall.

On the wall $y = 0$, two boundary conditions are appropriate. First we impose the usual condition $u = 0$; hence $f_1(0) = 0$ in (96). In addition, it follows from the definition of the shear stress that

$$\mu\left[\frac{\partial u}{\partial y}\right]_{y=0} = \tau_0 = \mu u_\tau \left[\frac{df_1(u_\tau y/\nu)}{dy}\right]_{y=0} = \tau_0 f_1'(0). \tag{97}$$

When the shear stress is related to the friction velocity by means of (93), f_1 must have a linear behavior as $y \to 0$, of the form

$$f_1(u_\tau y/\nu) \simeq u_\tau y/\nu. \tag{98}$$

Now let us consider the outer region of the boundary layer, where the Reynolds stress dominates the viscous stress. In this region we expect the parameter y/δ to be important, in conjunction with the outer "boundary condition" that $u = U$ when $y/\delta = 1$. This suggests the appropriate form for the outer approximation of (95), or the *velocity-defect law*

$$\frac{U-u}{u_\tau} = f_2(y/\delta). \tag{99}$$

The velocity difference has been used to impose the outer boundary condition in the form

$$f_2(1) = 0. \tag{100}$$

Moreover, equation (99) reflects the significant fact that in the outer region, momentum flux is balanced by the Reynolds stress and the former depends not on the absolute velocity u but on the velocity defect $U - u$.

To amplify briefly on the justification for the velocity-defect law, let us view the flow in a frame of reference moving with the free stream. Here the relative fluid velocity is $u_r = u - U$, which can be expressed in the non-dimensional form (99). In these terms, U now enters only in the boundary condition on the plate $y = 0$, where $u_r = -U$. However, in terms of the flow in the outer region, the plate is deeply embedded within the viscous sublayer; its effect on the momentum balance in the outer region is represented by the shear stress at the wall, and hence by the friction velocity u_τ. On this basis the relative velocity u_r *in* the outer region must depend only on u_τ and y/δ, as assumed in (99).

At this point two complementary expressions have been obtained for the velocity in the turbulent boundary layer; the inner approximation (96) is valid in the viscous sublayer, and the outer approximation (99) is valid in the main portion of the boundary layer where the Reynolds stress is dominant. Neither of these is a complete solution, of course, but they at least represent simplifications of the overall problem compared with the more general velocity (95) which depends on two parameters. Moreover, further progress can be made with the apparently bold assumption that there exists an *overlap region* where both approximations are valid simultaneously—that is, where

$$f_1(u_\tau y / \nu) = U / u_\tau - f_2(y / \delta). \tag{101}$$

This seems at first to be an unjustifiable assumption, since the inner expression f_1 has been derived by neglecting the Reynolds stress, whereas the outer function on the right side of (101) is assumed to hold if and only if the Reynolds stress is dominant. The principal justification for (101) is that it leads to conclusions that compare well with experiments, namely a logarithmic velocity profile that will be derived later. First, however, a mathematical justification for (101) can be found from the concept of *matched asymptotic expansions*. Thus, the inner approximation should hold when $y/\delta \ll 1$, and the outer approximation if $u_\tau y/\nu \gg 1$. Can both these inequalities hold simultaneously for some suitable values of the distance from the wall y? In other words, can an intermediate value of y be found such that $\nu/u_\tau \ll y \ll \delta$? Clearly this is possible, provided $\nu/u_\tau \ll \delta$ or

$u_\tau\delta/\nu \gg 1$. The latter is a Reynolds number, which represents the ratio of inertial forces to viscous forces in the outer portion of the boundary layer. By assumption this Reynolds number is large, and hence an overlap region exists where (101) is valid.

It is now a simple matter to show that the functions f_1 and f_2 must be logarithmic in the overlap region. Differentiating both sides of (101) with respect to y and multiplying the result by y, we obtain

$$(u_\tau y / \nu) f_1'(u_\tau y / \nu) = -(y / \delta) f_2'(y / \delta), \tag{102}$$

where a prime denotes the derivative with respect to the appropriate argument of each function. The left side of (102) is a function only of the Reynolds number $u_\tau y/\nu$, while the right side depends only on the ratio y/δ. Since these two variables are independent, the functions on both sides of the equation can be equal to each other if and only if they are equal to a constant, say $\boldsymbol{A}$. Thus,

$$f_1'(u_\tau y / \nu) = A(\nu / u_\tau y) \quad \text{and} \quad f_2'(y / \delta) = -A(\delta / y)$$

or, after integrating,

$$f_1 = A \log(u_\tau y / \nu) + C_1, \tag{103}$$

$$f_2 = -A \log(y / \delta) + C_2, \tag{104}$$

where C_1 and C_2 are constants of integration.

These equations are valid only in the overlap region. Indeed, it would be remarkable if (103) and (104) represented the complete solution throughout the boundary layer, for they have been derived from the boundary conditions alone; the Navier-Stokes equations and Reynolds stress have been used only for qualitative arguments. Thus the inner velocity function f_1 cannot satisfy the inner boundary condition, $u = 0$ on $y = 0$, with any choice of the constants A and C_1, and the outer function f_2 cannot satisfy the outer condition $u = U$ when $y = \delta$, unless $C_2 = 0$. The latter condition was invoked originally by Prandtl, but this does not permit a satisfactory empirical fit within the overlap region.

Thus we have arrived at the conclusion, originally due to Prandtl and von Kármán, that the mean velocity profile is logarithmic in the intermediate or overlap region. Much effort has gone into corroborating these formulas and determining the constants A, C_1, and C_2. Surveys of these investigations and more complete expositions of the above derivation are given by Schlichting (1968) and by Monin and Yaglom (1971). Figures 3.10 and 3.11, reproduced from the latter reference, show comparisons with experiments for the inner and outer functions respectively; the constants there have been chosen empirically to give the following equations in place of (103) and (104):

$$\frac{u}{u_\tau} = 2.5\log(u_\tau y / \nu) + 5.1, \qquad u_\tau y / \nu > 30, \tag{105}$$

$$\frac{U - u}{u_\tau} = -2.5\log(y / \delta) + 2.35, \qquad y / \delta < 0.15. \tag{106}$$

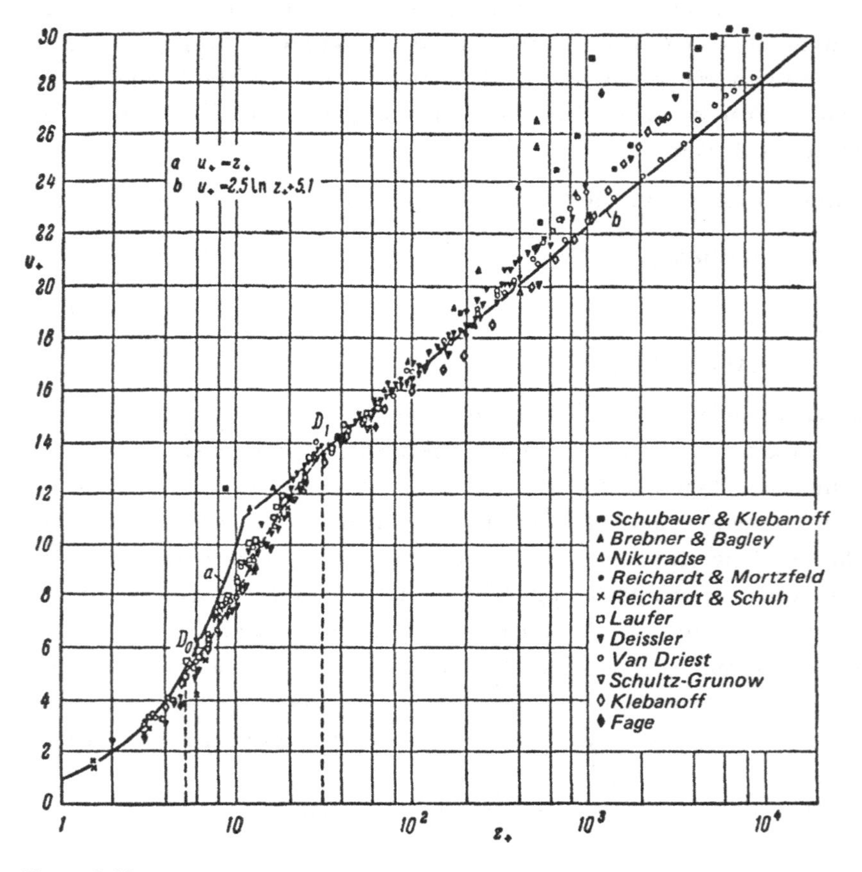

Figure 3.10
Universal dimensionless mean velocity profile of turbulent flow close to a smooth wall according to the data of tube-, channel- and boundary-layer measurements. Here $u_+ = u/u_\tau$ and $z_+ = u_\tau y/\nu$. (From Monin and Yaglom 1971)

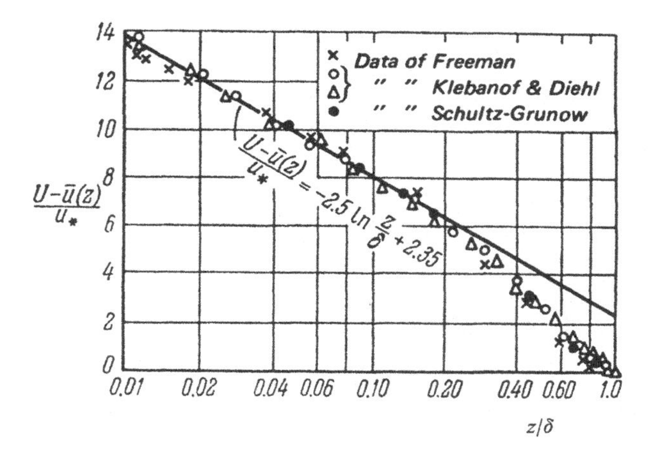

Figure 3.11
Logarithmic form of the velocity defect law for a boundary layer. Here $u_* = u_\tau$ and $z = y$. (From Monin and Yaglom 1971)

Natural logarithms are implied here and subsequently, except where explicitly stated in (115). Figures 3.10 and 3.11 both display the restriction that the logarithmic profiles are correct only in the intermediate region, and the respective domains of validity are given in (105–106). Figure 3.10 also shows the region of validity of the near-wall approximation (98), which is valid for $u_\tau y/\nu < 5$. The experimental data in figure 3.10 includes results not only for a flat plate boundary layer, but for various tubes and channels as well, since the inner velocity profile is the same for all these cases.

An equation for the friction velocity u_τ, and hence the shear stress on the wall, may be deduced by adding (105) and (106) to give the relation

$$\frac{U}{u_\tau} = A\log\frac{u_\tau\delta}{\nu} + (C_1 + C_2). \tag{107}$$

However, a second equation is needed to relate the two unknowns u_τ and δ. For this purpose, we recall the momentum equation (72), which was derived for a laminar boundary layer. It is possible to show that this same equation is valid for a turbulent boundary layer by including the Reynolds stress in the boundary-layer equations, provided the fluctuating components of the velocity satisfy the isotropic relation $\overline{u'^2} = \overline{v'^2}$. With this additional assumption, and deleting the pressure gradient term in (72), we get

$$\tau_0 = \rho U^2 d\theta / dx = \rho u_\tau^2, \tag{108}$$

where (93) has been used, and θ denotes the momentum thickness. The boundary-layer thicknesses can be calculated in terms of the two definite integrals

$$I = \int_0^1 f_2(y/\delta)d(y/\delta) \simeq \int_0^1 \left(\frac{U-u}{u_\tau}\right) d(y/\delta), \tag{109}$$

$$J = \int_0^1 [f_2(y/\delta)]^2 d(y/\delta) \simeq \int_0^1 \left(\frac{U-u}{u_\tau}\right)^2 d(y/\delta). \tag{110}$$

Except for the comparatively thin viscous sublayer, these approximations are valid, and the integrals in (109–110) must have universal constant values independent of the Reynolds number. Thus, from the definitions (55) and (57),

$$\delta^* / \delta = I(u_\tau / U), \tag{111}$$

$$\theta / \delta = I(u_\tau / U) - J(u_\tau / U)^2. \tag{112}$$

Equations (107), (108), and (112) give a total of three equations for the three unknowns u_τ, δ, and θ; these can be solved, without further assumptions, in terms of the constants that can be determined empirically. One of the principal results of this procedure is the *local* frictional-drag coefficient $c_f = \tau_0 \Big/ \frac{1}{2}\rho U^2$, which satisfies the implicit equation

$$\frac{1}{\sqrt{c_f}} = 2^{-1/2} A \log(R_x c_f) + C_3, \tag{113}$$

where $R_x = Ux/\nu$ is the local Reynolds number. The total drag coefficient

$$C_F = \frac{1}{l}\int_0^l c_f(x)dx \tag{114}$$

can be determined from a similar equation, and this is the basis for Schoenherr's flat-plate frictional-drag formula

$$\frac{1}{\sqrt{C_F}} = 1.79 \log(R_l C_F) = 4.13 \log_{10}(R_l C_F). \tag{115}$$

This important result is displayed in figure 2.3, which also shows that experimental confirmation exists over a broad range of Reynolds numbers.

The lower limit for this range of validity is where transition from laminar flow occurs, generally between $R_l = 2 \times 10^5$ and 2×10^6. The upper limit is undetermined; the maximum Reynolds number where experimental data exist is about 4.5×10^8. In this range of two to three decades, the Schoenherr formula (115) is apparently a satisfactory fit to the experimental data. Nevertheless, some uncertainty remains from the standpoint of ship-resistance predictions, insofar as large, high-speed ships operate at Reynolds numbers between 10^9 and 10^{10}. The Schoenherr formula also has been criticized at the lower limit of Reynolds numbers, where towing-tank tests are carried out on small ship models. This subject is discussed by Matheson and Joubert (1974) and by Granville (1975).

3.19 The 1/7-Power Approximation

A simpler velocity distribution, but one that has less empirical support and scientific motivation than the logarithmic profile, is the 1/7-power relation

$$u / u_\tau = 8.7(u_\tau y / \nu)^{1/7}. \tag{116}$$

Despite its shortcomings, this formula is useful as a basis for qualitative conclusions. For example, if we invoke the condition that $u = U$ when $y = \delta$ and solve for the shear stress on the boundary, it follows that

$$\tau_0 = \rho u_\tau^2 = 0.0227 \rho U^2 (U\delta / \nu)^{-1/4}. \tag{117}$$

A direct calculation of the momentum thickness gives

$$\begin{aligned} \theta = \int_0^\delta \frac{u}{U}(1 - u / U)dy &= \int_0^\delta (y / \delta)^{1/7}[1 - (y / \delta)^{1/7}]dy \\ &= (7/8 - 7/9)\delta = 0.0972\delta. \end{aligned} \tag{118}$$

Then from (108) and (117), it follows that δ must satisfy the differential equation

$$0.0972 \frac{d\delta}{dx} = 0.0227(U\delta / \nu)^{-1/4}. \tag{119}$$

Multiplying both sides by $\delta^{1/4}$, and integrating, we have

$$\delta^{5/4} = 0.292(\nu / U)^{1/4} x + C, \tag{120}$$

where the constant of integration must be determined from the initial thickness δ at the point of transition from laminar to turbulent flow. If the Reynolds number is sufficiently high, or if turbulence is induced near the leading edge by studs or other devices, then we can set $C = 0$ in (120) and obtain the relations

$$\delta = 0.373\, x R_x^{-1/5}, \tag{121}$$

$$\theta = 0.0363\, x R_x^{-1/5}, \tag{122}$$

$$\tau_0 = 0.0290 \rho U^2 R_x^{-1/5}, \tag{123}$$

$$\delta^* = 0.0467\, x R_x^{-1/5}. \tag{124}$$

Here $R_x = Ux/\nu$, and the last result for the displacement thickness δ^* follows from a straightforward integration from 0 to δ using (55).

By comparison with the laminar boundary layer, which increases in thickness with distance downstream at a rate proportional to $x^{1/2}$, the turbulent boundary layer has a faster rate of growth, proportional to $x^{4/5}$ according to the 1/7-th power assumption.

3.20 Roughness Effects on Turbulent Boundary Layers

So far we have assumed that the flat plate is *hydraulically smooth.* Marine vessels may be roughened either by features of their fabrication, such as welding seams or rivet heads, or by the inevitable marine fouling that accumulates in service. Thus, it is appropriate to discuss the effects of roughness and the extent to which a distinction can be made between rough and smooth surfaces.

To introduce roughness into the boundary-layer theory, we must define a scale of roughness, say the height k of the roughness elements. The significance of the roughness can be estimated then in terms of the magnitude of k relative to the dimensions of the boundary layer. If k is small compared to the height of the viscous sublayer, the plate is effectively smooth and the frictional drag independent of k. In the other extreme, if k is large compared to δ, the boundary layer will be unimportant by comparison to the form drag resulting from the roughness elements, which in turn will be essentially independent of the Reynolds number.

For a fixed scale of roughness, in terms of the characteristic length l of the plate, k/l = constant. In this case the frictional drag is identical to that of a smooth plate below a Reynolds number which depends on the roughness ratio; above this point, the frictional drag is effectively constant, as shown in figure 3.12. From this result we can infer that smoothness of a full-scale ship hull is much more important than smoothness of its model, for if the two had the same relative roughness or equal values of k/l, the effects of the roughness would increase with Reynolds number.

On the other hand, for constant values of the Reynolds number based on the roughness scale, Uk/ν, corresponding to k/δ = constant, there is a family of curves approximately parallel to the smooth curve. The latter are more representative of the ship roughness, insofar as the roughness scale of ship hulls of different lengths is effectively independent of the length. In this sense a *roughness allowance* can be approximated by a constant additive resistance coefficient such as 0.0004, as discussed in chapter 2.

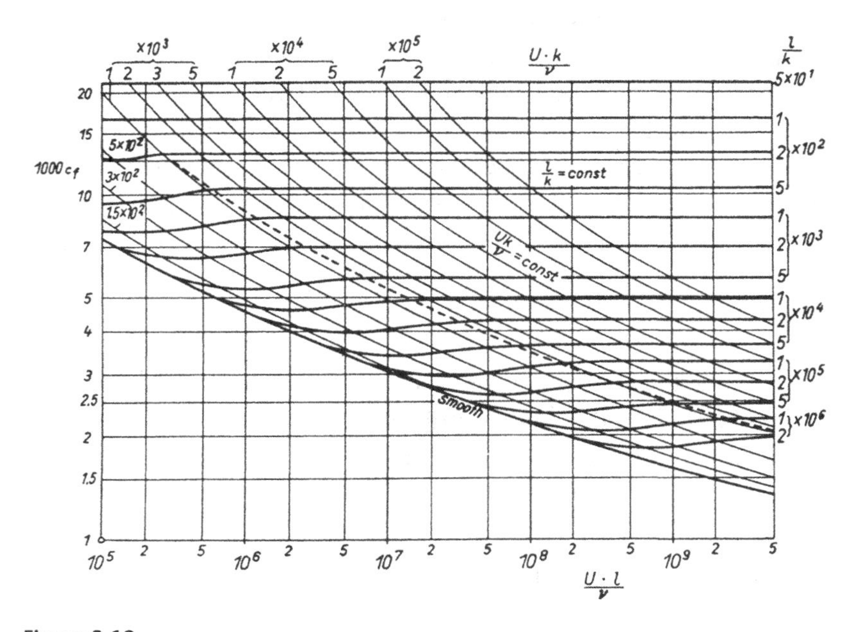

Figure 3.12

Resistance formula of sand-roughened plate; coefficient of total skin friction. (From Schlichting 1968)

3.21 Turbulent Boundary Layers: Closing Remarks

Semiempirical methods have been developed for analyzing the effects of a pressure gradient. A critique of this subject is given by Thompson (1967). Qualitatively, the effects of the pressure gradient are analogous to the case of a laminar boundary layer. In particular, negative and positive values of the pressure gradient in a streamwise direction along the body surface are respectively favorable and unfavorable for the stability of a thin boundary layer, as compared with the occurrence of separation. Turbulence serves the useful role of convecting more momentum from the free stream into the boundary layer, by comparison with the laminar case, and thus the separation point is delayed. In this connection, we recall the drag coefficient of a sphere, shown in figure 2.2, which is decreased by the effects of turbulence at the transition Reynolds number of approximately 400,000. Similar remarks apply to the development of separation or stall on lifting surfaces; even in the normal regime of small angles of attack, the pressure gradient can significantly affect the turbulent boundary layer.

Semiempirical extensions of the turbulent boundary-layer theory have been developed to account for the presence of external pressure gradients and of three-dimensional effects. Extensive discussion of these topics is given in the monograph by Cebeci and Smith (1974). Applications to the calculation of the viscous drag on a ship hull are discussed by Huang and von Kerczek (1972) and Kux (1974).

Problems

1. The Couette flow shown in figure 3.3 is a rough model for the flow in a lubricated bearing. Find the bearing friction per unit area, in terms of the viscosity of the lubricant and the gap width of the bearing. How does the bearing friction change if there is a pressure gradient in the direction of motion?

2. The simplest equation of motion for the spar buoy shown in figure 2.16 is $m\ddot{y} + \rho g S_0(y - \boldsymbol{A} \cos \omega t) = 0$ where $y(t)$ is the heave amplitude, S_0 is the waterplane area, and $\boldsymbol{A}$ is the wave amplitude. Use the solution in section 3.13 for sinusoidal motion of a flat plate to derive a linear damping coefficient proportional to the submerged area of the cylindrical portion of the spar buoy. Using this estimate of the damping coefficient, find the resonant

response for a spar buoy of diameter 1m and draft 5 m in waves of amplitude 0.5 m.

3. Modify the analysis of section 3.10 for the case where the upper boundary is a free surface with no stress. Include the component of the hydrostatic pressure acting in the direction of the flow and find the surface velocity of a fluid of depth 1cm flowing down an inclined plane of angle 1 degree. Based on this steady-state solution, what is the surface current in the Cape Cod Canal, where the depth is 10 m, the length is 10 km, and the maximum difference in elevation between the two ends is 3 m? Give several reasons why this estimate is exaggerated.

4. Reformulate the problem of the Cape Cod Canal as stated in problem 3, assuming that this motion is sinusoidal in time with period 12 hours and that the velocity is horizontal and a function only of the vertical coordinate. Show that the solution is of the form

$$u(y,t) = \mathrm{Re}[e^{i\omega t}(ae^{ky} + be^{-ky} + c)].$$

Determine the constants in this equation, the maximum velocity on the surface, and the phase shift between the current and the tidal difference at the two ends. (The actual current in this canal is about five knots.)

5. For a Reynolds number of 10^6, compute the displacement thickness and shear stress at the boundary for laminar (Blasius) and turbulent (1/7 power) boundary layers. Use these results to sketch the two velocity profiles, as was done in figure 3.9, but with the vertical scale (y) preserved in both curves.

6. For a 2 m ship model moving at 1.0 m/s, estimate the displacement thickness of the boundary layer at the stern, using the Blasius result for Laminar flow, and the 1/7 power law for turbulent flow. Compare the resulting values of δ^*/l with the corresponding turbulent result for a 200 m ship moving at 10 m/s.

7. Given the differences between salt and fresh water at the same temperature, how would the boundary-layer thickness and shear stress of a ship change as it moves from salt water into fresh water, if the ship's velocity and the water temperature are unchanged?

8. A common speed-measuring device for small boats is a paddlewheel that protrudes about 1cm from the hull surface and rotates about an axis flush

with the hull surface in response to the fluid moving past the hull. If such a paddlewheel is mounted 3 m downstream from the bow, estimate the linearity of this device over speeds ranging between 1 and 10 m/s.

9. Comment on the validity of the following statement: Barnacles and similar marine growth are unimportant on the surface of a ship, since the size of these organisms is negligible relative to the size of the ship. Using figures 2.11 and 3.12, estimate the effects of barnacles, of typical scale $k = 1$ mm, on the full-scale resistance of the *Lucy Ashton* at a speed of 15 knots. How much will this speed be reduced assuming equal total drag in the smooth and rough conditions?

References

Abramowitz, M., and I. A. Stegun, eds. 1964. *Handbook of mathematical functions.* Washington: U.S. Government Printing Office.

Bark, F. H., E. J. Hinch, and M. T. Landahl. 1975. Drag reduction in turbulent flow due to additives: A report on Euromech 52. *Journal of Fluid Mechanics* 68:129–138.

Batchelor, G. K. 1967. *An introduction to fluid dynamics.* Cambridge: Cambridge University Press.

Cebeci, T., and A. M. O. Smith. 1974. *Analysis of turbulent boundary layers.* New York: Academic.

Granville, P. S. 1975. *Drag and turbulent boundary layer of flat plates at low Reynolds numbers.* David W. Taylor Naval Ship Research and Development Center Report 4682.

Huang, T. T., and C. H. von Kerczek. 1972. Shear stress and pressure distribution on a surface ship model: Theory and experiment. In *Ninth symposium on naval hydrodynamics,* eds. R. Brard and A. Castera, pp. 1963–2010. Washington: U.S. Government Printing Office.

Kux, J. 1974. Three-dimensional turbulent boundary layers. In *Tenth symposium on naval hydrodynamics,* eds. R. D. Cooper and S. W. Doroff, pp. 685–703. Washington: U.S. Government Printing Office.

Landau, L. D., and E. M. Lifshitz. 1959. *Fluid mechanics. English translation.* London: Pergamon.

Lumley, J. L. 1969. Drag reduction by additives. *Annual Review of Fluid Mechanics* 1:367–384.

Matheson, N., and P. N. Joubert. 1974. A note on the resistance of bodies of revolution and ship forms. *J. Ship Res.* 18:153–168.

Monin, A. S., and A. M. Yaglom. 1971. *Statistical fluid mechanics. English translation.* Cambridge, Mass.: The MIT Press.

Schlichting, H. 1968. *Boundary layer theory. English translation.* 6th ed. New York: McGraw-Hill.

Thompson, B. G. J. 1967. *A critical review of existing methods of calculating the turbulent boundary layer.* Ministry of Aviation, Aeronautical Research Council Reports and Memoranda 3447.

White, F. M. 1974. *Viscous fluid flow.* New York: McGraw-Hill.

Yih, C. S. 1969. *Fluid mechanics.* New York: McGraw-Hill.

4 The Motion of an Ideal Fluid

Having obtained an understanding of viscous effects, we can justify their neglect in certain circumstances. Thus, when the boundary layer is thin, the bulk of the fluid is effectively inviscid or *ideal* and can be analyzed from the Navier-Stokes equations with the viscous stress tensor deleted. At first glance the resulting simplification of these equations is less apparent than might be expected from physical considerations. Ultimately, the mathematical simplifications are considerable, and a wide variety of inviscid flow problems are amenable to solution.

The fundamental equations are the continuity equation and Euler's equations, derived in chapter 3. Thus, the velocity vector $\mathbf{V} = (u_1, u_2, u_3)$ must satisfy the equations

$$\partial u_i / \partial x_i = 0, \tag{1}$$

and

$$\frac{\partial u_i}{\partial t} + u_j \frac{\partial u_i}{\partial x_j} = -\frac{1}{\rho}\frac{\partial p}{\partial x_i} + \frac{1}{\rho}F_i, \qquad i = 1,2,3. \tag{2}$$

In the second equation only the normal pressure stress is included, on the assumption that the fluid is inviscid.

Hereafter, the external force field F_i will be assumed to consist only of the gravitational force ρg which acts vertically downward. If x_2 is the vertical coordinate and is positive in the upward direction, then $F_i = (0, -\rho g, 0)$. Under these circumstances equation (2) can be written as

$$\frac{\partial u_i}{\partial t} + u_j \frac{\partial u_i}{\partial x_j} = -\frac{1}{\rho}\frac{\partial}{\partial x_i}(p + \rho g x_2). \tag{3}$$

In the simplest case of hydrostatic equilibrium, the velocity is zero (or a constant) and $p + \rho g\, x_2$ is a constant.

4.1 Irrotational Flows

To simplify these equations, the class of possible fluid motions must be restricted. For this purpose we define the *circulation* Γ as the integrated tangential velocity around any closed contour C in the fluid,

$$\Gamma = \int_C u_i dx_i. \tag{4}$$

Kelvin's theorem of the conservation of circulation states that for an ideal fluid acted upon by conservative forces (e.g., gravity) the circulation is constant about any closed material contour moving with the fluid. Physically, this happens because no shear stresses act within the fluid; hence it is impossible to change the rotation rate of the fluid particles. Thus, any motion that started from a state of rest at some initial time, will remain *irrotational* for all subsequent times, and the circulation about any material[1] contour will vanish.

To prove Kelvin's theorem, we consider the derivative of (4) with respect to time, $d\Gamma/dt$. Since the contour C moves with the fluid particles, or with velocity **V**, it follows that

$$\frac{d\Gamma}{dt} = \frac{d}{dt}\int_C u_i dx_i = \int_C \left(\frac{\partial}{\partial t} + u_j \frac{\partial}{\partial x_j}\right)(u_i dx_i). \tag{5}$$

Here the differential operator acting upon the integrand is the substantial derivative D/Dt defined in equation (3.19), since the contour of integration C is a material contour moving with the fluid particles. The resulting derivatives of the velocity components u_i are straightforward to compute, but some care is required for the differential element dx_i. Since C is a material contour, the differential element will itself be a function of time; to analyze its resulting distortion, we resort to the definition of the integral as the limit of a finite sum. Thus, if $x_i^{(n)}$ denotes the coordinates of the nth point along the contour C, say with $n = 1, 2, \ldots, N$, (5) can be replaced by

$$\frac{d\Gamma}{dt} = \lim_{N\to\infty} \sum_n \left(\frac{\partial}{\partial t} + u_j \frac{\partial}{\partial x_j}\right) u_i \delta x_i^{(n)}, \tag{6}$$

provided that as $N \to \infty$,

$$\delta x_i^{(n)} \equiv x_i^{(n+1)} - x_i^{(n)} \to 0.$$

Since the coordinates $x_i^{(n)}$ move with individual fluid particles, they must be functions of time. By definition the velocity components at the same points are given by

$$u_i^{(n)} = \partial x_i^{(n)} / \partial t. \tag{7}$$

Using the chain rule in (6), noting that the coordinates $x_i^{(n)}$ depend on time but not on the space coordinates, and finally using (7), we obtain

$$\frac{d\Gamma}{dt} = \lim_{N\to\infty} \sum_n \left[\left(\frac{\partial u_i}{\partial t} + u_j \frac{\partial u_i}{\partial x_j} \right) \left(x_i^{(n+1)} - x_i^{(n)} \right) + u_i \left(u_i^{(n+1)} - u_i^{(n)} \right) \right]. \tag{8}$$

In this form, the limit is given by the integrals

$$\frac{d\Gamma}{dt} = \int_C \left(\frac{\partial u_i}{\partial t} + u_j \frac{\partial u_i}{\partial x_j} \right) dx_i + \int_C u_i du_i, \tag{9}$$

or, using the equations of motion (3),

$$\frac{d\Gamma}{dt} = -\frac{1}{\rho} \int_C \frac{\partial}{\partial x_i} (p + \rho g x_2) dx_i + \frac{1}{2} \int_C d(u_i u_i). \tag{10}$$

The right side of equation (10) is the integral of a perfect differential over a closed contour and is, therefore, equal to zero. In effect, the integral can be integrated to give

$$-\frac{1}{\rho} \left[p + \rho g x_2 - \frac{1}{2} \rho u_j u_j \right]_A^B,$$

where B and A are the upper and lower limits of integration; since these points are identical, the difference $[\quad]_A^B = 0$.

Since $d\Gamma/dt = 0$, the circulation must be a constant. We shall assume hereafter that the fluid motion started from an initial state of rest, say when $t = -\infty$, so that $\Gamma = 0$ for all time and for all material contours within the fluid.

Finally, we invoke Stokes' theorem for a continuous differentiable vector $\mathbf{V}$,

$$\iint_S (\nabla \times \mathbf{V}) \cdot d\mathbf{S} = \int_C \mathbf{V} \cdot d\mathbf{x}, \tag{11}$$

where the surface integral is taken over any surface S bounded by the contour C. The line integral is by definition the circulation, which must equal zero for any material contour lying within the fluid. Thus, the surface integral must vanish for any surface S, situated in the fluid, and this can be true only if the integrand is identically zero. Hence the motion of the fluid is *irrotational*,

$$\nabla \times \mathbf{V} = 0, \tag{12}$$

or, in Cartesian components,

$$\frac{\partial u}{\partial y} - \frac{\partial v}{\partial x} = 0, \quad \frac{\partial v}{\partial z} - \frac{\partial w}{\partial y} = 0, \quad \frac{\partial w}{\partial x} - \frac{\partial u}{\partial z} = 0. \tag{12a}$$

The left side of (12) is the *vorticity*, or the curl of the velocity, and (12a) gives the Cartesian components of the vorticity. To say that the flow is irrotational is equivalent to saying that the vorticity is zero throughout the fluid. This conclusion is important because an irrotational vector can be represented as the gradient of a scalar. This statement is the result of Helmholtz' theorem in vector analysis, which states that any continuous and finite vector field can be expressed as the sum of the gradient of a scalar function ϕ plus the curl of a zero-divergence vector $\mathbf{A}$. The vector $\mathbf{A}$ vanishes identically if the original vector field is irrotational. The general proof of this theorem can be found in Morse and Feshbach (1953). For our present purposes, a simpler approach will suffice to show that if the velocity field is irrotational, it can be represented simply as the gradient of the scalar function ϕ, or the *velocity potential.*

4.2 The Velocity Potential

Here we wish to show that if the fluid motion is irrotational, the velocity vector $\mathbf{V}$ can be represented by the gradient of a scalar potential ϕ which will depend generally on the three space coordinates x_i and time t. Replacing the velocity by its potential may at first seem an unnecessary complication, since we can envisage the velocity and measure it with suitable instruments in the laboratory, whereas the velocity potential is no more than a mathematical abstraction. However, the velocity is a vector quantity with three unknown scalar components, whereas the velocity potential is a single scalar unknown from which all three velocity components may be computed.

Before proceeding to exploit the velocity potential, we must prove first that it is indeed a valid representation of the unknown velocity vector. Our proof will display clearly the importance of assuming that the motion is irrotational. In particular, let us consider the definite integral

$$\phi(\mathbf{x},t) = \int_{\mathbf{x}_0}^{\mathbf{x}} u_i dx_i, \tag{13}$$

where the lower limit is some arbitrary constant position $\mathbf{x}_0$ and the upper limit is the point $\mathbf{x} = (x_1, x_2, x_3)$. This integral is independent of the particular path of integration between the points $\mathbf{x}_0$ and $\mathbf{x}$, since the difference in the value of any two integrals, between the same two points, is equal to the circulation around the closed path from $\mathbf{x}_0$ to $\mathbf{x}$ along one path and back to $\mathbf{x}_0$ along the other path, which is equal to zero if the fluid motion is irrotational. Thus, the integration in (13) can be performed along any desired path. If we choose a path that approaches the point $\mathbf{x}$ along a straight line parallel to the x_1-axis, then along this final portion of the path of integration $u_i dx_i = u_1 dx_1$ so that

$$\frac{\partial \phi}{\partial x_1} = \frac{\partial}{\partial x_1} \int_{\mathbf{x}_0}^{\mathbf{x}} u_1 dx_1 = u_1. \tag{14}$$

The remaining portion of the integral, being a constant, does not contribute to the derivative. Applying a similar argument for the other two coordinates, we have in general that

$$u_i = \partial \phi / \partial x_i, \tag{15}$$

or

$$\mathbf{V} = \nabla \phi. \tag{15a}$$

Thus, one can assume the existence of a velocity potential, subject only to the requirement that the fluid motion is irrotational.

If (15) is substituted for the velocity vector in the continuity equation (1), it follows that

$$\frac{\partial}{\partial x_i} \frac{\partial \phi}{\partial x_i} = \frac{\partial^2 \phi}{\partial x_i \partial x_i} = 0, \tag{16}$$

or

$$\nabla^2\phi \equiv \frac{\partial^2\phi}{\partial x^2}+\frac{\partial^2\phi}{\partial y^2}+\frac{\partial^2\phi}{\partial z^2}=0. \tag{16a}$$

This is the *Laplace equation* which expresses conservation of fluid mass for potential flows and provides the governing partial differential equation to be solved for the function ϕ.

It can be confirmed that the fluid motion defined by the velocity (15) is irrotational, by recalling from vector analysis that the curl of a gradient is identically zero or simply by substituting (15) in (12).

4.3 Bernoulli's Equations

We have not yet exploited Euler's equations or conservation of momentum, except to prove Kelvin's theorem. Nor have we discussed the pressure, which is related to the fluid velocity by the three Euler equations (2).

In two important cases the Euler equations can be integrated to give an explicit equation for the pressure known as the *Bernoulli integral*, or simply Bernoulli's equation. The first result is perhaps the most familiar one from elementary courses in fluid mechanics and pertains to steady flow without assumption about the vorticity. The second case is irrotational flow, but it includes the possibility of unsteadiness. The two proofs are different and will be presented separately.

First we consider steady, but possibly rotational, flow. Here Euler's equations take the form

$$u_j\frac{\partial u_i}{\partial x_j}=-\frac{\partial}{\partial x_i}(p/\rho+gx_2). \tag{17}$$

Multiplying both sides by the velocity components u_i and summing over i gives

$$u_iu_j\frac{\partial u_i}{\partial x_j}=-u_i\frac{\partial}{\partial x_i}(p/\rho+gx_2). \tag{18}$$

The left side of this equation can be rewritten as

$$\frac{1}{2}u_j\frac{\partial}{\partial x_j}u_iu_i=\frac{1}{2}u_i\frac{\partial}{\partial x_i}u_ju_j. \tag{19}$$

In the last step, we have interchanged the two indices *i* and *j*; this has no effect on the summation over both. Combining the last two equations and collecting terms on one side gives the result

$$u_i \frac{\partial}{\partial x_i}\left(p/\rho + \frac{1}{2} u_j u_j + g x_2 \right) = 0. \tag{20}$$

The differential operator $u_i \partial/\partial x_i$ in (20) is the steady form of the substantial derivative, or the rate of change following a fluid particle with velocity **V**. Thus along any *streamline*, whose tangent is everywhere parallel to **V**, the quantity in parentheses in (20) is constant, and

$$p = -\frac{1}{2}\rho V^2 - \rho g x_2 + C, \tag{21}$$

where

$$V^2 = u_i u_i = u^2 + v^2 + w^2. \tag{22}$$

In this result the constant C may vary for different streamlines, and a steady motion independent of time has been assumed.

The alternative form of Bernoulli's equation, valid for unsteady irrotational flows, is more useful since the flow of an inviscid fluid is generally irrotational but may be unsteady. In this case we utilize the velocity potential, substituting (15) in the general form of Euler's equations (3),

$$\frac{\partial}{\partial t}\frac{\partial \phi}{\partial x_i} + \frac{\partial \phi}{\partial x_j}\frac{\partial}{\partial x_j}\frac{\partial \phi}{\partial x_i} = -\frac{1}{\rho}\frac{\partial}{\partial x_i}(p + \rho g x_2). \tag{23}$$

Direct differentiation shows that

$$\frac{\partial \phi}{\partial x_j}\frac{\partial}{\partial x_j}\frac{\partial \phi}{\partial x_i} = \frac{1}{2}\frac{\partial}{\partial x_i}\frac{\partial \phi}{\partial x_j}\frac{\partial \phi}{\partial x_j}. \tag{24}$$

Thus (23) can be rewritten in the form of a perfect differential

$$\frac{\partial}{\partial x_i}\left(\frac{\partial \phi}{\partial t} + \frac{1}{2}\frac{\partial \phi}{\partial x_j}\frac{\partial \phi}{\partial x_j} \right) = -\frac{1}{\rho}\frac{\partial}{\partial x_i}(p + \rho g x_2). \tag{25}$$

Integrating this with respect to the three space variables x_i gives

$$\frac{\partial \phi}{\partial t} + \frac{1}{2}\frac{\partial \phi}{\partial x_j}\frac{\partial \phi}{\partial x_j} = -\frac{1}{\rho}(p + \rho g x_2) + C(t). \tag{26}$$

The "constant" in (26) is independent of the space coordinates but may depend on time. This constant can be absorbed into the velocity potential by redefining the latter as

$$\phi' = \phi - \int^{t} C(t)dt. \tag{27}$$

Since the difference between ϕ and ϕ' is a function only of time, the gradient is not affected and either of these functions is an equally suitable velocity potential from the standpoint of (15). Substituting (27) in (26) leaves the form of Bernoulli's equation unchanged in terms of the velocity potential ϕ', but the constant $C(t)$ no longer appears. In other words, the value of $C(t)$ is related to a corresponding constant in the velocity potential which has no effect on the velocity vector, and $C(t)$ may be chosen arbitrarily. $C(t)$ can be set equal to zero and deleted entirely from Bernoulli's equation, or it can be set equal to some desired value such as the atmospheric pressure or other relevant ambient pressure. In the remainder of this chapter, we shall delete $C(t)$ entirely.

Both forms of Bernoulli's equation, (21) and (26), include in common the dynamic pressure $-\frac{1}{2}\rho V^2$ and the hydrostatic pressure $-\rho g x_2$. The former expresses that the pressure is reduced in regions of high velocity in accordance with the well-known Venturi effect. The hydrostatic term, anticipated in section 3.9, represents the simple increase in pressure with depth, equal to the static weight per unit area of the fluid above this point.

4.4 Boundary Conditions

The distinction between different types of fluid motions results from the conditions imposed on the boundaries of the fluid domain. Thus we must relate the physical conditions on these boundaries to appropriate mathematical statements.

As in chapter 3, two different types of boundary conditions must be discussed: a kinematic condition corresponding to a statement regarding the velocity of the fluid on the boundary and a dynamic boundary condition corresponding to a statement about the forces on the boundary. Unlike the viscous case, there are now fewer conditions to impose, because in the present case there are no shear stresses acting within the fluid.

A kinematic boundary condition is appropriate on any boundary surface with a specified geometry and position. In the simplest case the boundary may be fixed and rigid, as at the sides and bottom of a stationary container of fluid. In the more general instance, say of a rigid body moving with prescribed velocity U through the fluid, the velocity of this surface is nonzero. In either case the physically relevant boundary condition for the fluid flow at the boundary is that the normal component $\mathbf{V}\cdot\mathbf{n}$ of the fluid velocity must be equal to the normal velocity $\mathbf{U}\cdot\mathbf{n}$ of the boundary surface itself. In other words, no fluid can flow through the boundary surface. Expressed in terms of the velocity potential, this condition becomes

$$\partial\phi / \partial n = \mathbf{U}\cdot\mathbf{n}, \tag{28}$$

where $\partial/\partial n$ denotes the derivative in the direction of the unit normal $\mathbf{n}$ directed out of the fluid.

In the absence of viscous shear stresses, only the normal component of the fluid velocity is prescribed on the boundary. A tangential velocity difference between the fluid and the boundary surface is not only permitted but also expected. In a real fluid the conditions described are relevant immediately outside the boundary layer, as opposed to the actual body surface. However, the distinction between these two positions is not significant to the formulation of the inviscid flow problem, on the presumption that the boundary-layer thickness is very small.

If the normal velocity (28) is prescribed on all the boundary surfaces of a simply-connected fluid domain and the position of these surfaces is known, then the solution of Laplace's equation for the velocity potential is specified uniquely except for an additive constant. The proof of this statement is sufficiently simple to repeat here. Let us suppose that two solutions ϕ_1 and ϕ_2 both satisfy the same boundary condition (28) and Laplace's equation throughout the fluid. The difference, $\phi_3 = \phi_1 - \phi_2$, is a homogeneous solution with zero normal derivative on the boundary surface S and also a solution of Laplace's equation throughout the fluid volume $\mathcal{V}$. By a well-known consequence of the divergence theorem (3.5),

$$\begin{aligned}\iint_S \phi\frac{\partial\phi}{\partial n}dS &= \iint_S (\phi\nabla\phi)\cdot d\mathbf{S} = \iiint_{\mathcal{V}} \nabla\cdot(\phi\nabla\phi)d\mathcal{V} \\ &= \iiint_{\mathcal{V}} [\nabla\phi\cdot\nabla\phi + \phi\nabla^2\phi]d\mathcal{V}.\end{aligned} \tag{29}$$

Under the specified conditions, substituting the potential ϕ_3 gives

$$0 = \iiint_{\mathcal{V}} (\nabla\phi_3 \cdot \nabla\phi_3) d\mathcal{V}. \tag{30}$$

Since the integrand of this volume integral is positive definite, it must vanish everywhere throughout the fluid. Hence there is no fluid motion if there is no normal velocity on the boundaries, and the potential ϕ_3 can be at most a constant.

From this *uniqueness proof* it follows that the kinematic boundary condition (28) contains precisely the right amount of information regarding the fluid motion. To prescribe more information on the boundaries would be to overspecify the solution. On the other hand, there will be problems where the position and velocity of the boundary are unknown, in particular on the free surface where waves occur and the form of the wave motion, and free surface elevation, is not known *a priori*. In this instance the kinematic boundary condition (28) remains valid, but the velocity $\mathbf{U}$ of the boundary is unknown. Additional information must be provided, and the physically relevant condition is that the pressure on the free surface is equal to atmospheric pressure. Thus the dynamic boundary condition is based on Bernoulli's equation (26), with the pressure p prescribed. Further discussion of this situation is deferred until we discuss free-surface flows in chapter 6.

4.5 Simple Potential Flows

The simplest example of ideal fluid motion is the uniform stream, say with velocity U flowing in the x-direction. The velocity potential of this flow is clearly $\phi = Ux$. More generally the components of similar streaming flows in the y- and z-directions can be superposed to obtain the velocity potential $\phi = Ux + Vy + Wz$. This potential is a solution of Laplace's equation (16), and operating with the gradient (15) confirms that this is indeed the potential of a flow with velocity components (U, V, W).

This streaming flow will be disturbed by the presence of a body or bodies or of some other type of boundary geometry such as the variable form of a flow channel. From the mathematical viewpoint such a disturbance requires the introduction of *singularities*; the simplest of these are the source potential and its compatriot, the sink. Since these singularities are literally a source or sink of fluid, they violate the continuity condition and Laplace's equation locally at the point where they are located. This is permissible provided the singularity is situated within the body, or at most on the boundary surfaces, and is not allowed within the interior of the fluid.

The velocity potential of a source, situated at the origin, is

$$\phi = \frac{-m}{4\pi}\left(x^2 + y^2 + z^2\right)^{-1/2} = -\frac{m}{4\pi r}, \tag{31}$$

where r is the radial distance from the origin, or from the position of the source, and m is a constant. Since the source potential (31) depends only on the coordinate r, its gradient will be a vector in the radial direction, and hence the velocity field will consist of radial streamlines. It is straightforward to confirm that (31) is a solution of the Laplace equation (16) for $r \neq 0$; hence this radial flow satisfies the requirement of conservation of fluid mass everywhere except at the singular point $r = 0$.

If the source potential (31) is differentiated with respect to r to obtain the radial velocity and integrated over the surface of a sphere centered upon the origin, it follows that the rate of flux Q of fluid emitted from the source is precisely equal to m. This parameter is known as the *source strength*. If m is negative, the flux direction is reversed, and the singularity is called a *sink*. Mathematically, the distinction between a source and a sink is simply the sign of the strength m; generally we shall use the term "source" without distinction.

The streaming flow past a semi-infinite *half-body* can be developed by superposing the source potential (31) and a free stream so that

$$\phi = Ux - \frac{m}{4\pi}(x^2 + y^2 + z^2)^{-1/2}. \tag{32}$$

The resulting flow is axisymmetric about the x-axis, and the streamlines in the x-y plane are as shown in figure 4.1. Differentiation of (32) with respect to x indicates a *stagnation point* at $x = -(m/4\pi U)^{1/2}$, where $V = 0$. Here the flow is deflected around the source; thereafter the outer flow upstream of the stagnation point continues downstream, but with a permanent deflection from the x-axis due to the fluid emanating from the source. Although the inner flow does not correspond to the physical domain of the fluid, it is of interest because it reveals how the source serves to generate the body. Thus, fluid originally emitted from the source tends to oppose

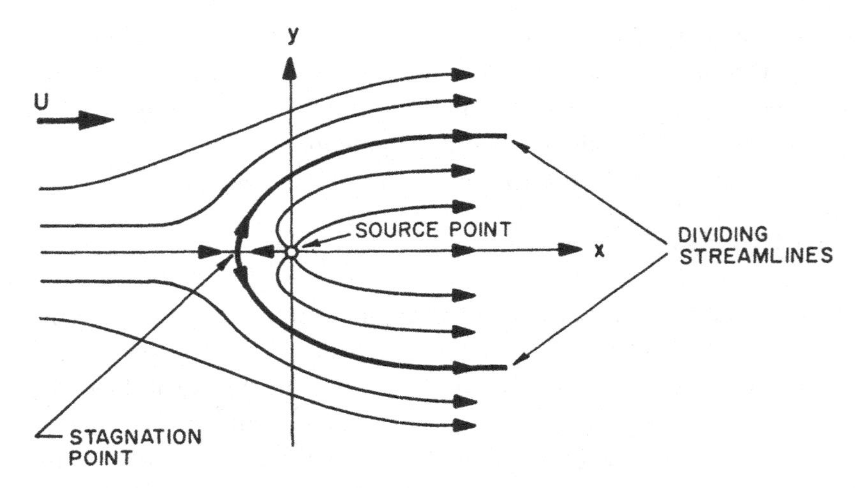

Figure 4.1
Streaming flow past a semi-infinite half-body generated by a point source at the origin. The body surface is axisymmetric about the *x*-axis and corresponds to the position of the dividing streamlines.

the incoming stream and produces the stagnation point, but ultimately all of the inner flow is diverted downstream to infinity. Since the rate of flux emitted by the source is m, and since far downstream this fluid must move with velocity U for the pressure to balance across the dividing streamline, the cross-sectional area of the dividing streamline far downstream is equal to m/U. The resulting half-body is semi-infinite in extent.

To represent the more practical situation where the body is closed and of finite length, we need to introduce not only a source but also a sink of opposite strength, located so that the fluid emitted by the source will be absorbed into the sink. The sink clearly must be located downstream of the source, and if these two singularities are situated symmetrically about the origin, at $x = \mp a$, the potential will be

$$\phi = Ux - \frac{m}{4\pi}[(x+a)^2 + y^2 + z^2]^{-1/2} + \frac{m}{4\pi}[(x-a)^2 + y^2 + z^2]^{-1/2}. \tag{33}$$

Differentiation with respect to x, with $y = z = 0$, reveals that the stagnation points are located at $x = \pm l/2$, where the length l is determined from the equation

$$(l^2/4 - a^2)^2 = 2al(m/4\pi U). \tag{34}$$

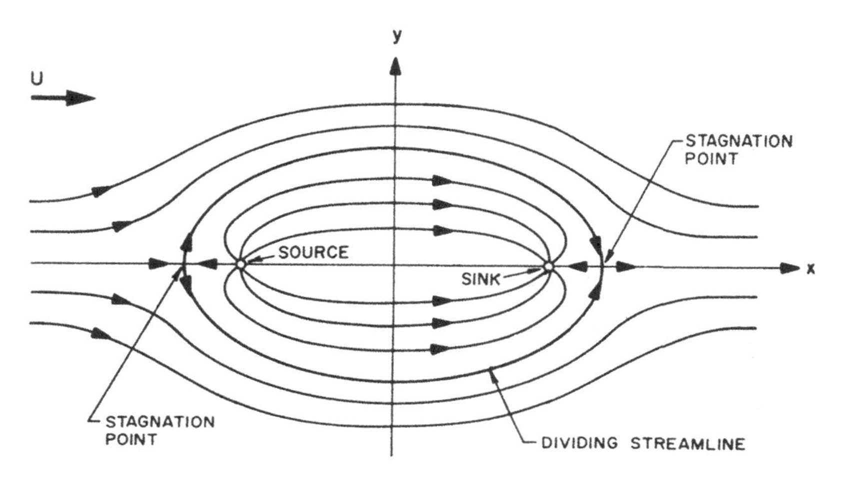

Figure 4.2
Streaming flow past a Rankine ovoid, or source-sink combination.

The streamlines associated with (33) define a *Rankine ovoid,* as shown in figure 4.2. The maximum radius b can be found from continuity, since the flux passing across the plane $x = 0$, inside a circle of radius b, will be equal to the flux emitted from the source. Thus, with $x = 0$,

$$2\pi\int_0^b \left[U + \frac{2am/4\pi}{(a^2 + R^2)^{3/2}}\right] R\,dR = m, \tag{35}$$

or

$$m = \pi U b^2 (1 + b^2/a^2)^{1/2}. \tag{36}$$

The resulting flow, shown in figure 4.2, is similar to that actually observed for streamlined axisymmetric bodies. From Bernoulli's equation one can compute the pressure distribution on the body; it will take a maximum value at the two stagnation points and a minimum at the central plane $x = 0$ where the velocity is a maximum.

We might proceed to construct more general axisymmetric bodies by distributing sources and sinks continuously along the body axis. This is a practical method for determining the flow characteristics of bodies of revolution, especially if they are relatively slender, as will be shown in chapter 7. Instead, let us focus our attention on the opposite extreme, where the separation between the two singularities shrinks to zero. To preserve a finite effect from the two singularities as they are brought together, their strengths m must increase at the same time, for otherwise they will cancel out in the limit when they coincide. Thus it is necessary to make the product $\mu = 2ma$ a constant, with the result

$$\begin{aligned}\phi &= Ux + \lim_{a \to 0} \frac{\mu}{8\pi a}\{[(x-a)^2 + y^2 + z^2]^{-1/2} - [(x+a)^2 + y^2 + z^2]^{-1/2}\} \\ &= Ux + \frac{\mu}{4\pi}\frac{\partial}{\partial a}\{[(x-a)^2 + y^2 + z^2]^{-1/2}\}_{a=0} \\ &= Ux + \frac{\mu x}{4\pi(x^2 + y^2 + z^2)^{3/2}}.\end{aligned} \tag{37}$$

The last term is called a *dipole* or *doublet*, and the constant μ is its *moment*. If we examine the resulting flow, from the combination of this dipole with the uniform stream, in a spherical coordinate system where $x = r\cos\theta$ and $(y^2 + z^2)^{1/2} = r\sin\theta$, (37) takes the form

$$\phi = Ur\cos\theta + \frac{\mu\cos\theta}{4\pi r^2}. \tag{38}$$

Since the derivative with respect to the radius r vanishes on $r = (\mu/2\pi U)^{1/3}$, (37) and (38) give the flow of a uniform stream, of velocity U, past a sphere of this radius.

Two-dimensional singularities can be constructed analogously. Thus, if the flow is independent of the z coordinate, or parallel to the x-y plane, the source and dipole potentials are

$$\frac{m}{2\pi}\log(x^2 + y^2)^{1/2} \quad \text{(source)},$$

$$\frac{\mu}{2\pi}\frac{x}{x^2 + y^2} \quad \text{(dipole)}.$$

Once again, a source will combine with a uniform stream to generate a half-body, a source and equal sink will generate an elongated closed body, and a point dipole will give the limiting and important case of the flow past a circle of radius $R = (\mu/2\pi U)^{1/2}$ in the form

$$\phi = Ux + \frac{\mu x}{2\pi(x^2 + y^2)}. \tag{39}$$

4.6 The Stream Function

In the preceding sections the concept of the velocity potential has been used exclusively, as a functional representation of the flow field, and justified on the basis that the fluid velocity field is irrotational. Alternatively, the velocity vector can be represented in terms of the *stream function,* whose existence is instead a consequence of the continuity equation. For any incompressible fluid, with or without viscosity, the velocity vector **V** must satisfy the condition that its divergence is equal to zero; a well-known result of vector analysis is that any divergence-free vector can be written in the form

$$\mathbf{V} = \nabla \times \boldsymbol{\Psi}, \tag{40}$$

where $\boldsymbol{\Psi}$ is the vector stream function. In general, this is not a very useful concept, since the new unknown function is also a vector. However, in the special cases of two-dimensional plane flow and three-dimensional axisymmetric flow, the vector stream function has only one component; thus it becomes a scalar unknown, with the same resulting simplification as the velocity potential.

First, let us consider two-dimensional plane motion in the x-y plane. The velocity component $w = 0$ and the stream function should be independent of z. These conditions can be met only if $\boldsymbol{\Psi} = (0,0,\psi)$ and the scalar stream function ψ determines the velocity components u and v in accordance with the relations

$$u = \partial\psi / \partial y, \qquad v = -\partial\psi / \partial x. \tag{41}$$

The axisymmetric case follows from analogous reasoning. Thus, if we introduce circular cylindrical coordinates (R,θ,x) and assume that the flow is independent of θ, it follows that the stream function vector consists only of a component in the θ-direction, which is independent of that coordinate. It is desirable in this case to define $\boldsymbol{\Psi} = (0,\ \psi/R,\ 0)$ so that the radial and axial velocity components are given by

$$u_R = -\frac{1}{R}\partial\psi / \partial x, \qquad u_x = \frac{1}{R}\partial\psi / \partial R. \tag{42}$$

(The form of the curl operator, in circular cylindrical coordinates, is given in Hildebrand, 1976, section 6.18.)

The derivation of the stream function can be supplemented by a physical argument. For plane flow in the x-y plane, we define the scalar, which is a measure of the flux across a contour C, as

$$\psi = \int_C (u dy - v dx). \tag{43}$$

The contour C is shown in figure 4.3. From continuity, (43) is independent of the choice of C; as in the analogous integral (13) defining the velocity potential, (43) depends only on the two endpoints, say $\mathbf{x}_0$ and $\mathbf{x}$, of C. It is clear that differentiation, with respect to the endpoint $\mathbf{x}$, gives the relations (41). In addition, the stream function is a constant along fluid streamlines, or contours everywhere tangent to the local velocity vector.

For axisymmetric flow, the flux-integral analogous to (43) is

$$Q = 2\pi \iint_{S(R,x)} (u_x R dR - u_R R dx), \tag{44}$$

where $S(R, x)$ is any surface bounded by the circle (R, x). The derivatives of (44) differ from (42) by the factor 2π; hence $Q = 2\pi\psi$, and we see that the stream function is equal to $(1/2\pi)$ times the flux passing through the circle (R, x). This property ensures that the surfaces ψ = constant will define the stream surfaces of the flow field.

As an example of the use, of the axisymmetric stream function, we consider again the flow past a Rankine ovoid, or source-sink combination, for which the velocity potential is given by equation (33). In polar coordinates, the potential (33) is

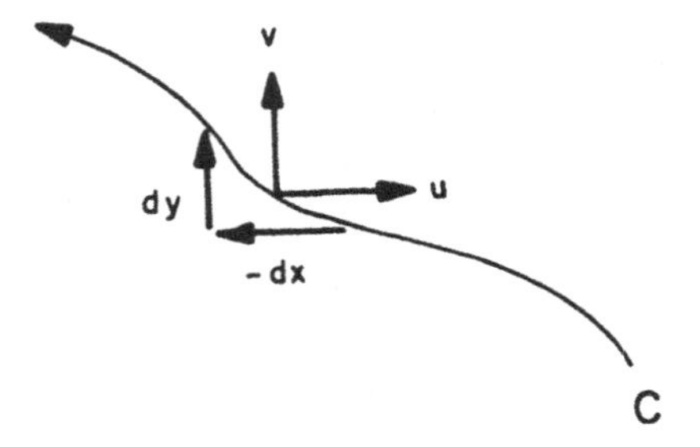

Figure 4.3
Contour of integration in equation (43). The fluid flux across a differential element of this contour is $u\,dy - v\,dx$. A positive flux is defined as moving across C to the right, with respect to an observer moving along C.

$$\phi(R,\theta,x) = Ux - \frac{m}{4\pi}[(x+a)^2 + R^2]^{-1/2} + \frac{m}{4\pi}[(x-a)^2 + R^2]^{-1/2}. \tag{45}$$

From (44), we can compute the stream function as

$$\begin{aligned}\psi(R,x) &= \int_0^R \frac{\partial \phi}{\partial x} R'\,dR' = \frac{\partial}{\partial x}\int_0^R \phi R' dR' \\ &= \frac{\partial}{\partial x}\left\{\frac{1}{2}Ux R^2 - \frac{m}{4\pi}[(x+a)^2 + R^2]^{1/2} + \frac{m}{4\pi}[(x-a)^2 + R^2]^{1/2}\right\} \\ &= \frac{1}{2}UR^2 - \frac{m(x+a)}{4\pi[(x+a)^2 + R^2]^{1/2}} + \frac{m(x-a)}{4\pi[(x-a)^2 + R^2]^{1/2}}.\end{aligned} \tag{46}$$

In deriving (46), we have overlooked two contributions that cancel out. Thus, for the stream function to remain continuous and single valued, as a measure of the flux passing through the surface S defined in (44), this surface must not pass through the source points at $x = \pm a$. The contribution from ignoring this restriction is equal to the flux of these sources, $\pm m$; this is canceled by the ignored contribution to the integral in (46) from the lower limit at $R = 0$.

The first term in (46) gives the stream function of the axial stream; the second and third terms give the stream functions of a source at $x = -a$ and a sink at $x = +a$, respectively. For $|x| > a$, the axis $R = 0$ corresponds to the streamline $\psi = 0$, and since the stagnation points are included on this line, it follows that $\psi = 0$ is the equation of the dividing streamlines that pass around the body surface. Hence the equation of the body surface is given by

$$0 = \frac{2\pi U R^2}{m} - \frac{(x+a)}{[(x+a)^2 + R^2]^{1/2}} + \frac{(x-a)}{[(x-a)^2 + R^2]^{1/2}}. \tag{47}$$

In particular, setting $x = 0$ gives equation (36) for the maximum radius of the body, and it is apparent that the stream function was used implicitly in (36).

4.7 The Complex Potential

In the case of plane two-dimensional flows, analytic functions of a complex variable can be used to represent the velocity field. This allows us to exploit a wide variety of mathematical tools.

Let us assume that the flow depends only on the coordinates x and y, which are taken to be the real and imaginary parts of the complex variable $z = x + iy$. We define the *complex potential* $F(z)$ to be

$$F(z) = \phi + i\psi, \tag{48}$$

where ϕ is the velocity potential and ψ the stream function. Since the two velocity components can be determined by differentiating either of these real functions, it follows that

$$u = \partial\phi / \partial x = \partial\psi / \partial y, \qquad v = \partial\phi / \partial y = -\partial\psi / \partial x,$$

or the real and imaginary parts of F satisfy the *Cauchy-Riemann equations.* Therefore the complex potential F is an analytic function of the complex variable z, and its derivative is

$$\frac{dF}{dz} = u - iv. \tag{49}$$

Conversely, any analytic function defines a complex potential, and its derivative is a plausible velocity field. Some simple analytic functions with corresponding velocity fields are listed in table 4.1.

An example of particular interest in two dimensions is the corner flow $F(z) = z^n$, which gives the flow interior to a sector of included angle π/n. This can be verified by using polar coordinates, $z = re^{i\theta}$, and writing the complex potential as $F(z) = r^n e^{in\theta}$. Since the imaginary part is zero on $\theta = 0$ and $\theta = \pi/n$, these are streamlines of the flow. Special cases of particular interest are $n = 2$, or the flow interior to a right angle, and $n = 1/2$ which gives the exterior flow around the edge of a semi-infinite flat plate. Both cases are illustrated in figure 4.4. A significant difference between these is the local behavior near the origin. The case $n = 2$ has a stagnation point. This is

Table 4.1
Simple Examples of Complex Potentials and the Corresponding Velocity Fields

$F(z)$	$u - iv$	Type of Flow
Uz	U	Stream
$\log z$	$1/z$	Source
$1/z$	$-1/z^2$	Dipole
$Uz + A/z$	$U - A/z^2$	Stream of velocity U flowing past a circle of radius $(A/U)^{1/2}$
z^n	$n\, z^{n-1}$	Corner flow within a sector of angle π/n.

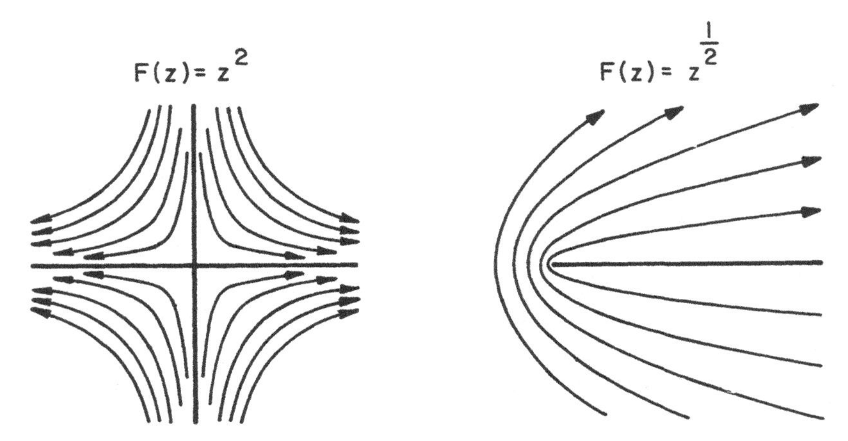

Figure 4.4
Simple complex potentials and the corresponding streamlines.

characteristic of interior corner flows, where the included angle is less than 180°, or where the exponent n is greater than one. On the other hand, for $n = 1/2$ an infinite velocity occurs at the sharp edge, and this is characteristic of all exterior corner flows where $1/2 < n < 1$.

4.8 Conformal Mapping

A particularly useful facet of the application of analytic function theory to two-dimensional flows is the use of conformal mapping techniques. Here, the objective is to map a complicated geometry onto a simpler one, replacing the fluid flow and body profile in the physical z-plane by a fictitious flow past a hypothetical profile in, say, the ζ-plane, where ζ is related to z by a known analytic function.

One of the simplest examples is the corner flow considered in the preceding section. A corner with included angle θ can be mapped onto a straight line in the ζ-plane by the transformation $\zeta = z^{\pi/\theta}$, as shown in figure 4.5. In the ζ-plane the flow past the straight line is simply a uniform stream; hence the complex potential is $F(\zeta) = \zeta$. Inverting the mapping procedure, it follows that

$$F(z) = z^{\pi/\theta}. \tag{50}$$

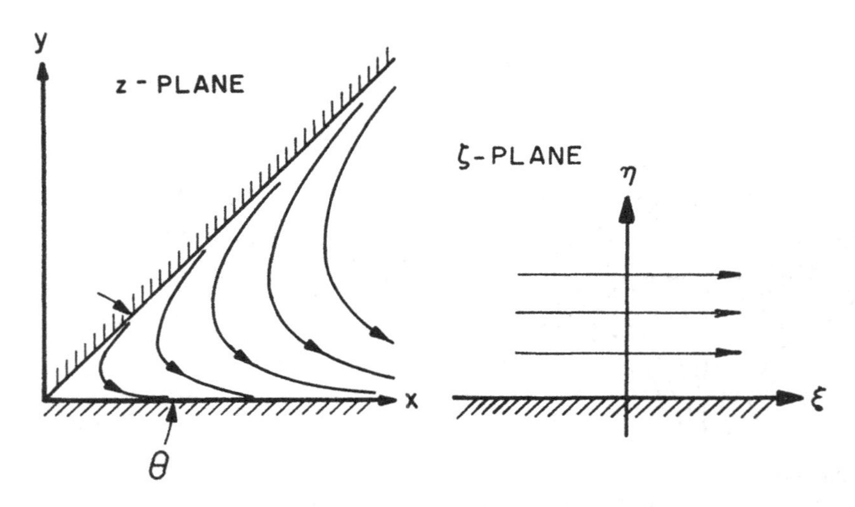

Figure 4.5
The complex potential $F(z) = z^{\pi/\theta}$, and the corresponding conformal mapping onto the plane $\zeta = z^{\pi/\theta}$.

This answer agrees with the result shown in table 4.1, which was deduced by the indirect process of first choosing a potential and subsequently determining the flow geometry.

As a more interesting example of conformal mapping, let us generalize the complex potential for the flow past a circle. Replacing the potential by the variable ζ we are led to consider the mapping

$$\begin{aligned}\zeta &= Uz + A/z = Ure^{i\theta} + \frac{A}{r}e^{-i\theta} \\ &= (Ur + A/r)\cos\theta + i(Ur - A/r)\sin\theta,\end{aligned} \tag{51}$$

where $z = re^{i\theta}$ has been substituted. Equation (51) defines a family of confocal ellipses in the ζ-plane; the major semiaxis $Ur + A/r$ and the minor semiaxis $Ur - A/r$ coincide with the real and imaginary axes respectively. In particular, the circle $r = (A/U)^{1/2}$ in the z-plane is mapped onto the segment between $\pm\, 2(AU)^{1/2}$ in the ζ-plane. For convenience we shall set $A = U = 1/2$, so that the circle of unit radius in the z-plane is mapped onto the straight line $(-1, 1)$ in the ζ-plane. Physically we can suppose that this segment corresponds with a flat plate, which is the conformal mapping of the circle. In this special case (51) reduces to the mapping

$$\zeta = \frac{1}{2}(z + 1/z). \tag{52}$$

Solving the resulting quadratic equation for z gives

$$z = \zeta + (\zeta^2 - 1)^{1/2}. \tag{53}$$

Here the branch of the square root is chosen so that $(\zeta^2 - 1)^{1/2} \simeq \zeta$ as $\zeta \to \infty$; hence $z \simeq 2\zeta$ which is consistent with the corresponding limit of (52) as $z \to \infty$.

With the above preliminaries we now are ready to move back and forth between the z-plane and ζ-plane and to relate the flow past the unit circle to the flow past a flat plate. The simplest solution to start with is the complex potential $F(\zeta) = \zeta$, corresponding to a uniform stream in the ζ-plane parallel to the plate. Since the plate does not disturb the flow in this particular case, the solution is complete. Transforming this complex potential back to the z-plane by means of (53) gives the potential for the flow past the unit circle:

$$F(z) = \zeta = \frac{1}{2}(z + 1/z). \tag{54}$$

Note that the stream velocity at infinity in the z-plane is parallel to the x-axis, of magnitude 1/2. Thus the conformal mapping has enabled us to recover the solution for the flow past a circle in terms of the simple uniform-stream solution past a flat plate at zero incidence.

To proceed further, we rotate the stream velocity by 90 degrees. In the z-plane the situation is unchanged, except that the flow is now vertical past the circle. But in the ζ-plane the flow is normal to the flat plate. At this point the latter solution is unknown, but the solution in the z-plane is known, and this can be mapped into the ζ-plane. First, if z is replaced by iz, the complex potential (54) becomes

$$F(z) = \frac{1}{2} i(z - 1/z). \tag{55}$$

Note that, from (49), the streaming flow at infinity is positive downward with magnitude equal to 1/2. Transforming to the ζ-plane requires that (53) be substituted in (55). Thus the complex potential for flow normal to the flat plate is given by

$$\begin{aligned} F(\zeta) &= \frac{1}{2} i[\zeta + (\zeta^2 - 1)^{1/2}] - \frac{1}{2} i[\zeta + (\zeta^2 - 1)^{1/2}]^{-1} \\ &= i(\zeta^2 - 1)^{1/2}. \end{aligned} \tag{56}$$

In the ζ-plane the complex velocity is $F'(\zeta)$, and at infinity (56) gives a downward streaming velocity of unit magnitude past the horizontal plate. The complex potential can be multiplied by any real constant to change the value of the stream velocity; for the general case of flow at normal incidence with stream velocity U, past a flat plate of width $2a$, the appropriate generalization of (56) is

$$F(\zeta) = iU(\zeta^2 - a^2)^{1/2}. \tag{57}$$

Differentiating (57) gives the complex velocity

$$u - iv = \frac{iU\zeta}{(\zeta^2 - a^2)^{1/2}}. \tag{58}$$

As in the case of the flow around a semi-infinite flat plate, the velocity is infinite at the sharp edges with a square-root singularity. More generally one can expect the singular behavior of the flow near a sharp corner to depend only on the local shape near the corner.

Since the general conformal mapping (51) maps concentric circles in the z-plane onto confocal ellipses in the ζ-plane, the flow past an ellipse can be obtained in a similar manner.

Certain fundamental characteristics of conformal mapping transformations that we have taken for granted should be stated explicitly. The following three properties apply to any conformal transformation given by an analytic function:

1. The mapping is one-to-one, so that to each point in the physical domain, there is one and only one corresponding point in the mapped domain.
2. Closed curves map to closed curves.
3. Angles are preserved between the intersections of any two lines in the physical domain and in the mapped domain.

The last property obviously is violated in the case of the flat plate, at the two ends of the plate, because the mapping function is not analytic at these points. This feature is exploited in the *Schwarz-Christoffel transformation,* where polygonal profiles of arbitrary form are mapped onto a straight line, utilizing products of corner mappings analogous to (50). An extensive treatment of conformal mapping is given by Milne-Thomson (1968).

Finally, the Riemann mapping theorem, which states that an arbitrary closed profile can be mapped onto the unit circle, enables the solution for the flow past any practical two-dimensional body to be obtained by conformal transformation. However, the practical significance of the Riemann mapping theorem is diminished by the numerical complexity of mapping arbitrary body profiles, and generally speaking other techniques are equally useful under these circumstances. Principal among the alternatives is the representation of the body by a distribution of singularities on the body surface. Before discussing this approach, however, we will first discuss a three-dimensional method more akin to conformal mapping.

4.9 Separation of Variables

Aside from guessing solutions indirectly, the simplest technique for solving three-dimensional problems is *separation of variables*. Laplace's equation is *separable* in thirteen coordinate systems including rectangular, circular cylindrical, elliptic cylindrical, parabolic cylindrical, spherical, conical, parabolic, prolate spheroidal, oblate spheroidal, ellipsoidal, paraboloidal, bispherical, and toroidal coordinates. The features of these are given in detail by Morse and Feshbach (1953). The cylindrical coordinate systems are generalizations from two dimensions where there is a close relation with complex-variable methods. The essential idea is to assume a solution for the velocity potential as the product of three functions, each depending separately on one of the three coordinates. If the coordinate system is separable, Laplace's equation will reduce to a system of three ordinary differential equations for these three functions.

To illustrate this method, let us return to the problem of a uniform stream flowing past a sphere. After the introduction of spherical coordinates (r, θ, α), which are related to the Cartesian coordinates by

$$x = r\cos\theta,$$
$$y = r\sin\theta\cos\alpha,$$
$$z = r\sin\theta\sin\alpha,$$

Laplace's equation takes the form

$$\frac{1}{r^2}\frac{\partial}{\partial r}\left(r^2\frac{\partial\phi}{\partial r}\right)+\frac{1}{r^2\sin\theta}\frac{\partial}{\partial\theta}\left(\sin\theta\frac{\partial\phi}{\partial\theta}\right)+\frac{1}{r^2\sin^2\theta}\frac{\partial^2\phi}{\partial\alpha^2}=0. \tag{59}$$

Assume that the uniform stream at infinity is in the positive x-direction and that the sphere is centered about the origin; then the velocity potential will be independent of the coordinate α. The resulting boundary-value problem for $\phi(r, \theta)$ is stated as follows:

$$\frac{1}{r^2}\frac{\partial}{\partial r}\left(r^2\frac{\partial\phi}{\partial r}\right)+\frac{1}{r^2\sin\theta}\frac{\partial}{\partial\theta}\left(\sin\theta\frac{\partial\phi}{\partial\theta}\right)=0, \qquad r>r_0, \tag{60}$$

$$\partial\phi/\partial r=0, \qquad \text{on } r=r_0, \tag{61}$$

$$\phi\to Ur\cos\theta, \qquad \text{as } r\to\infty. \tag{62}$$

Here $r = r_0$ is the radius of the sphere, (60) states that the velocity potential satisfies the Laplace equation in the fluid domain, (61) is the kinematic boundary condition on the sphere, and (62) states that the flow tends to a uniform stream at large radial distances from the sphere.

This problem is solved, by the method of separation of variables, by assuming a solution of (60–62) in the form

$$\phi(r,\theta)=R(r)\Theta(\theta). \tag{63}$$

Then, from (60),

$$r^2\phi^{-1}\nabla^2\phi=\frac{1}{R}\frac{d}{dr}(r^2R')+\frac{1}{\Theta\sin\theta}\frac{d}{d\theta}(\sin\theta\Theta')=0, \tag{64}$$

where a prime denotes the derivative with respect to the appropriate variable. The first term on the right side of (64) depends only on r, and the second only on θ; if the two are to be equal and opposite for all values of r and θ, they must each be a constant. Thus R and Θ satisfy the ordinary second-order differential equations

$$\frac{d}{dr}(r^2R')-RC=0, \tag{65}$$

$$\frac{d}{d\theta}(\sin\theta\Theta')+C\Theta\sin\theta=0. \tag{66}$$

Solutions of the second equation are the Legendre functions

$$P_n(\cos\theta), \quad \text{where } C=n(n+1),$$

and the Legendre functions of the second kind $Q_n(\cos\theta)$. If we require the solution to be regular for $0\le\theta\le\pi$, only the first functions can be allowed, and n must then be zero or a positive integer. The first equation now takes the form

$$\frac{d}{dr}(r^2R') - Rn(n+1) = 0, \tag{67}$$

and solutions are clearly given by

$$R = r^n, \qquad R = 1/r^{n+1}.$$

Thus, with A_n and B_n a set of undetermined constants, the most general solution for the velocity potential $\phi(r, \theta)$ or (60) is

$$\phi = \sum_{n=0}^{\infty} P_n(\cos\theta)[A_n r^n + B_n / r^{n+1}]. \tag{68}$$

The Legendre functions $P_n(x)$, with n an integer, are polynomials in x; the first three of these are

$$P_0(x) = 1, \qquad P_1(x) = x, \qquad P_2(x) = \frac{1}{2}(3x^2 - 1).$$

These polynomials are orthogonal in the interval $-1 < x < +1$, and the term with $n = 1$ in (68) is the only member of this series that can be proportional to $\cos\theta$ as required by the condition (62) at infinity. Thus, the remaining coefficients for $n > 1$ must vanish in (68). The solution of (60–62) is therefore given by the potential

$$\phi = U\cos\theta\left(r + \frac{1}{2}r_0^3 \middle/ r^2\right). \tag{69}$$

This potential is recognized as the superposition of a dipole and a uniform stream; cf. equation (38).

The term in (68) proportional to A_0 is a constant, and in (69) this has been set equal to zero. The corresponding term B_0/r is a source, which is ruled out here because the sphere has constant volume. An alternative problem would be the pulsation of a spherical bubble of gas within the fluid, where the source term would be the appropriate solution.

For polar coordinates (r, θ) in two dimensions, the analog to (68) is the expansion

$$\phi = \sum_{n=1}^{\infty} \cos n\theta[A_n r^n + B_n / r^n] + A_0 + B_0 \log r \tag{70}$$

together with a similar series with $\cos n\theta$ replaced by $\sin n\theta$. This is equivalent to expressing the complex potential in its most general form as a Laurent series plus a logarithmic source term.

A common feature of (68) and (70) is that two sets of independent radial functions arise, one singular at the origin and the other at infinity. The latter is appropriate for *internal* flows, where the fluid is contained within a spherical or circular container with prescribed normal velocity on the boundary. However, for external flows in a fluid domain that extends to infinity radially, the solutions of Laplace's equation proportional to r^n are ruled out, with the exception of a uniform free stream obtained when $n = 1$. On the other hand, for external flows the entire set of solutions singular at $r = 0$ is admissible, since the fluid domain is exterior to this point. Individual terms in this set are known as *multipoles;* the source and dipole are the first two members. (The next two are *quadrupoles* and *octapoles*.) The multipole of order n can be derived by differentiating the source potential n times.

The method of separation of variables can be applied, using the same scheme with more complicated coordinate systems, to determine the flow past spheroids and ellipsoids and more generally to solve problems where the boundary geometry coincides with one of the coordinate surfaces of the separable systems listed above. Details of these problems and the resulting solutions are discussed by Lamb (1932) and by Morse and Feshbach (1953). The solutions obtained are analogous in form to (68) and (70), with undetermined coefficients that must be found from the boundary conditions of the problem, but with more complicated special functions representing the spatial dependence. Several applications of this technique in the field of ship hydrodynamics are included in the collected works of Havelock (1963).

4.10 Fixed Bodies and Moving Bodies

In the previous sections, streaming flows have been considered where the body is fixed and the fluid at infinity has a uniform constant velocity. Except for the change in reference frame, these are identical to corresponding problems where the body moves with constant velocity through otherwise undisturbed fluid, provided the relative velocity between the body and the fluid at infinity is the same in both cases. However, the boundary conditions must be changed if the fluid velocity is referred consistently to a fixed reference frame, and the two solutions will differ by the free-stream potential.

Table 4.2
Comparison of Flows for Circle and Sphere of Radius r_0

Horizontal Streaming Flow Past a Fixed Body		
	Circle	**Sphere**
Boundary Condition on Body	$\partial\phi/\partial r = 0$	$\partial\phi/\partial r = 0$
Boundary Condition at Infinity	$\phi \to Ux = Ur\cos\theta$	$\phi \to Ux = Ur\cos\theta$
Velocity Potential	$\phi = U(r + r_0^2/r)\cos\theta$	$\phi = U(r + \frac{1}{2}r_0^3/r^2)\cos\theta$
Horizontal Translation of Body with Fluid at Rest at Infinity		
	Circle	**Sphere**
Boundary Condition on Body	$\partial\phi/\partial r = U\cos\theta$	$\partial\phi/\partial r = U\cos\theta$
Boundary Condition at Infinity	$\phi \to 0$	$\phi \to 0$
Velocity Potential	$\phi = -(Ur_0^2/r)\cos\theta$	$\phi = -(\frac{1}{2}Ur_0^3/r^2)\cos\theta$

To illustrate this comparison, table 4.2 shows the boundary conditions and velocity potentials for a circle and a sphere when the fluid streams past the fixed body, as well as when the fluid is at rest with the body moving. The stream velocity in the former case is $+U$ in the $+x$-direction; the body velocity in the second case is denoted by the same symbol, with the convention that a positive velocity of the body is in the $+x$-direction. Thus the relative velocities differ in sign, and for this reason the dipole moments of the corresponding solutions are equal in magnitude, but of opposite signs.

When the body moves with an unsteady motion, the same *kinematic* conclusions apply to the velocities of the body and fluid, with $U(t)$ a function of time. In this instance, however, the problems in each frame of reference will differ from the *dynamic* standpoint; this difference will be examined in section 4.17.

4.11 Green's Theorem and Distributions of Singularities

Earlier we showed that certain bodies could be represented by suitable combinations of a source and sink or by a point dipole. Given the effect of these singularities on the surrounding fluid, it seems plausible that a larger number of them could be used to represent more complicated body shapes. In the most general case, one might envisage a continuous volume distribution of singularities within the body, extending out to its surface.

Alternatively, from the standpoint of the flow exterior to the body, a distribution on the body surface alone might suffice. This latter speculative conjecture turns out to be correct, and in the process of establishing it a number of useful results will be obtained.

For reasons that will become clear later, let us consider two solutions of Laplace's equation in a volume $\mathcal{V}$ of fluid bounded by a closed surface S. Denoting these two potentials by ϕ and φ and applying the divergence theorem we get

$$\begin{aligned}\iint_S \left[\phi\frac{\partial\varphi}{\partial n} - \varphi\frac{\partial\phi}{\partial n}\right] dS &= \iiint_{\mathcal{V}} \nabla\cdot(\phi\nabla\varphi - \varphi\nabla\phi)d\mathcal{V} \\ &= \iiint_{\mathcal{V}} (\phi\nabla^2\varphi + \nabla\phi\cdot\nabla\varphi - \varphi\nabla^2\phi - \nabla\varphi\cdot\nabla\phi)d\mathcal{V} \\ &= 0.\end{aligned} \tag{71}$$

This important result is a form of *Green's theorem* that will be utilized later in a variety of contexts. Here we shall consider the consequence of replacing φ by the potential of a source.

In the following analysis it is convenient to set the source strength $m = -1$. Of more significance is the position of this unit source, which must be carefully specified. We shall define the *source point* $\boldsymbol{\xi} \equiv (\xi, \eta, \zeta)$ as the position of the source in the coordinates $\mathbf{x} = (x, y, z)$. For a unit source the potential at the *field point* $\mathbf{x}$ is given by

$$\varphi = 1/4\pi r = (1/4\pi)[(x-\xi)^2 + (y-\eta)^2 + (z-\zeta)^2]^{-1/2}. \tag{72}$$

In accordance with the principle of reciprocity, the value of (72) is unchanged if the source point and field point are interchanged. By the same token, (72) is a solution of Laplace's equation with respect to $\boldsymbol{\xi}$ as well as $\mathbf{x}$. Thus, in utilizing this source potential in Green's theorem (71) we can integrate with respect to either coordinate system. Mathematically the results are equivalent, but different physical interpretations follow.

In the subsequent derivation we shall perform the surface integration of (71) with respect to the coordinates of the source point $\boldsymbol{\xi}$. This requires that the potential ϕ and the normal derivative $\partial\phi/\partial n$ be defined with respect to (ξ, η, ζ) by a simple change of the dummy variable of integration. Physically we shall integrate over a continuous distribution of sources and normal dipoles that are located on the surface S, with a fixed value of the field point $\mathbf{x}$.

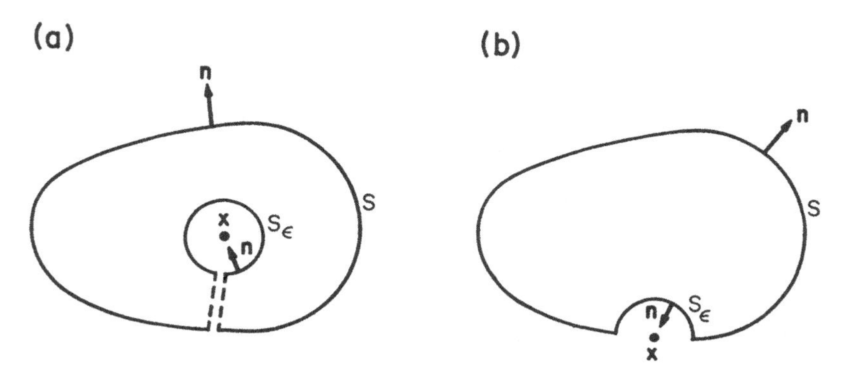

Figure 4.6
Surfaces of integration for Green's theorem. In (a) the field point is interior to S, surrounded by a small spherical surface S_ε; in (b) the field point is on the boundary surface S and S_ε is a hemisphere.

Substituting (72) in (71) requires caution, for the source potential does not satisfy the Laplace equation at the singular point $r = 0$, and thus (71) is not valid when the source point is situated within $\mathcal{V}$. This difficulty can be circumvented by surrounding the source point by a small sphere of radius $r = \varepsilon$ with surface S_ε as shown in figure 4.6 (a). Then $S + S_\varepsilon$ is a closed[2] surface surrounding the volume of fluid interior to S but exterior to S_ε; within this volume, (72) is regular. Thus, (71) can be replaced by

$$\frac{1}{4\pi}\iint_{S+S_\varepsilon}\left[\phi\frac{\partial}{\partial n}\frac{1}{r}-\frac{1}{r}\frac{\partial\phi}{\partial n}\right]dS=0, \tag{73}$$

or

$$\frac{1}{4\pi}\iint_{S}\left[\phi\frac{\partial}{\partial n}\frac{1}{r}-\frac{1}{r}\frac{\partial\phi}{\partial n}\right]dS=-\frac{1}{4\pi}\iint_{S_\varepsilon}\left[\phi\frac{\partial}{\partial n}\frac{1}{r}-\frac{1}{r}\frac{\partial\phi}{\partial n}\right]dS. \tag{74}$$

In the limit $\varepsilon \to 0$, the contribution from the integral over S_ε in (74) can be evaluated under the assumption that the velocity potential and its normal derivative on S_ε are both regular. The area of S_ε is $4\pi r^2$, while the normal derivative of $1/r$ is $-\partial/\partial r(1/r) = 1/r^2$. (Note that the normal is positive when pointing out of the fluid volume and hence in the direction opposite to r on S_ε.) Thus the first term in the integrand on the right side of (74) is singular in proportion to $1/r^2$, and when it is multiplied by the area $4\pi r^2$, a finite limit will result as $\varepsilon \to 0$, whereas the weaker singularity of the second term will give no contribution. For sufficiently small ε, the potential ϕ may be assumed constant and taken outside of the integral sign, so that the final limiting value of the right-hand side of (74) becomes

$$-\frac{1}{4\pi}\phi(x,y,z)\iint_{S_\varepsilon}\frac{\partial}{\partial n}\frac{1}{r}dS=-\phi(x,y,z). \tag{75}$$

Thus, if (x, y, z) is *inside* S,

$$\phi(x,y,z)=-\frac{1}{4\pi}\iint_S\left[\phi\frac{\partial}{\partial n}\frac{1}{r}-\frac{1}{r}\frac{\partial\phi}{\partial n}\right]dS. \tag{76}$$

Equation (76) can be regarded as a representation of the velocity potential in terms of a normal dipole distribution of moment ϕ and a source distribution of strength $-\partial\phi/\partial n$ distributed over the boundary surface S. Thus, in general, the flow can be represented by a suitable distribution of dipoles and sources on the boundary surface S. By further manipulation with Green's theorem, one can obtain alternative integrals involving sources or dipoles only, as shown by Lamb (1932). (See problem 13.)

If the point (x, y, z) is situated *on* the surface S, the surface S_ε is chosen to be a small hemisphere that indents the original surface S inside the source point, as in figure 4.6(b). The contribution from this hemisphere is just half that given by (75), so that (76) is replaced by

$$\phi(x,y,z)=-\frac{1}{2\pi}\iint_S\left[\phi\frac{\partial}{\partial n}\frac{1}{r}-\frac{1}{r}\frac{\partial\phi}{\partial n}\right]dS. \tag{77}$$

Here, the surface integral must be defined to exclude the immediate vicinity of the singular point, i.e., the locally-plane infinitesimal surface bounded by the intersection of S and the hemisphere S_ε. This situation is analogous to a principal value integral, except that in (77) the precise shape of the excluded infinitesimal area is not important.

Finally, when the point (x, y, z) is situated *outside* S, (71) is valid without modification and the left-hand side of (76) or (77) is replaced by zero. Similar results hold in two dimensions, with the source potential $1/r$ replaced by $\log r$ and with the surface integrals in the above relations replaced by line integrals; in this case the factors 4π and 2π are multiplied by $-1/2$, corresponding to the difference in the radial flux between the singularities $1/r$ and $\log r$.

Equation (77) is frequently used for constructing the velocity potential due to the motion of a ship hull or other moving body. Generally, the normal derivative $\partial\phi/\partial n$ is known on the body, so that (77) is an integral equation for the determination of the unknown potential; it may be solved by approximations or numerical techniques.

For a body moving in an otherwise infinite and unbounded fluid, the appropriate *closed* surface S enclosing the fluid volume must include both the body surface S_B and an additional control surface S_C surrounding the body. It then can be argued that $S_B + S_C$ together form a closed surface surrounding the fluid volume $\mathcal{V}$ or, at least, that portion of $\mathcal{V}$ within a finite distance of the body. The contribution from S_C to the integral in (77) can be estimated from the expansion of a general three-dimensional velocity potential in spherical harmonics, (69). For large spherical radius r, the potential due to the presence of the body is of order r^{-1}, and $\partial\phi/\partial n$ of order r^{-2}. Thus the integrand of (77) is of order r^{-3}, whereas on S_C the differential area dS will increase in proportion to the square of the distance from the origin. It then follows that the contribution to (77) from S_C is of order r^{-1}, and vanishes in the limit where this surface is an infinite distance from the body. Therefore, for a body moving in an unbounded fluid, the integral in (77) can be taken simply over the surface of the body S_B.

In many problems, however, the body may move in a fluid bounded by other boundaries, such as the free surface, the fluid bottom, or possibly lateral boundaries such as canal walls. In each of these cases additional boundary conditions are imposed, and there is often a computational advantage in solving (77) if the source potential is modified to satisfy the same boundary conditions as ϕ. In this context the *Green function*

$$G(x,y,z;\xi,\eta,\zeta) = 1/r + H(x,y,z;\xi,\eta,\zeta), \tag{78}$$

where $\boldsymbol{H}$ is any function that satisfies the Laplace equation, can be substituted for the source potential in (76–77) since (71) is valid for the contribution from the regular function $\boldsymbol{H}$. Thus, with the Green function defined by (78), we can state that

$$\iint_S \left(\phi\frac{\partial G}{\partial n} - G\frac{\partial \phi}{\partial n}\right) dS = \begin{Bmatrix} 0 \\ -2\pi\phi(x,y,z) \\ -4\pi\phi(x,y,z) \end{Bmatrix}, \tag{79}$$

for (x, y, z) outside, on, or inside the closed surface S. The regular function $\boldsymbol{H}$ can be chosen to suit any additional boundary conditions that may be imposed. If a function $\boldsymbol{H}$ can be found with the property that

$$\partial G / \partial n = 0 \tag{80}$$

on the boundary surfaces of the fluid, the unknown term in the integrand of (79) vanishes. With this choice of the Green function, (79) provides an explicit solution for the potential in terms of the prescribed normal velocity on the boundaries. This Green function corresponds to the velocity potential of a source, in the presence of the appropriate fixed boundary surfaces of the problem. Unfortunately, this source potential is not known, except for some very simple body geometries.

One type of body geometry for which the source potential is known is the thin nonlifting planar surface that can be associated with a symmetrical thin hydrofoil at zero angle of attack and with a ship hull of small beam. For these situations the body surface is to first approximation a flat disc, and the source potential $1/r$ itself satisfies a condition of zero normal velocity on the disc, provided the source is situated on the disc. Thus, thin bodies of this type can be represented hydrodynamically by a center-plane distribution of simple sources, of strength proportional to the normal velocity on the body surface, provided only that the flow is symmetrical with respect to the center plane. This approximation forms the basis for two important topics: the use of a source distribution to represent the thickness effects of thin wings or hydrofoils, and the Michell theory for the wave resistance of thin ships.

4.12 Hydrodynamic Pressure Forces

One of the primary reasons for studying the fluid motion past a body is our desire to predict the forces and moments acting on the body due to the dynamic pressure of the fluid. Thus, we wish to consider the six components of the force and moment vectors, which are represented by the integrals of the pressure over the body surface, or

$$\mathbf{F} = \iint_{S_B} p\mathbf{n}\, dS, \tag{81}$$

$$\mathbf{M} = \iint_{S_B} p(\mathbf{r} \times \mathbf{n})\, dS. \tag{82}$$

Here, the normal vector $\mathbf{n}$ is taken to be positive when pointing out of the fluid volume and hence into the body. Substituting for the dynamic pressure from Bernoulli's equation, we get

$$\mathbf{F} = -\rho \iint_{S_B} \left[\frac{\partial \phi}{\partial t} + \frac{1}{2} \nabla \phi \cdot \nabla \phi \right] \mathbf{n}\, dS, \tag{83}$$

$$\mathbf{M} = -\rho \iint_{S_B} \left[\frac{\partial \phi}{\partial t} + \frac{1}{2} \nabla \phi \cdot \nabla \phi \right] (\mathbf{r} \times \mathbf{n}) dS. \tag{84}$$

Equations (83) and (84) can be recast by using Gauss' theorem,

$$\iiint_{\mathcal{V}} \nabla f \, d\mathcal{V} = \iint_{S} f\mathbf{n}\, dS, \tag{85}$$

together with the transport theorem in equation (3.11). First let us surround the body by a *fixed* control surface S_C, exterior to the body surface S_B as shown in figure 4.7. Thus $S_B + S_C$ forms a closed surface, enclosing a fluid volume $\mathcal{V}(t)$. The rate of change of the fluid momentum in this volume is

$$\begin{aligned} \rho \frac{d}{dt} \iint_{S_B+S_C} \phi \mathbf{n}\, dS &= \rho \frac{d}{dt} \iiint_{\mathcal{V}(t)} \nabla \phi \, d\mathcal{V} \\ &= \rho \iiint_{\mathcal{V}(t)} \nabla \left[\frac{\partial \phi}{\partial t} \right] d\mathcal{V} + \rho \iint_{S_B+S_C} \nabla \phi (\mathbf{U} \cdot \mathbf{n}) dS \\ &= \rho \iint_{S_B+S_C} \left[\frac{\partial \phi}{\partial t} \mathbf{n} + \nabla \phi (\mathbf{U} \cdot \mathbf{n}) \right] dS, \end{aligned} \tag{86}$$

where first Gauss' theorem and then the transport theorem have been used.

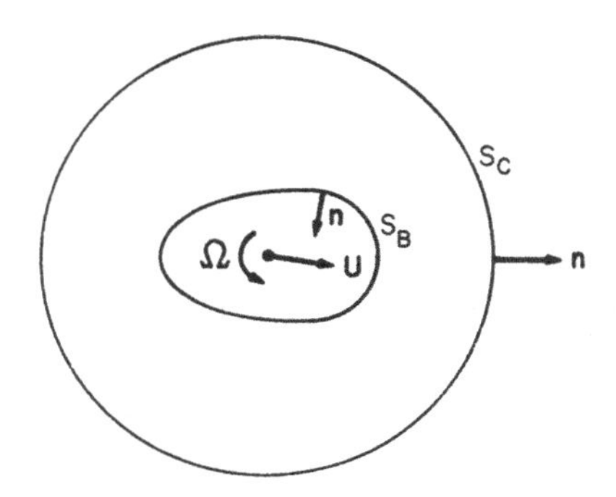

Figure 4.7
Fixed control surface S_C surrounding the moving body surface S_B.

On the fixed control surface S_C, $\mathbf{U}\cdot\mathbf{n} = 0$, and the time-derivative and surface integration can be interchanged. As a result, the first and last surface integrals in (86) may be equated separately over S_C and hence over S_B. Using the boundary condition $\partial\phi/\partial n = \mathbf{U}\cdot\mathbf{n}$ on the body surface gives

$$\rho\frac{d}{dt}\iint_{S_B}\phi\mathbf{n}\,dS = \rho\iint_{S_B}\left[\frac{\partial\phi}{\partial t}\mathbf{n} + \frac{\partial\phi}{\partial n}\nabla\phi\right]dS. \tag{87}$$

Adding (87) and (83) gives the hydrodynamic pressure force in the alternative form

$$\mathbf{F} = -\rho\frac{d}{dt}\iint_{S_B}\phi\mathbf{n}\,dS + \rho\iint_{S_B}\left(\partial\phi/\partial n\nabla\phi - \frac{1}{2}\nabla\phi\cdot\nabla\phi\mathbf{n}\right)dS. \tag{88}$$

Moreover, using the divergence and Gauss theorems,

$$\begin{aligned}\iint_{S_B+S_C}\left[\frac{\partial\phi}{\partial n}\nabla\phi - \frac{1}{2}\nabla\phi\cdot\nabla\phi\mathbf{n}\right]dS &= \iiint_{\mathcal{V}(t)}\left[\frac{\partial}{\partial x_j}\left(\frac{\partial\phi}{\partial x_j}\nabla\phi\right) - \frac{1}{2}\nabla\left(\frac{\partial\phi}{\partial x_j}\right)^2\right]d\mathcal{V} \\ &= \iiint_{\mathcal{V}(t)}\nabla\phi\frac{\partial^2\phi}{\partial x_j\partial x_j}d\mathcal{V} = \iiint_{\mathcal{V}(t)}\nabla\phi\nabla^2\phi d\mathcal{V} = 0.\end{aligned} \tag{89}$$

It follows that the second integral in (88) may be replaced by the negative of the same integral over S_C to give the desired expression for the force

$$\mathbf{F} = -\rho\frac{d}{dt}\iint_{S_B}\phi\mathbf{n}\,dS - \rho\iint_{S_C}\left[\frac{\partial\phi}{\partial n}\nabla\phi - \mathbf{n}\frac{1}{2}\nabla\phi\cdot\nabla\phi\right]dS. \tag{90}$$

A physical interpretation of this result can be inferred from the fact that the second member of (86) is the rate of change of fluid momentum enclosed between S_B and S_C, which must be equal to the sum of the external pressure, forces acting on $S_B + S_C$ plus the rate of flux of momentum across S_C. Thus

$$\begin{aligned}\mathbf{F}_B + \mathbf{F}_C + \rho\iint_{S_C}\frac{\partial\phi}{\partial n}\nabla\phi dS &= -\rho\frac{d}{dt}\iiint_{\mathcal{V}(t)}\nabla\phi d\mathcal{V} \\ &= -\rho\frac{d}{dt}\iint_{S_B}\phi\mathbf{n}\,dS - \rho\iint_{S_C}\frac{\partial\phi}{\partial t}\mathbf{n}\,dS.\end{aligned} \tag{91}$$

If Bernoulli's equation is used as in (83) to evaluate F_C, (90) follows directly.

Similarly, the relation

$$\iiint \nabla \times \mathbf{Q}\, d\mathcal{V} = \iint (\mathbf{n} \times \mathbf{Q})\, dS \tag{92}$$

can be used to derive from (84) an alternative expression for the moment in the form

$$\mathbf{M} = -\rho \frac{d}{dt} \iint_{S_B} \phi(\mathbf{r} \times \mathbf{n})\, dS - \rho \iint_{S_C} \mathbf{r} \times \left[\frac{\partial \phi}{\partial n} \nabla \phi - \frac{1}{2} \nabla \phi \cdot \nabla \phi \mathbf{n} \right] dS. \tag{93}$$

Equations (90) and (93) express the hydrodynamic pressure force and moment acting on the body surface S_B in an alternative form to the direct pressure integrals (83–84). By comparison with the latter, (90) and (93) involve a simpler integral over the body surface but an additional integral over the control surface surrounding the body. At first glance, the advantage of the new equations is not apparent—indeed they might appear more complicated. In practice, however, the flexibility afforded by the freedom to choose the control surface in an optimum manner often simplifies the computation of the force and moment. This is the principal justification for the analysis required to derive (90) and (93).

By virtue of (89), the control surface S_C could be placed within the body, in deriving (90) from (88), and similarly in (93). The only restriction here is that S_C must remain exterior to any singularities within the body, but it may be arbitrarily close to these singularities. On this basis one can derive the *Lagally theorem* for the force and moment, in terms of products of the singularity strengths and gradients of the velocity potential at the points of location of the singularities. Details are given by Cummins (1957), and by Landweber and Yih (1956), and reviewed by Milne-Thomson (1968).

Conversely, the evaluation of the integrals over S_C may be simplified by removing this control surface to the *far field,* where the details of the flow past the body are unimportant. This procedure will be followed in the next section, where we analyze the case of a rigid body moving in an otherwise infinite and unbounded fluid. For this particular problem, the integrals over S_C are simplified to the ultimate extent where they vanish!

4.13 Force on a Moving Body in an Unbounded Fluid

Now let us consider the hydrodynamic force and moment acting on a rigid body, which moves in an arbitrary manner in an otherwise unbounded fluid. In this case the contributions from the integrals over the control surface S_C, in (90) and (93), can be estimated in an analogous manner to that employed in connection with the corresponding integral in Green's theorem. In particular, for large spherical radius r, the potential[3] due to the body motions will tend to zero at a rate proportional to r^{-2}, and the gradient will vanish as r^{-3}. Let S_C be a sphere of large radius; since the surface element dS is proportional to r^2, the contribution from the integrals over S_C in (90) and (93) will be of order r^{-4} and r^{-3}, respectively. As r tends to infinity, these integrals will vanish, and the force and moment are given by the simple expressions

$$\mathbf{F} = -\rho \frac{d}{dt} \iint_{S_B} \phi n \, dS, \tag{94}$$

$$\mathbf{M} = -\rho \frac{d}{dt} \iint_{S_B} \phi (\mathbf{r} \times \mathbf{n}) \, dS. \tag{95}$$

These results imply that the total momentum of the fluid is equal to ρ times the integral in (94), and the moment of momentum is similarly proportional to the integral in (95).

In the special but important case of *steady translation* of the body, the integral in (94) will not depend on time, and hence we obtain D'Alembert's "paradox." This states that no hydrodynamic force acts on a body moving with constant translational velocity in an infinite, inviscid, and irrotational fluid. A moment may exist in this case, however. To verify this, we note that problems of steady translation are independent of time only when viewed in a reference frame moving with the body. If the body moves with velocity U_1 parallel to the x_1-axis, the appropriate moving coordinate system is

$$x_1' = x_1 - U_1 t, \qquad x_2' = x_2, \qquad x_3' = x_3;$$

the potential

$$\phi'(x_1', x_2', x_3') \equiv \phi(x_1' + U_1 t, x_2', x_3', t) \tag{96}$$

will be independent of time. On the body surface S_B, the unit normal also will be independent of time, but not the cross-product

$$\begin{aligned} \mathbf{r} \times \mathbf{n} &= (x_2 n_3 - x_3 n_2, x_3 n_1 - x_1 n_3, x_1 n_2 - x_2 n_1) \\ &= \mathbf{r}' \times \mathbf{n} + (0, -U_1 t n_3, U_1 t n_2). \end{aligned} \tag{97}$$

More generally, if $\mathbf{U}$ is the body velocity,

$$\mathbf{r} \times \mathbf{n} = \mathbf{r}' \times \mathbf{n} + t(\mathbf{U} \times \mathbf{n}). \tag{98}$$

Hence, for steady translation, the moment (95) is given by

$$\mathbf{M} = -\rho \iint_{S_B} \phi(\mathbf{U} \times \mathbf{n})\, dS = -\rho \mathbf{U} \times \iint_{S_B} \phi \mathbf{n}\, dS, \tag{99}$$

and it is perpendicular to the velocity vector of the body. This moment is equal to zero if the body is symmetrical with respect to the two directions normal to **U**. Thus, for example, a submarine in steady forward motion with port-starboard symmetry may experience only a pitching moment (bow up or bow down).

Now let us consider the most general case—unsteady motion with six degrees of freedom. If the translational velocity is $\mathbf{U}(t)$ and the body is rotating with angular velocity $\boldsymbol{\Omega}(t)$ about an origin that moves with the body, the velocity potential must satisfy the boundary condition

$$\frac{\partial \phi}{\partial n} = \mathbf{U} \cdot \mathbf{n} + \Omega \cdot (\mathbf{r}' \times \mathbf{n}) \tag{100}$$

on the body surface. Here $\mathbf{r}'$ is the radius vector from the center of rotation. It is convenient to define the six velocity components by the redundant notation

$$U = (U_1, U_2, U_3),$$
$$\boldsymbol{\Omega} = (\Omega_1, \Omega_2, \Omega_3) \equiv (U_4, U_5, U_6).$$

Thus, U_1, U_2, and U_3 denote the three components of translational velocity (surge, heave, and sway), and $\Omega_1 = U_4$, $\Omega_2 = U_5$, and $\Omega_3 = U_6$ denote the corresponding rotational velocity components (roll, yaw, and pitch).

The boundary condition (100) suggests that the total potential be expressed as the sum

$$\phi = U_i \phi_i. \tag{101}$$

Here the summation convention applies, with $i = 1, 2, \ldots, 6$. Physically, each ϕ_i represents the velocity potential due to a body motion with unit velocity in the ith mode. Note that the dimensions of ϕ are $\mathbf{L}^2\mathbf{T}^{-1}$, whereas those of **U** and $\boldsymbol{\Omega}$ are $\mathbf{LT}^{-1}$ and $\mathbf{T}^{-1}$, respectively. Therefore the potentials ϕ_i in (101) are not dimensionally homogeneous; the units are **L** and $\mathbf{L}^2$, respectively, for $i = 1, 2, 3$, and $i = 4, 5, 6$.

The potential defined by (101) satisfies the body boundary condition (100), provided each component ϕ_i satisfies the corresponding condition

$$\frac{\partial \phi_i}{\partial n} = n_i, \qquad i = 1,2,3, \tag{102}$$

$$\frac{\partial \phi_i}{\partial n} = (\mathbf{r}' \times \mathbf{n})_{i-3}, \quad i = 4,5,6. \tag{103}$$

The boundary-value problem for each ϕ_i is completed by the usual requirements that $\nabla^2 \phi_i = 0$ in the fluid volume and $\phi_i = O(r^{-2})$, or smaller, as $r \to \infty$. These six potentials, when expressed in body-fixed coordinates, will depend only on the body geometry via the boundary conditions (102) and (103), and do not depend on the velocities U_i or time. Thus, the linear decomposition (101) isolates the time dependence of the velocity potential, and substitution in (94) yields the hydrodynamic force expression

$$\mathbf{F} = -\rho \frac{d}{dt} U_i(t) \iint_{S_B} \phi_i(x_1', x_2', x_3') \mathbf{n} \, dS. \tag{104}$$

This last integral will depend on time, because of the effect of body rotation on the vector $\mathbf{n}$. From vector analysis this contribution is

$$\frac{d\mathbf{n}}{dt} = \boldsymbol{\Omega} \times \mathbf{n}, \tag{105}$$

since $\boldsymbol{\Omega}$ is the rotational velocity vector of the body-fixed coordinates. From (104) it follows that

$$\mathbf{F} = -\rho \dot{U}_i \iint_{S_B} \phi_i \mathbf{n} \, dS - \rho U_i \boldsymbol{\Omega} \times \iint_{S_B} \phi_i \mathbf{n} \, dS. \tag{106}$$

The corresponding result for the moment can be derived from (95) after decomposition of the position vector r in the form

$$\mathbf{r} = \mathbf{r}_0(t) + \mathbf{r}'. \tag{107}$$

Here $\mathbf{r}_0$ is the vector from the origin of the fixed reference system to the point $\mathbf{r}' = 0$ which moves with the body, and thus $d\mathbf{r}_0/dt = \mathbf{U}$. On the other hand, $(\mathbf{r}' \times \mathbf{n})$ is a vector fixed with respect to the body, and thus its time-derivative is $\boldsymbol{\Omega} \times (\mathbf{r}' \times \mathbf{n})$ as in (105). Substituting (107) in (95) and evaluating the time derivative, the moment can be expressed in the form

$$\mathbf{M} = \mathbf{r}_0 \times \mathbf{F} - \rho U_i \mathbf{U} \times \iint_{S_B} \phi_i \mathbf{n}\, dS - \rho \dot{U}_i \iint_{S_B} \phi_i (\mathbf{r}' \times \mathbf{n})\, dS$$
$$- \rho U_i \boldsymbol{\Omega} \times \iint_{S_B} \phi_i (\mathbf{r}' \times \mathbf{n})\, dS. \tag{108}$$

The first term in (108) is the moment due to the force **F** acting at the body origin **r**' = 0; therefore, the remaining terms in this expression give the moment about the same point.

At this stage the time derivatives in (94–95) have been evaluated with respect to the space-fixed reference frame **r**. This reference frame loses its significance if the force and moment are redefined with respect to the body-fixed coordinates. In fact, the need to discriminate between these two coordinate systems can be avoided simply by setting $\mathbf{r}_0 = 0$, with the assumption that the two coordinate systems coincide at the particular instant of time under consideration. However, this artifice is not essential, and we shall proceed hereafter without assumption as to the origin of the space-fixed coordinates.

An indicial notation will be adopted for the components of the force and moment, with respect to the body-fixed reference frame. These will be denoted by F_j and M_j respectively. The subsequent representation of cross-products is facilitated by using the *alternating tensor* ε_{jkl}, which is equal to +1 if the indices are in cyclic order (123, 231, 312), equal to −1 if the indices are acyclic (132, 213, 321), and equal to zero if any pair of the indices are equal. With this notation, the *j*th component of a vector cross-product can be written in the indicial notation as

$$(\mathbf{A} \times \mathbf{B})_j = \varepsilon_{jkl} A_k B_l, \tag{109}$$

where the summation convention is implied for *k* and *l*.

With the conventions noted above (106) may be rewritten in the indicial form

$$F_j = -\rho \dot{U}_i \iint_{S_B} \phi_i n_j\, dS - \rho \varepsilon_{jkl} U_i \Omega_k \iint_{S_B} \phi_i n_l\, dS. \tag{110}$$

Similarly, from (108), the components of the moment about the body-fixed origin are given by

$$M_j = -\rho \dot{U}_i \iint_{S_B} \phi_i (\mathbf{r}' \times \mathbf{n})_j\, dS - \rho \varepsilon_{jkl} U_i \Omega_k \iint_{S_B} \phi_i (\mathbf{r}' \times \mathbf{n})_l\, dS$$
$$- \rho \varepsilon_{jkl} U_i U_k \iint_{S_B} \phi_i n_l\, dS. \tag{111}$$

It is apparent that the force and moment depend on the body shape and on the potentials ϕ_i only in terms of the integrals shown in (110–111); in view of the boundary conditions (102–103), these integrals can be written in the alternative forms

$$\iint_{S_B} \phi_i n_j \, dS = \iint_{S_B} \phi_i \frac{\partial \phi_j}{\partial n} dS, \tag{112}$$

$$\iint_{S_B} \phi_i (\mathbf{r}' \times \mathbf{n})_j \, dS = \iint_{S_B} \phi_i \frac{\partial \phi_{j+3}}{\partial n} dS. \tag{113}$$

The importance of these quantities in (110–111) suggests the definition of the *added-mass* tensor

$$m_{ji} = \rho \iint_{S_B} \phi_i \frac{\partial \phi_j}{\partial n} dS \tag{114}$$

since, with this substitution in (110–111), the latter equations can be expressed in the form

$$F_j = -\dot{U}_i m_{ji} - \varepsilon_{jkl} U_i \Omega_k m_{li}, \tag{115}$$

$$M_j = -\dot{U}_i m_{j+3,i} - \varepsilon_{jkl} U_i \Omega_k m_{l+3,i} - \varepsilon_{jkl} U_i U_k m_{li}. \tag{116}$$

Here the three indices j, k, l take on the values 1, 2, 3, whereas the index i is used to denote the six components of the velocity potential in accordance with (101).

These expressions (115–116) for the force and moment depend only on the body velocity and acceleration components and the added-mass coefficients m_{ij}. No surface integrals are involved, other than those in (114) which define the added-mass coefficients; so from a computational standpoint, (115) and (116) are much simpler than the pressure integrals (81) and (82). Indeed, this is the principal justification for the extensive analysis just performed. The added-mass coefficients depend only on the body geometry; they can be regarded as the most important hydrodynamic characteristics of the body, except in the case of steady translation where (115) predicts no force and the viscous drag is clearly more important.

It may be worthwhile to review briefly the assumptions implicit in (115–116). First, the fluid is ideal and irrotational, and viscous forces are neglected. Second, the body has been taken to be rigid and of constant volume; its motions are defined by three translational and three rotational velocity components. Third, except for the presence of the body itself, the fluid is unbounded and of infinite extent.

For a variety of problems in marine hydrodynamics, the second assumption is the most acceptable of the three. We have shown that the assumption of irrotational flow is consistent with the assumption of an ideal fluid motion; however, important modifications are required for lifting surfaces with sharp trailing edges and associated shed vorticity. The relative importance of the inertial forces and moments (115–116) by comparison to viscous forces and moments is discussed in section 2.11. Finally, with respect to the third assumption, (115–116) will have to be modified in cases where the fluid domain is bounded by other boundaries. The presence of a free surface is the most important and troublesome of these possibilities, but simpler problems also occur, as, for example, when the body moves in proximity to a plane rigid wall. The latter situation can be dealt with by the *method of images,* an approach which will be described briefly in the final section of this chapter.

4.14 General Properties of the Added-Mass Coefficients

As the name implies, there is an important analogy of the added-mass coefficients m_{ij} with the body mass and moments of inertia. In section 4.16 we shall demonstrate this analogy by showing that the force and moment due to the body mass are of precisely the same form as (115–116). Thus from the physical standpoint the added-mass coefficients represent the amount of fluid accelerated with the body. However, the added-mass coefficients for translation generally differ depending on the direction of the body motion, as opposed to Newton's equation $F = ma$ where the body mass m is independent of the direction of the acceleration a. Also, in the absence of body symmetry, the cross-coupling coefficients such as m_{12}, m_{13}, and m_{23} are non-zero, implying that the hydrodynamic force differs in direction from the acceleration.

The added mass can be interpreted as a particular volume of fluid particles that are accelerated with the body. Strictly, however, the particles of fluid adjacent to the body will accelerate to varying degrees, depending on their position relative to the body. In principle, every fluid particle will accelerate to some extent as the body moves, and the added mass is a weighted integration of this entire mass.

One convenient feature of the added-mass coefficients is their symmetry, $m_{ij} = m_{ji}$. To confirm this property, Green's theorem (71) is applied to the potentials ϕ_i and ϕ_j over the closed surface $S_B + S_C$. The contribution from the external surface S_C will vanish as $r \to \infty$, as noted above. Hence, it follows that

$$\iint_{S_B} \left[\phi_i \frac{\partial \phi_j}{\partial n} - \phi_j \frac{\partial \phi_i}{\partial n} \right] dS = 0, \tag{117}$$

or

$$m_{ij} = m_{ji}. \tag{118}$$

Thus, there are twenty-one independent coefficients, and this number is reduced substantially if the body is symmetrical about one or more axes.

A simple relation exists between the added-mass coefficients and the kinetic energy of the fluid. From (114) and the divergence theorem,

$$\begin{aligned} m_{ij} &= \rho \iint_{S_B} \phi_j \frac{\partial \phi_i}{\partial n} dS = \rho \iiint_{\mathcal{V}} \nabla \cdot (\phi_j \nabla \phi_i) d\mathcal{V} \\ &= \rho \iiint_{\mathcal{V}} (\nabla \phi_i \cdot \nabla \phi_j) d\mathcal{V} \end{aligned} \tag{119}$$

since $\nabla^2 \phi_i = 0$. Here, the vanishing of ϕ at infinity has been invoked to omit the surface integral at infinity, and the volume integrals are over the entire fluid volume. The total velocity potential in the fluid is given by (101), and the kinetic energy of the *fluid* is

$$\begin{aligned} T &= \frac{1}{2} \rho \iiint (\nabla \phi \cdot \nabla \phi) d\mathcal{V} = \frac{1}{2} \rho \iiint (U_i \nabla \phi_i) \cdot (U_j \nabla \phi_j) d\mathcal{V} \\ &= \frac{1}{2} \rho U_i U_j \iiint \nabla \phi_i \cdot \nabla \phi_j d\mathcal{V}. \end{aligned} \tag{120}$$

Combining this result with (119) gives

$$T = \frac{1}{2} U_i U_j m_{ij}. \tag{121}$$

Therefore the added-mass coefficients are the constants in a quadratic equation for $2T$, in terms of the body velocities. This result can be obtained from physical arguments by equating the change in fluid energy to the work done in accelerating the body, but for rotational modes (i = 4, 5, 6) the rate of rotation of coordinates must be accounted for, as in equations (105) and (107).

The third and final property of the added-mass coefficients considered here is a relation between the translation added-mass coefficients and the dipole moment describing the fluid motion at large distances from the body. For sufficiently large radial distances r from the body, the disturbance due to the body can be written in the form

$$\phi = \frac{A_0}{r} + \mathbf{A} \cdot \nabla \left[\frac{1}{r} \right] + O(1/r^3). \tag{122}$$

Here, A_0 is the net source strength, $\mathbf{A}$ the dipole moment, and neglected terms correspond to higher-order multipoles (quadrupoles, etc.). If the body is rigid, there can be no net flux out of the body surface S_B, and hence from continuity the source strength $A_0 = 0$. For symmetrical rotational modes the dipole moment will vanish, and thus we restrict ourselves here to the translations ($i = 1, 2, 3$), where it can be assumed that

$$\phi_i \simeq A_{ij} \frac{\partial}{\partial x_j} \frac{1}{r}, \quad \text{as } r \to \infty. \tag{123}$$

Thus, the far-field disturbance is identical to that of a dipole or a combination of dipoles. We recall the case of a sphere (38) where from symmetry $A_{ij} = 0$ for $i \neq j$, and where (123) is exact for all r. In effect, translation of an arbitrary three-dimensional body will cause a far-field fluid motion qualitatively similar to a sphere, and the details of the body shape are unimportant in the far field.

An equation for the dipole moment A_{ij} can be obtained from Green's theorem. Using (76), we may write the potential at a point exterior to the body in the form

$$\begin{aligned} \phi(x,y,z) = \frac{1}{4\pi} \iint_{S_B} \Big\{ & [(x-\xi)^2 + (y-\eta)^2 + (z-\zeta)^2]^{-1/2} \frac{\partial \varphi}{\partial n} \\ & - \phi \frac{\partial}{\partial n} [(x-\xi)^2 + (y-\eta)^2 + (z-\zeta)^2]^{-1/2} \Big\} dS, \end{aligned} \tag{124}$$

where the first term is a distribution of sources and the second a distribution of normal dipoles. If $r = [x^2 + y^2 + z^2]^{1/2}$ is large compared to the body dimensions and terms of order r^{-3} are neglected, we can write from Taylor series expansions

$$[(x-\xi)^2 + (y-\eta)^2 + (z-\zeta)^2]^{-1/2} \simeq \frac{1}{r} - \boldsymbol{\xi} \cdot \nabla \left[\frac{1}{r} \right]$$

and

$$\frac{\partial}{\partial n}[(x-\xi)^2+(y-\eta)^2+(z-\zeta)^2]^{-1/2} \simeq -\mathbf{n}\cdot\nabla\left[\frac{1}{r}\right].$$

Thus, the potential at large distances is

$$\phi=\frac{1}{4\pi}\left[\iint\frac{\partial\phi}{\partial n}dS\right]\left[\frac{1}{r}\right]-\frac{1}{4\pi}\left[\iint\left(\xi\frac{\partial\phi}{\partial n}-\phi\mathbf{n}\right)dS\right]\cdot\nabla\left[\frac{1}{r}\right], \tag{125}$$

which may be compared with (122). The first term vanishes if there is no net flux through the body surface, and

$$\phi=-\frac{1}{4\pi}\left[\iint_{S_B}\left(\xi\frac{\partial\phi}{\partial n}-\phi\mathbf{n}\right)dS\right]\cdot\nabla\left[\frac{1}{r}\right]\equiv\mathbf{A}\cdot\nabla\left[\frac{1}{r}\right]. \tag{126}$$

Thus, the coefficients A_{ij} in (123) may be expressed by the integrals

$$A_{ij}=-\frac{1}{4\pi}\iint_{S_B}\left(x_j\frac{\partial\phi_i}{\partial n}-\phi_i n_j\right)dS. \tag{127}$$

Moreover, recalling the body boundary conditions (102), equation (127) may be replaced by

$$\begin{aligned}A_{ij}&=-\frac{1}{4\pi}\iint\left(x_j n_i-\phi_i\frac{\partial\phi_j}{\partial n}\right)dS\\&=\frac{1}{4\pi}(\forall\delta_{ij}+m_{ij}/\rho),\quad i,j=1,2,3.\end{aligned} \tag{128}$$

Here, $\forall$ is the body volume, m_{ij} the added-mass coefficient as defined by (112), and δ_{ij} is the Kroenecker delta function

$$\delta_{ij}=\begin{cases}0 & \text{if } i\neq j;\\ 1 & \text{if } i=j.\end{cases}$$

Thus, from equation (128), the far-field disturbance and associated dipole moment A_{ij} are directly related to the body volume and the added-mass coefficients m_{ij}. From the symmetry of m_{ij} it follows that $A_{ij}=A_{ji}$. Finally, we note without rederivation the analogous results for two-dimensional motion,

$$\phi_i\simeq A_{ij}\frac{\partial}{\partial x_j}\log r, \tag{129}$$

and

$$A_{ij} = -\frac{1}{2\pi}(S\delta_{ij} + m_{ij} / \rho), \tag{130}$$

where S is the cross-sectional area.

4.15 The Added Mass of Simple Forms

The simplest examples of added-mass coefficients are those for a circular two-dimensional cylinder, and in three dimensions for a sphere. For the circular section, the dipole moment is obtained from equation (39), and comparison with (129) gives

$$A_{ij} = -R^2\delta_{ij}. \tag{131}$$

Here R is the cylinder radius, and the minus sign is inserted because the streaming flow is positive in (39); hence the body velocity relative to the fluid is $-U$. From (130) it follows that

$$m_{ij} = \pi\rho R^2\delta_{ij} = \rho S\delta_{ij}. \tag{132}$$

Thus, the added mass of a circular cylinder is precisely equal to the displaced mass of fluid. For a sphere, equation (38) yields the dipole moment

$$A_{ij} = \tfrac{1}{2}R^3\delta_{ij}. \tag{133}$$

The minus sign in this equation is canceled by the minus sign associated with the derivative in (123). From (128) it follows that

$$m_{ij} - \tfrac{2}{3}\pi\rho R^3\delta_{ij} = \tfrac{1}{2}\rho\forall\delta_{ij}, \tag{134}$$

so that for a sphere, the added mass is half of the displaced mass. This reduced added-mass effect, by comparison with the two-dimensional case, can be explained by stating that the sphere obstructs less fluid, since the fluid can flow around the sphere in all meridional planes.

In many situations involving elongated or cylindrical bodies, such as slender pilings, ship hulls, and mooring cables, the three-dimensional added-mass coefficients can be approximated by a *strip theory* synthesis, in which the flow at each section is assumed to be locally two dimensional. Thus, the simpler two-dimensional added-mass coefficients are of practical importance. The case of a circular cylinder is given by (132), and the corresponding results for an elliptical cylinder can be derived in a similar manner, taking advantage of the conformal transformation (51) to map the ellipse onto a circle. The added-mass coefficients of an ellipse are closely related to those of a circle, as shown in table 4.3. For acceleration parallel to either of the principal axes of the ellipse, the added mass is equal to that of a circle having the same radius as the semiaxis normal to the motion. The same results hold for a flat plate, or an ellipse of vanishing minor axis, where the added mass for motion normal to the plate is equal to the displaced mass of the circumscribed circle.

The last result can be confirmed from the complex potential (57), either by direct integration over the plate using (114) or by computing the effective dipole moment and using (130). For acceleration in the plane of the plate there is no added-mass force, in the absence of viscous effects. Similarly, the added moment of inertia of the circle is zero, but nonzero for all other cases.

Table 4.3
Added-Mass Coefficients for Various Two-Dimensional Bodies.

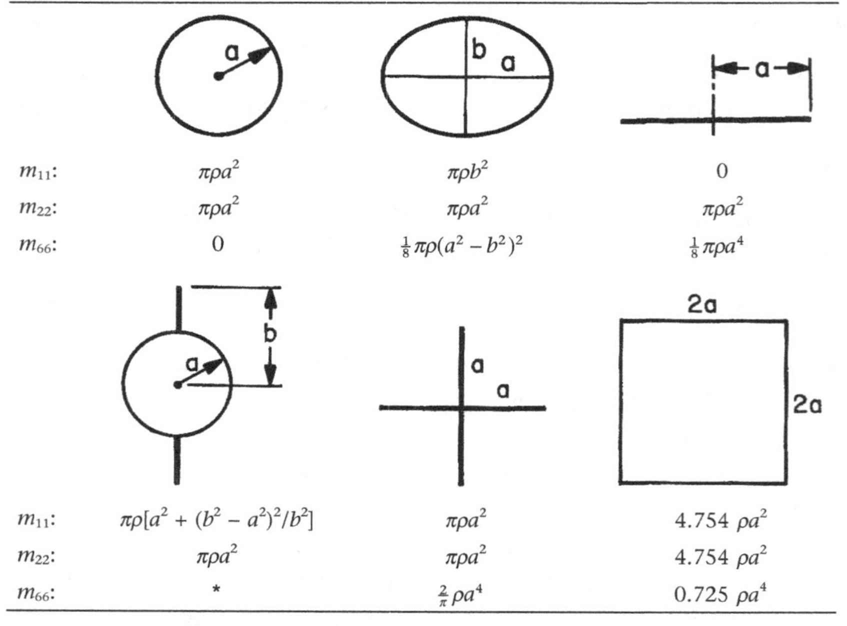

	Circle	Ellipse	Plate
m_{11}:	$\pi\rho a^2$	$\pi\rho b^2$	0
m_{22}:	$\pi\rho a^2$	$\pi\rho a^2$	$\pi\rho a^2$
m_{66}:	0	$\frac{1}{8}\pi\rho(a^2 - b^2)^2$	$\frac{1}{8}\pi\rho a^4$

	Finned circle	Crossed fin	Square
m_{11}:	$\pi\rho[a^2 + (b^2 - a^2)^2/b^2]$	$\pi\rho a^2$	$4.754\ \rho a^2$
m_{22}:	$\pi\rho a^2$	$\pi\rho a^2$	$4.754\ \rho a^2$
m_{66}:	*	$\frac{2}{\pi}\rho a^4$	$0.725\ \rho a^4$

*For the finned circle the added moment of inertia is given by the formula $m_{66} = \rho a^4(\pi^{-1}\csc^4\alpha[2\alpha^2 - \alpha\sin 4\alpha + \frac{1}{2}\sin^2 2\alpha] - \pi/2)$, where $\sin\alpha = 2ab/(a^2 + b^2)$ and $\pi/2 < \alpha < \pi$. The derivations of this result, and of m_{66} for the crossed fin and square, are given by Newman (1978). Derivations of the remaining coefficients in this table are given or cited by Kennard (1967).

Table 4.3 lists the added-mass coefficients for several two-dimensional bodies including the circle, ellipse, flat plate, finned circle, crossed flat plates, and square. The symmetry of these bodies is such that the remaining added-mass coefficients are zero. Thus in these cases there is no *coupling* between the three principal modes of motion of the body. The results shown in table 4.3 can be derived by appropriate conformal mappings, some of which are illustrated in the problems listed at the end of this chapter.

In three dimensions, ellipsoids are the most general bodies where comparable analytical results are available. Since there are three planes of symmetry, the only nonzero added-mass coefficients are the six values of m_{ij} where $i = j$. These will depend on the two nondimensional ratios defining the shape of the ellipsoid; appropriate graphs may be found in Kochin, Kibel, and Roze (1964).

A simpler three-dimensional body is the spheroid, or ellipsoid of revolution. If the coordinate x_1 is chosen to coincide with the axis of revolution, with a the semilength of this axis, and b the radius at the equatorial plane $x_1 = 0$, three nonzero coefficients must be considered, including the longitudinal added mass m_{11}, the lateral added mass $m_{22} = m_{33}$, and the rotational added mass $m_{55} = m_{66}$. These will depend on the ratio b/a or the beam-length ratio of the body.

Graphs of the three added-mass coefficients for a spheroid are shown in figure 4.8. The upper half of this figure shows the coefficients nondimensionalized in terms of the mass and moment of inertia of the displaced volume of fluid. In the limiting case where $b/a \to 0$, or a slender spheroid, the longitudinal added-mass coefficient tends to zero in a manner analogous to the flat plate moving in its own plane. On the other hand, the corresponding limits for the lateral added-mass coefficient and added moment of inertia are both equal to 1.0, as one would predict using the strip theory and the added mass of a circular cylinder. For increasing values of b/a, the longitudinal added mass becomes significant, whereas the lateral added-mass coefficient decreases because of the three-dimensional effects. The case $b/a = 1$ coincides with the sphere, where the translational added-mass equals one half of the displaced mass, and the rotational coefficients vanish. For $b/a > 1$, the spheroid is flattened, or *oblate*. For $b/a \to \infty$, corresponding to a circular disc, the coefficients based on displaced mass and mass-moment are degenerate; to avoid this difficulty the coefficients are renormalized in the lower half of figure 4.8.

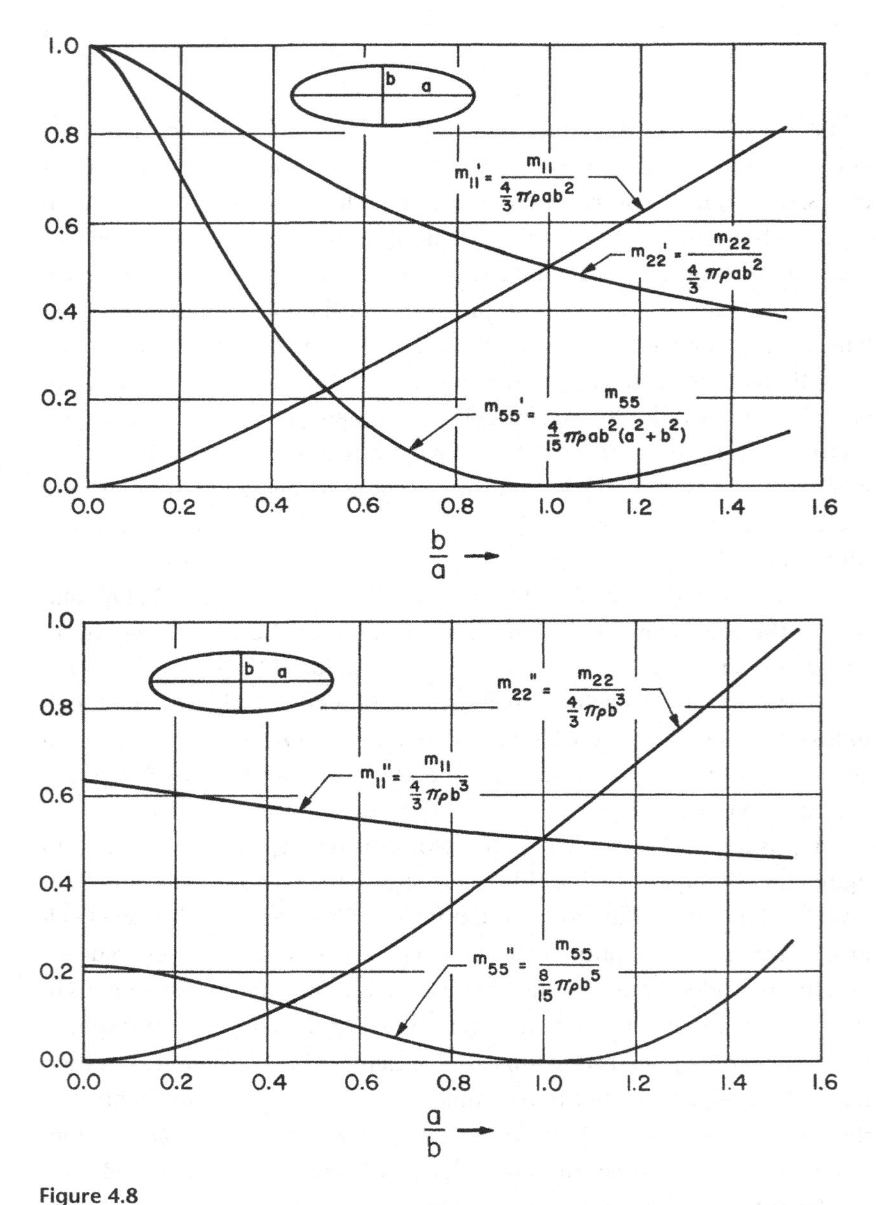

Figure 4.8
Added-mass coefficients for a spheroid, of length $2a$ and maximum diameter $2b$. The added mass m_{11} corresponds to longitudinal acceleration, m_{22} to lateral acceleration in the equatorial plane, and m_{55} denotes the added moment of inertia, for rotation about an axis in the equatorial plane. In the upper figure, the coefficients are non-dimensionalized with respect to the mass and moment of inertia of the displaced volume of the fluid, and in the lower figure with respect to the same quantities for a sphere of radius b.

4.16 The Body-Mass Force

It is useful to note here the similarity between Newton's equations, for the force and moment associated with the inertia of the body mass, and the corresponding hydrodynamic pressure force and moment. This analogy is of interest for its own sake, and the force and moment due to the body mass will be required in chapters 6 and 7 to derive the equations of motion for unrestrained vessels.

If ρ_B denotes the mass-density of the body, which is a function of position, the inertial force associated with this mass is given by the rate of change of linear momentum, in the form

$$\mathbf{F} = \frac{d}{dt}\iiint_{\mathcal{V}_B} \rho_B(\mathbf{U} + \boldsymbol{\Omega} \times \mathbf{r}')\,d\mathcal{V}. \tag{135}$$

Here $\mathcal{V}_B$ denotes integration over the body volume. Similarly, from conservation of angular momentum, the moment is given by

$$\mathbf{M} = \frac{d}{dt}\iiint_{\mathcal{V}_B} \rho_B \mathbf{r} \times (\mathbf{U} + \boldsymbol{\Omega} \times \mathbf{r}')\,d\mathcal{V}. \tag{136}$$

This pair of integrals, for the linear and angular momentum of the body, may be compared with the surface integrals (94–95) which give the corresponding momentum of the fluid.

To take advantage of the subsequent reduction of (94–95), a decomposition of the body velocity can be made, corresponding to the gradient of (101) for the fluid velocity. For this purpose we define six vectors $\mathbf{b}_j$ such that

$$\mathbf{U} + \Omega \times \mathbf{r}' = U_j \mathbf{b}_j. \tag{137}$$

Equating the factors of the six velocity components U_j in (137), it follows that

$$(\mathbf{b}_1, \mathbf{b}_2, \mathbf{b}_3) = (\mathbf{i}, \mathbf{j}, \mathbf{k}), \tag{138}$$

$$(\mathbf{b}_4, \mathbf{b}_5, \mathbf{b}_6) = (\mathbf{i}, \mathbf{j}, \mathbf{k}) \times \mathbf{r}', \tag{139}$$

where $(\mathbf{i}, \mathbf{j}, \mathbf{k})$ are the unit vectors parallel to (x, y, z).

The hydrodynamic pressure force and moment are related by (115–116) to the added-mass coefficients m_{ij}, which can be defined in terms of the kinetic energy of the fluid volume by (119). Thus, by analogy, the coefficients of the body-mass matrix can be defined as

$$M_{ij} = \iiint_{\mathcal{V}_B} \rho_B(\mathbf{b}_i \cdot \mathbf{b}_j)\, d\mathcal{V}. \tag{140}$$

From the definitions (138–139), the matrix of body-inertia coefficients follows in the more conventional form

$$M_{ij} = \begin{Bmatrix} m & 0 & 0 & 0 & mz_G & -my_G \\ 0 & m & 0 & -mz_G & 0 & mx_G \\ 0 & 0 & m & my_G & -mx_G & 0 \\ 0 & -mz_G & my_G & I_{11} & I_{12} & I_{13} \\ mz_G & 0 & -mx_G & I_{21} & I_{22} & I_{23} \\ -my_G & mx_G & 0 & I_{31} & I_{32} & I_{33} \end{Bmatrix}. \tag{141}$$

Here the body mass is

$$m = \iiint_{\mathcal{V}_B} \rho_B\, d\mathcal{V}, \tag{142}$$

the vector position of the center of gravity is

$$\mathbf{x}_G = \frac{1}{m} \iiint_{\mathcal{V}_B} \rho_B \mathbf{x}\, d\mathcal{V}, \tag{143}$$

and the moments of inertia can be defined by

$$I_{ij} = \iiint_{\mathcal{V}_B} \rho_B[\mathbf{x}' \cdot \mathbf{x}' \delta_{ij} - x_i' x_j']\, d\mathcal{V}, \tag{144}$$

where δ_{ij} is the Kroenecker delta function.

With these definitions and M_{ij} substituted for m_{ij}, equations (115–116) can be used to evaluate the force and moment (135–136). Alternatively, the total force including the body inertia and fluid pressure can be evaluated from (115–116), in terms of the *virtual-mass* coefficients $(M_{ij} + m_{ij})$. Thus the body behaves in the fluid as it would in a vacuum, but the body-mass matrix is replaced by the virtual-mass.

4.17 Force on a Body in a Nonuniform Stream

Some of the preceding results can be extended to a body in a nonuniform stream. The fundamental assumption is that the nonuniformity of the stream is *slowly varying*, relative to the length scale of the body. A practical and important example of this situation is the case where the body is acted upon by an incident wave field, with the restriction that the body is small compared to the wavelength.

Let us assume that the fluid motion in the absence of the body is described by a velocity potential $\Phi(x, t)$. This may be unsteady in time, but the length scale characterizing changes in Φ, say $\Phi / |\nabla \Phi|$, is assumed large compared to the dimensions of the body. To treat the simplest case, we assume that the orientation of the body, with respect to the stream, is such that the stream velocity at the body is parallel to the x-axis with local velocity $U = \partial \Phi / \partial x$, and that the body symmetry is such that with this orientation of the coordinates, $m_{12} = m_{13} = 0$.

From the kinematic viewpoint, the disturbance of the stream due to the body depends only on the relative flow between the body and the fluid. Thus, with the assumptions noted above, the total potential including the body disturbance is given by

$$\phi = \Phi + (U_1 - U)\phi_1. \tag{145}$$

Here U_1 is the velocity of the body, and ϕ_1 the corresponding velocity potential, as defined by (101). Only the velocity U_1 is included here since, with the symmetry noted above, there is no interaction between the remaining body velocities and the nonuniformity of the stream.

The pressure force F_x acting on the body in the x-direction may be computed by substituting (145) in (90):

$$\begin{aligned} F_x = &-\rho \frac{d}{dt} \iint_{S_B} [\Phi + (U_1 - U)\phi_1] n_x \, dS \\ &- \rho \iint_{S_C} \left\{ \left[\frac{\partial \Phi}{\partial n} + (U_1 - U)\frac{\partial \phi_1}{\partial n} \right] \left[\frac{\partial \Phi}{\partial x} + (U_1 - U)\frac{\partial \phi_1}{\partial x} \right] \right. \\ &\left. - \frac{1}{2} n_x [\nabla \Phi + (U_1 - U)\nabla \phi_1]^2 \right\} dS. \end{aligned} \tag{146}$$

The integral over the body surface S_B can be evaluated using Gauss' theorem for the term involving Φ and (112) for the term involving ϕ_1. Since $U = \partial \Phi / \partial x$ is effectively constant over the length scale of S_B, it follows that

$$\begin{aligned} \iint_{S_B} [\Phi + (U_1 - U)\phi_1] n_x \, dS &= -\iiint_{V_B} \frac{\partial \Phi}{\partial x} dV + \frac{1}{\rho} m_{11}(U_1 - U) \\ &\simeq \frac{1}{\rho} m_{11} U_1 - U(\forall + m_{11} / \rho). \end{aligned} \tag{147}$$

Thus the first term in (146) contributes a force equal to

$$m_{11}\left[\frac{dU}{dt}-\frac{dU_1}{dt}\right]+\rho\forall\frac{dU}{dt}.$$

Here dU/dt represents the value of $\partial^2\Phi(\mathbf{x},t)/\partial x\partial t$ at the body.

The integral over the control surface S_C can be divided into three separate contributions, associated respectively with the stream potential Φ, the body potential ϕ_1, and cross-terms involving both. Since $\nabla^2\Phi = 0$ throughout the entire region interior to S_C, including the interior of S_B, (89) can be applied without S_B to show that there is no contribution from the terms involving Φ alone. Moreover, by the argument leading to (94), there is no contribution to the control-surface integral from ϕ_1 by itself, and only the cross-terms remain to be considered.

For the contribution from the cross-terms, a straightforward extension of (89) shows that the control surface S_C can be chosen arbitrarily, so long as it is exterior to the body. We shall locate this surface in the far field of the body, where the dipole approximation (123) is valid for the potential. If we use (128) for the dipole moment and consider only the cross-terms, the last integral in (146) takes the form

$$\begin{aligned}(U_1-U)&\iint_{S_C}\left[\frac{\partial\Phi}{\partial n}\frac{\partial\phi_1}{\partial x}+\frac{\partial\Phi}{\partial x}\frac{\partial\phi_1}{\partial n}-n_x\nabla\Phi\cdot\nabla\phi_1\right]dS\\&=\frac{1}{4\pi}(\forall+m_{11}/\rho)(U_1-U)\iint_{S_C}\left[\frac{\partial\Phi}{\partial n}\frac{\partial^2}{\partial x^2}\left(\frac{1}{r}\right)+\frac{\partial\Phi}{\partial x}\frac{\partial^2}{\partial n\partial x}\left(\frac{1}{r}\right)\right.\\&\left.\quad-n_x\nabla\Phi\cdot\nabla\frac{\partial}{\partial x}\left(\frac{1}{r}\right)\right]dS.\end{aligned}\tag{148}$$

The right-hand side of (148) can be expressed in the form

$$\begin{aligned}&\frac{1}{4\pi}(\forall+m_{11}/\rho)(U_1-U)\iint_{S_C}\left[\frac{\partial^2\Phi}{\partial x^2}\frac{\partial}{\partial n}\left(\frac{1}{r}\right)-\frac{1}{r}\frac{\partial}{\partial n}\frac{\partial^2\Phi}{\partial x^2}+(\nabla\times\mathbf{A})\cdot\mathbf{n}\right]dS\\&=-(\forall+m_{11}/\rho)(U_1-U)\left[\frac{\partial^2\Phi}{\partial x^2}\right]_{r=0},\end{aligned}\tag{149}$$

where the components of the vector A are

$$\begin{aligned}A_x&=0,\\A_y&=-\frac{\partial\Phi}{\partial z}\frac{\partial}{\partial x}\left[\frac{1}{r}\right]+\frac{1}{r}\frac{\partial^2\Phi}{\partial x\partial z}-\frac{\partial\Phi}{\partial x}\frac{\partial}{\partial z}\left[\frac{1}{r}\right],\\A_z&=\frac{\partial\Phi}{\partial y}\frac{\partial}{\partial x}\left[\frac{1}{r}\right]-\frac{1}{r}\frac{\partial^2\Phi}{\partial x\partial y}+\frac{\partial\Phi}{\partial x}\frac{\partial}{\partial y}\left[\frac{1}{r}\right].\end{aligned}$$

Here Laplace's equation has been used, for Φ and its derivatives as well as for the source potential; the right-hand side of (149) follows by using Green's theorem in the form (76) and noting from Stokes' theorem that the integral of $(\nabla \times \mathbf{A}) \cdot \mathbf{n}$ vanishes over the closed surface S_C.

Replacing $\partial^2\Phi/\partial x^2$ at the body by $\partial U/\partial x$, we obtain the total pressure force on the body in the form

$$F_x = (\rho \forall + m_{11})\left[\frac{\partial U}{\partial t} + (U - U_1)\frac{\partial U}{\partial x}\right] - m_{11}\frac{dU_1}{dt}. \tag{150}$$

This is the desired expression for the force acting on a body that moves with velocity $U_1(t)$ in the presence of a slowly-varying nonuniform stream of velocity $U(x, t)$. Significantly, this force is not simply a function of the relative velocity $U - U_1$. The factor of the added-mass coefficient in (150) can be written as a function of the relative velocity, but there is an additional "buoyancy force" proportional to the displaced volume of the body.

If Euler's equations (2) are rewritten in a coordinate system translating with the body velocity U_1 the quantity in square brackets in (150) can be identified with the pressure gradient $\partial p/\partial x$ of the nonuniform stream, in the absence of the body, and (150) can be rewritten in the form

$$F_x = -(\forall + m_{11} / \rho)(\partial p / \partial x) - m_{11}\frac{dU_1}{dt}. \tag{151}$$

Thus, the force due to the nonuniform stream is equal to the pressure gradient associated with this stream, multiplied by the *effective volume* $\forall + m_{11} / \rho$. Note that analogies based on Archimedian hydrostatics would overlook the contribution from the added-mass coefficient!

From (150) one can derive an equation of motion for the body, by equating F_x to the product of body mass m and acceleration dU_1/dt. In particular, if the body is neutrally buoyant with $m = \rho\forall$, equilibrium of the forces will occur if $U_1 = U$. Thus the body will be carried along as if it were a fluid particle.

The more general case where the added-mass tensor m_{ij} contains off-diagonal elements was derived originally by G. I. Taylor, using an energy argument due to Kelvin. This approach is outlined by Lamb (1932).

4.18 The Method of Images

When a body moves in proximity to a plane rigid boundary, such as a horizontal or vertical wall, the interaction between the two must be accounted for. Formally, this is done by imposing the boundary condition $\partial\phi/\partial n = 0$ on the wall and attempting to solve the resulting boundary-value problem for the prescribed motion of the body. A convenient approach is to replace the wall by an *image body,* symmetrically disposed on the opposite side of the wall with a suitably prescribed symmetric motion to ensure that the boundary condition on the wall is satisfied. In this sense body-wall interactions are closely related to the interactions between two adjacent and identical bodies. For example, the interaction between a ship and an adjacent vertical canal bank is identical to the interaction of the same ship hull with a catamaran twin-hull.

If a two-dimensional source is situated at a point $(0, b)$ above the x-axis and an image source is situated at $(0, -b)$, the resulting complex source potential is simply

$$F = \frac{m}{2\pi}\log(z^2 + b^2). \tag{152}$$

Differentiation gives a dipole plus its image, in the x-direction,

$$F = \frac{\mu}{2\pi}\left(\frac{1}{z - ib} + \frac{1}{z + ib}\right), \tag{153}$$

but the corresponding vertical dipole is obtained by differentiation with respect to the source point b as opposed to the vertical coordinate y.

From (153), the flow past a pair of circular cylinders of radius a can be approximated by

$$F = Uz + \frac{Ua^2}{z - ib} + \frac{Ua^2}{z + ib}. \tag{154}$$

This solution is not exact, because the flow associated with the image dipole will distort the streamlines moving past the first dipole, but this effect will be small if the ratio a/b is small. The corresponding result for a pair of spheres, or for a single sphere moving near a wall, is readily obtained.

This procedure can be extended to a body or source situated between two parallel walls, say at $y = \pm\frac{1}{2}b$. In this case, to satisfy the boundary conditions on both walls simultaneously, it is necessary to use an infinite row of images, at $y = \pm b, \pm 2b, \pm 3b, \ldots$. In two-dimensional problems, the resulting infinite series for sources and dipoles can be summed in closed form, or the solutions can be guessed by noting that these must be periodic functions of y with period b. Thus, the complex source potential is given by

$$F = \frac{m}{2\pi} \log \sinh \frac{\pi z}{b}, \tag{155}$$

and the flow past a (distorted) circle is given by the superposition of a free stream and dipole, in the form

$$F = Uz + \pi U(a^2 / b) \coth \pi z / b. \tag{156}$$

These problems are discussed by Lamb (1932), who notes that the distortion of the circle is negligible provided its diameter is less than half of the space between the walls.

The problem of a vertical flat plate, situated midway between two parallel horizontal walls, can be solved in closed form, as outlined by Sedov (1965). The added-mass coefficient in this case is given by the expression

$$m_{11} = -2\rho \frac{b^2}{\pi} \log \cos \frac{\pi a}{b}, \tag{157}$$

where $2a$ is the breadth of the plate. This result is exact, and the two limiting cases are a single flat plate in an infinite fluid when $a/b \to 0$ and a completely blocked channel with infinite added mass when $a/b \to 1/2$. Similar results for rectangular profiles are given in numerical form by Flagg and Newman (1971).

Wall corrections for wind tunnels and water tunnels can be derived in this manner. For three-dimensional motion in a rectangular tunnel, a doubly infinite array of image bodies or singularities is required.

Problems

1. Show that the velocity potentials defined by the real parts of the complex potentials in figures 4.4 and 4.5 satisfy the correct boundary conditions on the boundaries of each flow. Assuming each of these flows is steady, what is the corresponding expression for the pressure throughout the fluid and along the boundaries?

2. Show that the maximum tangential velocity for the streaming flow past a circle is twice the free-stream value, and for a sphere is one and one half the free-stream value. What is the corresponding result for a Rankine ovoid at the equatorial plane? How does this compare with the sphere as the distance between the source and sink goes to zero? What is the opposite limit as the separation distance becomes large?

3. For a Rankine half-body generated by the combination of the uniform stream $\phi = Ux$ and a three-dimensional source at the origin, find the equation of the dividing streamline in terms of the stream function. Show that the distance from the source to the stagnation point is one half of the limiting radius far downstream and that the body's radius in the plane $x = 0$ is the geometric mean of these two lengths. Show that the tangential velocity at $x = 0$ exceeds the free-stream velocity by a factor $(5/4)^{1/2}$. Sketch the tangential velocity as a function of position along the body surface.

4. A circular piston of radius a moving with axial velocity U is mounted flush in the plane of a rigid baffle extending to infinity in all radial directions in the plane $x = 0$. The fluid occupies the region $x > 0$. State the complete boundary-value problem for this situation. Using Green's theorem, give an appropriate source distribution representing this flow. What is the limiting form of the velocity potential at large distances from the piston? (Assume that the departure of the piston face from the plane $x = 0$ is negligible.)

5. Use the complex potential (57) to derive the added-mass coefficient of a flat plate, using (114). Show that the same result can be obtained from (130).

6. Show that the analytic function

$$F'(z) = u - iv = -(\tfrac{1}{2}a^2 - z^2)(a^2 - z^2)^{-1/2} - iz$$

satisfies the correct boundary condition for the rolling flat plate shown in table 4.3 and behaves correctly at infinity. Using equation (114) and integrating by parts, verify that

$$m_{66} = -2\rho \int_{-a}^{a} x\phi(x,0+)\,dx = -\rho \int_{-a}^{a} (a^2 - x^2)\,F'(x)\,dx = \frac{\pi}{8}\rho a^4.$$

7. Two common approximations for the added-mass coefficient m_{11} or m_{22} of two-dimensional bodies are (1) the displaced mass and (2) the displaced

mass of a circle with the same projected width normal to the direction of acceleration. Compare these for the bodies shown in table 4.3. Which approximation is, in general, the most accurate?

8. A naive ocean engineer, infatuated with strip theory and ignorant of three-dimensional effects, is required to compute the added-mass coefficients for a circular disc of radius b and negligible thickness. Using the correct result shown in figure 4.8, show that m_{11} will be overestimated by a factor of about 1.6, and the error for the added moment of inertia depends on whether the "strips" are parallel or normal to the axis of rotation.

9. By resolving the translation of a flat plate into components normal and tangential to its plane, show that if it is inclined at an angle θ from the x-axis, the added-mass coefficients are

$$m_{11} = \pi\rho a^2 \sin^2\theta$$
$$m_{22} = \pi\rho a^2 \cos^2\theta$$
$$m_{12} = m_{21} = \frac{\pi}{2}\rho a^2 \sin 2\theta$$

What is the added moment of inertia?

10. Show that the added mass coefficients m_{11} and m_{22} in table 4.3 for a crossed fin and square are unchanged, if the body is rotated through an arbitrary angle, and that $m_{12} = m_{21} = 0$.

11. Show that the conformal transformation (51), with $U = 1$, maps the circle of radius r_0 onto an ellipse in the ζ-plane, with semi-axes $r_0 + A/r_0$ and $r_0 - A/r_0$. Using the complex potential for the flow past a circle, determine the effective dipole moments for the ellipse, and use (130) to derive the added-mass coefficients m_{11} and m_{22}.

12. From qualitative estimates of the relative magnitudes of the added mass coefficients and from (116), show that a streamlined body is generally unstable when moving with constant velocity parallel to its long axis. Use the same argument to show that leaves falling from trees will generally be horizontal.

13. The velocity potentials for *internal* motions within the interior of a closed surface S, enclosing a volume of fluid, can be defined as in (101–103). Show that the solutions of the three internal translational potentials are given simply by $\phi_i = x_i$, whereas the solutions for rotational motions are

more complicated. By writing Green's theorem for the region exterior to the surface S, show that if ϕ_I and ϕ_E are solutions of Laplace's equation in the interior and exterior domains, respectively, then the exterior potential can be expressed as a distribution of sources on S, of strength $(\partial/\partial n)(\phi_I - \phi_E)$, provided ϕ_I satisfies the boundary condition $\phi_I = \phi_E$ on S. Show that the exterior potential can also be represented as a dipole distribution normal to S, of moment $\phi_I - \phi_E$, provided $\partial\phi_I/\partial n = \partial\phi_E/\partial n$ on S. Use the latter representation to derive (128).

14. Compare the relative magnitudes of the forces acting on a stationary prolate spheroid, with length-diameter ratio equal to 5.0, in the presence of an accelerating stream that is (1) parallel and (2) perpendicular to the longitudinal axis of the spheroid.

15. Using equation (90), derive an integral expression for the attraction force that occurs when a body moves with steady translation parallel to an infinite plane rigid wall, in terms of the pressure force on the wall. Show that this force always acts on the body in the direction toward the wall. If the body is sufficiently far from the wall so that the potential can be approximated by the dipole (123) plus an image dipole on the opposite side of the wall, show that the attraction force is given by

$$F = \frac{3\pi}{4}\frac{\rho U^2 A_{11}^2}{d^4} = \frac{3}{64\pi}\frac{\rho U^2}{d^4}(\forall + m_{11}/\rho)^2,$$

provided $A_{12} = A_{13} = 0$. Here d is the distance from the body axis to the wall. Derive from this equation and (133) the special case

$$F = \frac{3\pi}{16}\rho U^2 R^6 / d^4$$

for a sphere of radius R, and

$$F = \frac{3}{64\pi}\rho U^2 \forall^2 / d^4$$

for a slender prolate spheroid moving in the longitudinal direction.

References

Cummins, W. E. 1957. The force and moment on a body in a time-varying potential flow. *J. Ship Res.* 1 (1): 7–18.

Flagg, C., and J. N. Newman. 1971. Sway added-mass coefficients for rectangular profiles in shallow water. *J. Ship Res.* 15:257–265.

Havelock, T. H. 1963. *Collected papers. ONR/ACR-103.* Washington: U. S. Government Printing Office.

Hildebrand, F. B. 1976. *Advanced calculus for applications.* 2nd ed. Englewood Cliffs, N.J.: Prentice-Hall.

Kennard, E. H. 1967. *Irrotational flow of frictionless fluids, mostly of invariable density.* David Taylor Model Basin Report 2229.

Kochin, N. E., I. A. Kibel, and N. V. Roze. 1964. *Theoretical Hydromechanics. English translation of.* 5th Russian ed. New York: Wiley.

Lamb, H. 1932. *Hydrodynamics.* 6th ed. Cambridge: Cambridge University Press. Reprinted 1945, New York: Dover.

Landweber, L., and C. S. Yih. 1956. Forces, moments, and added masses for Rankine bodies. *Journal of Fluid Mechanics* 1:319–336.

Milne-Thomson, L. M. 1968. *Theoretical hydrodynamics.* 5th ed. New York: Macmillan.

Morse, P. M., and H. Feshbach. 1953. *Methods of theoretical physics,* 2 vols. New York: McGraw-Hill.

Newman, J. N. 1978. The added moment of inertia of two-dimensional cylinders. *J. Ship Res.* 23:1–8.

Sedov, L. I. 1965. *Two dimensional problems in hydrodynamics and aerodynamics. English translation.* New York: Interscience.

5 Lifting Surfaces

One of the most striking applications of fluid mechanics is the lifting surface that supports the flight of birds, airplanes, and hydrofoil boats. Similar effects are involved in the actions of control surfaces such as rudders, yacht sails and keels, and screw propellers. Thus, there are many applications of lifting surfaces in marine hydrodynamics; a proper understanding of their actions must preface an intelligent approach to their design and use.

Typically, a lifting surface is a thin streamlined body that moves through the surrounding fluid at a small angle of attack, with a resultant hydrodynamic *lift* force generated in the direction normal to the forward movement. We shall suppose that the lifting surface is primarily oriented in the x-z plane and moving with constant velocity U in the positive x-direction, as shown in figure 5.1. The lift force L is the component of the hydrodynamic force parallel to the y-axis; the drag force D acts in the negative x-direction.

The principal geometric dimensions of the lifting surface are its *chord length* l and *span* s, measured parallel to the x- and z-axes respectively. For a rectangular planform, l is a constant, but in general the chord $l(z)$ will depend on the transverse coordinate. The ratio of the span to the mean chord is the aspect ratio A, given by

$$A = s^2 / S, \tag{1}$$

where S is the projected area of the lifting surface on the plane $y = 0$.

The aspect ratio is an important measure of the effect of three-dimensionality. A large aspect ratio implies that the flow will be nearly independent of the transverse coordinate z, and a two-dimensional strip-theory approach is valid. This situation is exemplified by older aircraft wings, as well as by some hydrofoil configurations and deep narrow centerboards on

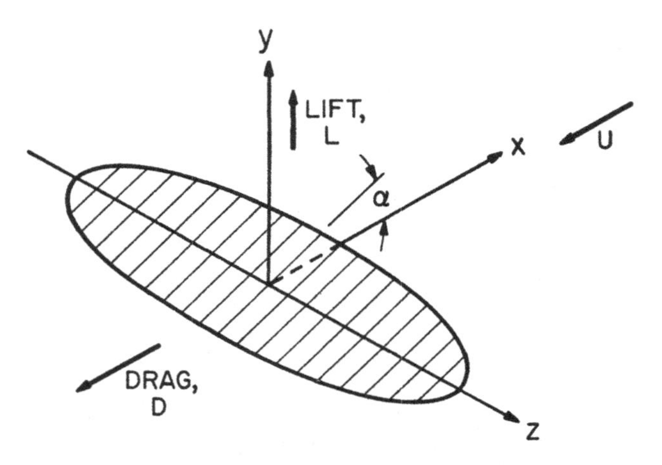

Figure 5.1
Three-dimensional lifting surface.

small sailing craft. On the other hand, if the aspect ratio is smaller, as in the case of most rudders and control surfaces, three-dimensional flow effects will be important and must be included in our analysis.

Lifting surfaces of the type described are *planar.* Excluded from this category is the screw propeller illustrated in figure 2.8. The results to be derived for planar foils are qualitatively applicable to the blades of a propeller. In particular, the two-dimensional theory can be used in a strip-theory sense to describe the flow past each section of a propeller blade, and the three-dimensional flow past a propeller will include trailing vortex sheets similar to those existing in the planar case. The principal distinctions are that the propeller blades and associated vortex sheets are roughly helical, and the blades affect each other in a manner analogous to a *cascade* of planar foils operating in proximity. Detailed descriptions of propeller theory are given by Cox and Morgan (1972) and Kerwin (1973).

The case of two-dimensional flow past a lifting surface or hydrofoil will be considered first to provide a quantitative understanding of lifting surfaces of high aspect ratio, as well as qualitative understanding of the more general case. Subsequently, we shall consider the more complicated three-dimensional case of finite aspect ratio planar lifting surfaces. Throughout our discussion we will assume that viscous effects are confined to a thin boundary layer along the surface of the hydrofoil and that the foil surface is itself thin and situated at a small angle of attack relative to the incident flow. Thus, in the jargon of aerodynamics, we shall restrict our discussion to the case of *linearized thin-wing theory.*

5.1 Two-Dimensional Hydrofoil Theory

Let us consider the steady flow of fluid past a thin streamlined section asymmetrical about $y = 0$. This asymmetry may be due either to a small angle of attack a between the axis of the hydrofoil and the x-axis, or to the existence of curvature or *camber* of the hydrofoil, or to a combination of these two effects. Under these circumstances the streamlines will appear roughly as shown in figure 5.2. In particular, viscous effects are confined to a thin boundary layer adjacent to the surface of the foil, and the flow will leave the trailing edge in a smooth tangential manner.

For viscous effects to be confined to a thin boundary layer, not only must the Reynolds number be reasonably large but also the body must be sufficiently thin and streamlined that separation does not occur. If the body is not thin or if the angle of attack is too large, the flow will be typified by the sketches shown in figure 5.3. Thus, there are practical limits to both the body shape and the angle of attack.

The additional assumption that the flow leaves the trailing edge in a smooth tangential manner (as shown in figure 5.2) is fundamental to the theory of lifting surfaces. Another possibility that can be envisaged is shown in figure 5.4; this involves an infinite velocity around the trailing edge and a stagnation point on the upper (or possibly the lower) surface of the foil. In fact, the streamlines as drawn in figure 5.4 correspond to a potential flow without circulation, whereas the expected situation shown in figure 5.2

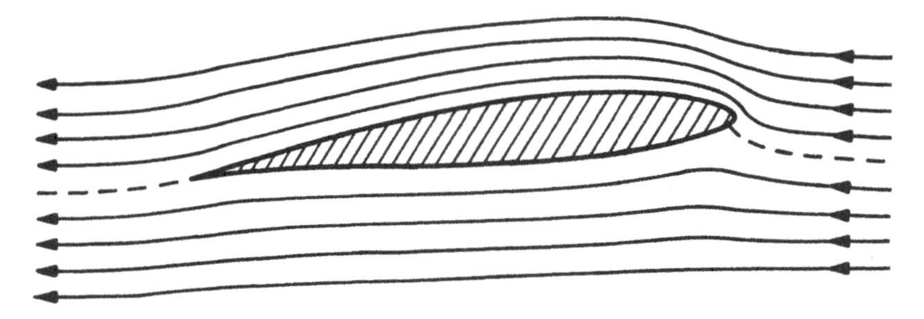

Figure 5.2
Assumed flow past foil.

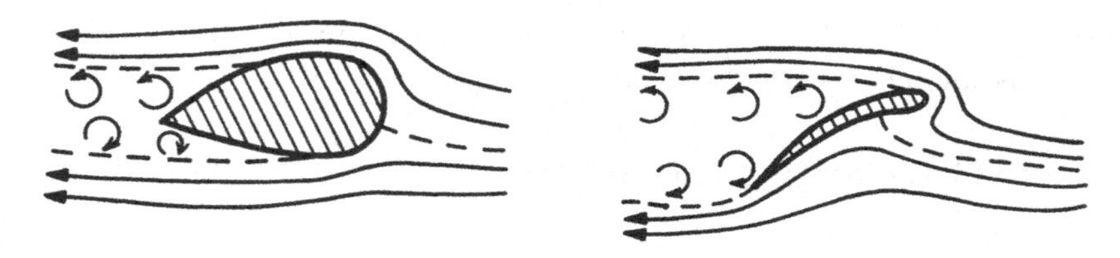

Figure 5.3
Separated flow due to a bluff body (left) or excessive angle of attack (right).

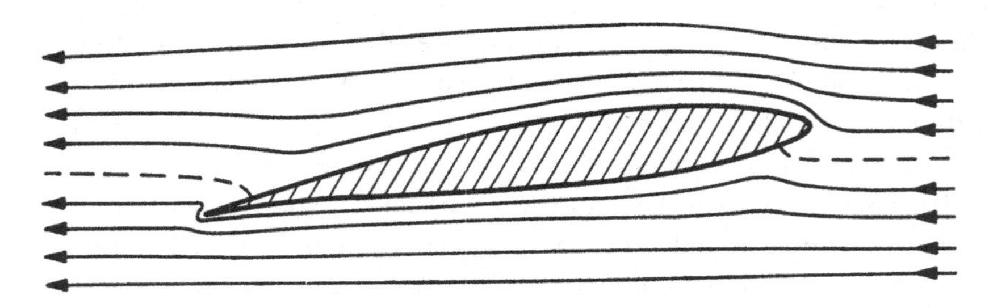

Figure 5.4
Flow past foil without circulation.

is obtained from that in figure 5.4 by adding a positive counterclockwise circulation around the foil, sufficient to move the stagnation point back to the trailing edge. This may seem to contradict Kelvin's theorem that the circulation is constant about any material contour always containing the same fluid particles, since this constant must equal zero if the fluid motion started from a state of rest.

This apparent contradiction can be resolved by noting that a material contour surrounding the foil cannot be related to another that is initially upstream, since the foil will "pierce" the fluid on the upstream contour. The contour that is initially upstream can only be related to a subsequent contour that surrounds the foil and a portion of its wake, as shown in figure 5.5, and the circulation about this contour will indeed remain zero. A more illuminating contour is one sufficiently large to surround the foil initially and at all subsequent times, as shown in figure 5.6. Here the circulation must be zero, and the circulation about the foil must, therefore, be canceled by an equal and opposite *starting vortex* shed from the trailing edge into the wake during the initial acceleration. The shedding of this starting vortex can be observed by accelerating a flat plate in a kitchen sink. If the plate is

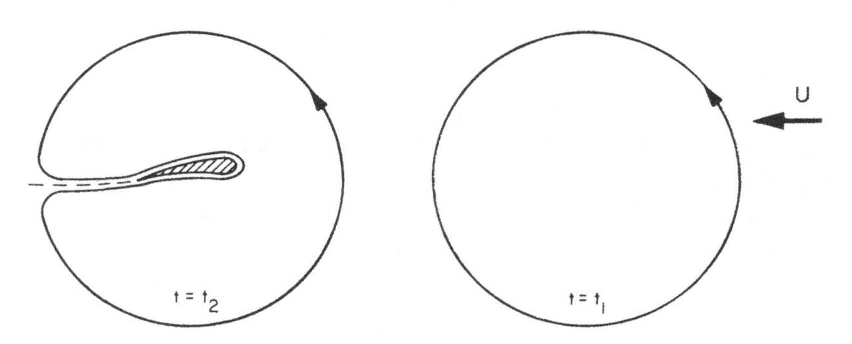

Figure 5.5
The material contour shown here is initially upstream of the foil, at time $t = t_1$. A short time later ($t = t_2$), it has surrounded the foil and a portion of the wake (- - - - -). From Kelvin's theorem, and the initial conditions upstream, the circulation around this contour is zero for all times.

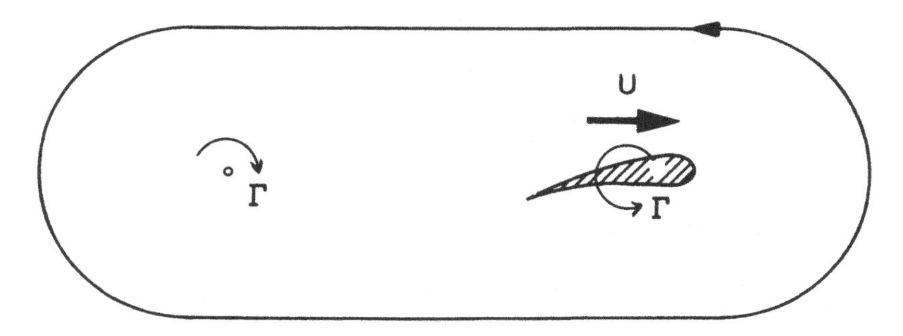

Figure 5.6
The material contour shown here is sufficiently large to surround the foil at its original position of rest, and subsequently after acceleration to a steady velocity U. The circulation about this contour is zero, so that the foil must shed a starting vortex of strength Γ, equal and opposite to the circulation of the foil.

at right angles to the flow, vortices will be shed from both edges, but if it is at a small angle to the flow, only the trailing edge will shed a vortex.

If our interest is in the steady-state hydrofoil problem, the starting vortex can be disregarded, since it will be situated infinitely far downstream and, in reality, will be dissipated by viscous diffusion. Thus, for the two-dimensional steady-state problem, it is reasonable to expect irrotational flow throughout the surrounding fluid, but with a net circulation about the foil. This circulation is essential to the development of a lift force since without it there can be no force acting on a body in a steady-state translation. In this respect, the smooth local flow at the trailing edge is vital to the development of the desired lift force.

The assumption of smooth tangential flow at the trailing edge is imposed mathematically by the *Kutta condition* requiring the velocity at the trailing edge to be finite. This extra condition is added to the conventional statement of the boundary-value problem, which may now be formulated. The conditions on the fluid flow are that the velocity vector should be equal to the free-stream velocity $-U\mathbf{i}$ at infinity, tangential to the surface of the foil, and finite at the trailing edge.

It is convenient to subtract the free-stream component and utilize the perturbation velocity components (u, v) with the understanding that the total velocity vector of the two-dimensional flow is $(u - U, v)$. If the perturbation velocity potential $\phi(x, y)$ is defined such that

$$(u,v) = \nabla\phi, \tag{2}$$

the following boundary-value problem results:

$$\nabla^2\phi = 0, \quad \text{throughout the fluid,} \tag{3}$$

$$\partial\phi / \partial n = Un_x, \quad \text{on the foil,} \tag{4}$$

$$\nabla\phi < \infty, \quad \text{at the trailing edge,} \tag{5}$$

$$\nabla\phi \to 0, \quad \text{at infinity.} \tag{6}$$

5.2 Linearized Two-Dimensional Theory

The boundary condition (4) on the foil can be simplified under the assumption that the foil is thin and nearly horizontal. Thus we now assume that the vertical coordinates of the upper and lower foil surfaces, $y = y_u$ and $y = y_l$, respectively, are both much smaller than the chord length l and that the slopes $y_u'(x)$ and $y_l'(x)$ are small compared to one. For convenience, we will take the origin of the coordinate system so that the leading and trailing edges are situated at $x = \pm\frac{1}{2}l$, respectively, as shown in figure 5.7.

The boundary condition (4) on the foil can be expressed in a more convenient form by utilizing the substantial derivative (3.19). Thus, following the fluid particles on the upper surface of the foil, where $y - y_u(x) = 0$,

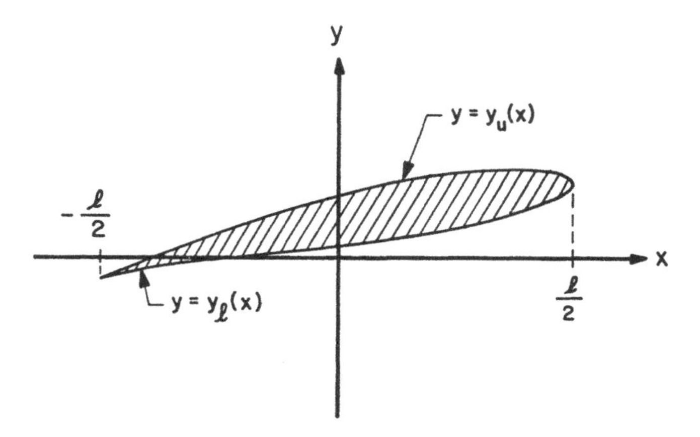

Figure 5.7
Notation for two-dimensional foil.

$$0 = \frac{D}{Dt}[y - y_u(x)] = \mathbf{V} \cdot \nabla[y - y_u(x)], \tag{7}$$

or

$$\frac{\partial \phi}{\partial y} = y_u'(x)\left[\frac{\partial \phi}{\partial x} - U\right], \quad \text{on } y = y_u(x). \tag{8}$$

A similar boundary condition applies on the lower surface $y = y_l$.

The boundary condition (8) shows that the vertical velocity component, and hence the perturbation potential ϕ, will be roughly proportional to the slope of the foil. For sufficiently small values of y_u', the magnitude of the free-stream velocity $-U$ will be much larger than the horizontal component of the perturbation velocity; in that case (8) can be approximated by

$$\partial \phi / \partial y \simeq -U y_u'(x), \quad \text{on } y = y_u(x), \tag{9}$$

which is valid provided $|\partial\phi/\partial x| \ll U$. Equation (9) is a *first-order* approximation to (8), in the sense that the perturbation velocity is proportional to the slope y_u', and the neglected term $y_u'(\partial\phi / \partial x)$ will be of second order in the slope.

The boundary condition on the foil can be further simplified, consistent with the first-order approximation (9). Since the vertical coordinates of the foil surface differ from zero by a small quantity, the exact surface of the foil may be collapsed onto the *cut* $(-l/2, l/2)$ of the real axis, with the upper surface of the foil corresponding to the upper surface $y = 0+$ of the cut, and conversely for the lower surface. Since singularities may exist on or inside the exact foil, these will be transferred to the cut in a corresponding manner. Then, provided $\partial\phi/\partial y$ is regular outside the foil, the difference between its value on the exact surface $y = y_u$ and the upper side of the cut $y = 0+$ will be a second-order quantity, i.e., from a Taylor series expansion,

$$(\partial\phi / \partial y)_{y=y_u} = (\partial\phi / \partial y)_{y=0+} + y_u(\partial^2\phi / \partial y^2)_{y=0+} + \ldots \simeq (\partial\phi / \partial y)_{y=0+}. \tag{10}$$

Combining the approximations (9) and (10) gives the first-order linearized boundary condition on the upper surface of the foil, in the form

$$\frac{\partial\phi}{\partial y} = -Uy_u'(x), \quad \text{on } y = 0+, -l/2 < x < l/2. \tag{11}$$

An identical argument gives the corresponding boundary condition

$$\frac{\partial\phi}{\partial y} = -Uy_l'(x), \quad \text{on } y = 0-, -l/2 < x < l/2, \tag{12}$$

on the lower surface. These linearized boundary conditions effectively replace the foil by a suitable distribution of the vertical velocity on the corresponding cut of the x-axis.

Before attempting to solve this linearized boundary-value problem, let us consider the corresponding first-order approximation for the hydrodynamic pressure force and moment. From the steady form of Bernoulli's equation (4.26), the dynamic pressure in the fluid is given by

$$p - p_\infty = -\frac{1}{2}\rho(\mathbf{V}\cdot\mathbf{V} - U^2) = -\frac{1}{2}\rho(u^2 + v^2 - 2uU). \tag{13}$$

Neglecting the nonlinear terms of second-order in the perturbation velocity gives

$$p - p_\infty \simeq \rho u U. \tag{14}$$

Thus, the linearized pressure is simply proportional to the horizontal perturbation velocity component $u = \partial\phi/\partial x$.

The vertical lift force L acting on the foil is obtained, to the same order of approximation, by integrating the jump in pressure across the cut. Thus, it follows that

$$L = \oint(p - p_\infty)dx = \rho U \oint u\, dx = \rho U \Gamma, \tag{15}$$

where Γ is the total circulation around the foil as defined by (4.4). Equation (15) is a derivation, for linearized thin foils, of the more general *Kutta-Joukowski theorem.* This theorem states that for any two-dimensional body, moving with constant velocity in an unbounded inviscid fluid, the hydrodynamic pressure force is directed normal to the velocity vector and is equal to the product of the fluid density, body velocity, and the circulation about the body (see problem 3).

The hydrodynamic moment M about the z-axis is given by the corresponding integral

$$M = \rho U \oint ux\, dx. \tag{16}$$

Now, to solve the boundary-value problem for the thin two-dimensional hydrofoil, it is convenient to decompose the velocity potential into even and odd functions of y:

$$\phi(x,y) = \phi_e(x,y) + \phi_o(x,y), \tag{17}$$

$$\phi_e(x,y) = \phi_e(x,-y) = \frac{1}{2}[\phi(x,y) + \phi(x,-y)], \tag{18}$$

$$\phi_o(x,y) = -\phi_o(x,-y) = \frac{1}{2}[\phi(x,y) - \phi(x,-y)]. \tag{19}$$

The boundary conditions on $y = 0\pm$ take the form

$$\frac{\partial \phi_e}{\partial y} = \mp \frac{1}{2} U(y_u' - y_l'), \quad \text{on } y = 0\pm, \tag{20}$$

$$\frac{\partial \phi_o}{\partial y} = -\frac{1}{2} U(y_u' + y_l'), \quad \text{on } y = 0\pm. \tag{21}$$

Since the operator $\partial/\partial y$ is odd, $\partial\phi_e/\partial y$ is odd and $\partial\phi_o/\partial y$ is even with respect to y.

The even and odd potentials correspond to two distinctly different physical problems. From (20), ϕ_e is the potential for a symmetrical strut, of thickness $y_u - y_l$, at zero angle of attack. On the other hand, from (21), ϕ_o is the potential of an asymmetric flow, past an arc of zero thickness defined by the curve $y = \frac{1}{2}(y_u + y_t)$, or the *mean-camber line.* Both problems are illustrated in figure 5.8.

The original problem has now been decomposed into two parts, one representing the effects of thickness and the other representing the effects of

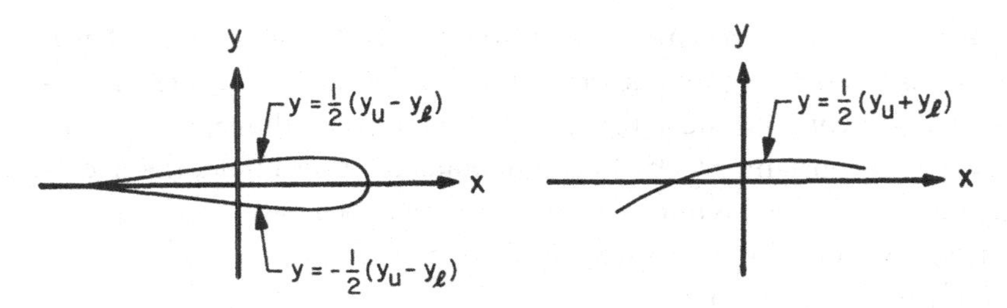

Figure 5.8
Definitions of thickness and lifting problems.

camber and angle of attack. Because the pressure distribution is symmetric in the thickness problem, there can be no lift force or moment in this case; hence thickness has no direct effect on the lift and moment. Thickness has practical effects only when modifications of the pressure distribution affect separation or cavitation. In particular, the effect of thickness will be important near the leading edge, where the mean-camber line contains a sharp edge, not present in the real foil with finite leading-edge radius. Thus, in the flow past the mean-camber line we must anticipate an infinite velocity at the leading edge which will not occur in practice because of thickness effects, but which, if it did occur, would be devastating from the standpoints of separation and cavitation. It is therefore permissible, in the lifting problem of the mean-camber line, to ignore an infinite velocity at the leading edge, on the premise that this singularity will be removed by the effects of thickness. An entirely different situation will occur at the sharp trailing edge; an infinite velocity here is prevented by viscous effects in a real fluid and by the Kutta condition in the ideal-fluid problem.

Since the thickness and lifting problems can be separated, and since the lift force and moment are independent of the thickness, we will examine primarily the solution of the lifting problem. The corresponding solution of the thickness problem, in terms of a source distribution, will be described briefly in section 5.6.

5.3 The Lifting Problem

Here we focus our attention on the flow past the mean-camber line and the resulting lift force and moment. It is convenient to denote the vertical coordinate of the mean-camber line by $\eta = \frac{1}{2}(y_u + y_l)$. The corresponding boundary condition on the cut is

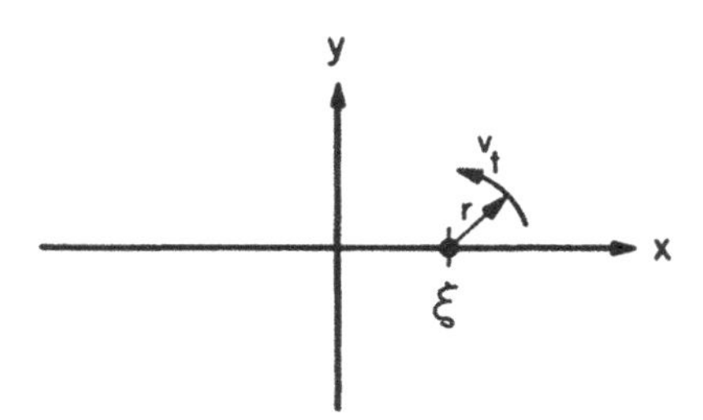

Figure 5.9
Tangential velocity induced by a point vortex around a circle of radius r.

$$v = \frac{\partial\phi}{\partial y} = -U\eta'(x). \tag{22}$$

From (22), the vertical velocity across the cut is continuous, but a discontinuity in the horizontal velocity component u must be anticipated. This is emphasized by the linearized form of Bernoulli's equation (14) and the Kutta-Joukowski theorem (15), where the presence of a lift force requires a discontinuity in the horizontal velocity component and a nonzero circulation about the foil section. These considerations suggest a distribution of vortices along the cut, for such a distribution will possess a nonzero circulation so long as the total integrated strength of the vortices is not zero. Moreover, a vortex distribution can be constructed to give any desired discontinuity of the horizontal velocity on the cut.

The velocity potential of a point vortex, of circulation γ situated at the point $x = \xi$ on the x-axis, is

$$\phi = \operatorname{Im}\frac{\gamma}{2\pi}\log(x-\xi+iy) = \frac{\gamma}{2\pi}\tan^{-1}\left(\frac{y}{x-\xi}\right). \tag{23}$$

The corresponding velocity components are

$$u = \frac{\partial\phi}{\partial x} = \frac{\gamma}{2\pi}\frac{-y}{(x-\xi)^2+y^2}, \tag{24}$$

$$v = \frac{\partial\phi}{\partial y} = \frac{\gamma}{2\pi}\frac{(x-\xi)}{(x-\xi)^2+y^2}. \tag{25}$$

Thus, the vortex induces a tangential velocity, $v_t = \gamma/2\pi r$, where r is the radial distance from the vortex, as shown in figure 5.9.

If vortices are distributed along the x-axis between the leading and trailing edges of the foil with local circulation density $\gamma(\xi)$, the resulting velocity components are given by the integrals

$$u(x,y) = -\frac{1}{2\pi}\int_{-l/2}^{l/2}\frac{\gamma(\xi)y\,d\xi}{(x-\xi)^2+y^2}, \tag{26}$$

$$v(x,y) = \frac{1}{2\pi}\int_{-l/2}^{l/2}\frac{\gamma(\xi)(x-\xi)\,d\xi}{(x-\xi)^2+y^2}. \tag{27}$$

To show that this vortex distribution can be used to solve the linearized lifting problem, it is necessary to examine the limiting values of the velocity components defined by (26) and (27) as the point (x, y) approaches the foil surface. These two limits must be considered separately.

First, let us consider the limiting value of the horizontal velocity, as defined by (26), when $y = \pm\varepsilon$, where ε is positive and $\varepsilon \ll l$. Then

$$u(x,\pm\varepsilon) = \mp\frac{1}{2\pi}\int_{-l/2}^{l/2}\frac{\gamma(\xi)\varepsilon}{(x-\xi)^2+\varepsilon^2}d\xi. \tag{28}$$

As $\varepsilon \to 0$ the integrand tends to zero, except for the point $\xi = x$, where the integrand tends to infinity. In this limit, the factor multiplying γ behaves like a Dirac delta function, extracting the value of $\gamma(x)$ at the point $\xi = x$. More specifically, since the integrand vanishes for $\xi \neq x$, (28) can be replaced by

$$\begin{aligned} u(x,\pm\varepsilon) &\simeq \mp\frac{1}{2\pi}\gamma(x)\int_{-l/2}^{l/2}\frac{\varepsilon}{(x-\xi)^2+\varepsilon^2}d\xi \\ &= \mp\frac{1}{2\pi}\gamma(x)\left[\tan^{-1}\left(\frac{\xi-x}{\varepsilon}\right)\right]_{-l/2}^{l/2} \\ &\simeq \mp\frac{1}{2}\gamma(x), \end{aligned} \tag{29}$$

where the approximations are valid for sufficiently small values of ε/l. Thus, the limiting values of the horizontal velocity component on the upper and lower surfaces of the foil are given by the last line of (29), and there will exist a discontinuity in this velocity component precisely equal to the local vortex strength $\gamma(x)$. This confirms our hypothesis that the lifting problem for the mean-camber line can be represented by a vortex distribution. To relate the local vortex strength γ to the geometry of the foil, we must invoke the boundary condition (22) on the surface of the foil; hence we must consider the limiting form of the vertical velocity component.

From (27) the vertical velocity component on the cut is given by the integral

$$v(x,0\pm) = -\frac{1}{2\pi}\fint_{-l/2}^{l/2} \frac{\gamma(\xi)}{\xi - x} d\xi, \qquad -\frac{1}{2}l < x < \frac{1}{2}l, \tag{30}$$

where the integral is to be interpreted in the sense of the Cauchy principal value, or

$$\fint_a^b \frac{f(\xi)}{\xi - x} d\xi \equiv \lim_{\varepsilon \to 0} \left\{ \fint_a^{x-\varepsilon} \frac{f(\xi)}{\xi - x} d\xi + \int_{x+\varepsilon}^b \frac{f(\xi)}{\xi - x} d\xi \right\}. \tag{31}$$

(The legitimacy of this interpretation is confirmed in section 5.6.)

If the geometry of the mean-camber line $\eta(x)$ is prescribed, the vertical velocity $v(x, 0)$ can be computed from the boundary condition (22). In this case, (30) is an *integral equation* since the unknown function γ appears in the integrand. We will see in section 5.7 that the general solution of this integral equation is given by the expression

$$\gamma(x) = \frac{2}{\pi[(l/2)^2 - x^2]^{1/2}} \left\{ \fint_{-l/2}^{l/2} \frac{v(\xi,0)[(l/2)^2 - \xi^2]^{1/2}}{\xi - x} d\xi + \tfrac{1}{2}\Gamma \right\}, \tag{32}$$

where the total circulation about the foil is

$$\Gamma = \int_{-l/2}^{l/2} \gamma(\xi)\, d\xi. \tag{33}$$

At this stage Γ is unknown and the solution (32) is not unique, even though it has been chosen to satisfy the boundary condition (22) on the foil. This is not surprising, since we have not yet imposed the Kutta condition at the trailing edge. The general case with arbitrary circulation will include an infinite velocity at the trailing edge, as in the case shown in figure 5.4, differing from the correct solution in figure 5.2 by the appropriate circulation.

In general, the vortex strength γ, and hence the horizontal velocity component on the foil, will be infinite both at the leading and trailing edges as a result of the square-root infinities in (32). This can be avoided at the trailing edge only if Γ is chosen so that the quantity in braces in (32) vanishes at $x = -\frac{1}{2}l$. Imposing the Kutta condition results in the following integral for the circulation:

$$\Gamma = -2\int_{-l/2}^{l/2} v(\xi, 0)\left[\frac{\frac{1}{2}l - \xi}{\frac{1}{2}l + \xi}\right]^{1/2} d\xi. \tag{34}$$

This is a particularly useful result because the lift is given by (15), in terms of the circulation, and substituting (34) in (15) gives an equation for the lift as

$$L = 2\rho U^2 \int_{-l/2}^{l/2} \frac{d\eta}{d\xi}\left[\frac{\frac{1}{2}l - \xi}{\frac{1}{2}l + \xi}\right]^{1/2} d\xi. \tag{35}$$

Here the boundary condition (22) has been utilized to substitute the slope of the mean camber line for the vertical velocity component. Alternatively, the nondimensional lift coefficient is

$$C_L = \frac{L}{\frac{1}{2}\rho U^2 l} = \frac{4}{l}\int_{-l/2}^{l/2} \frac{d\eta}{d\xi}\left[\frac{\frac{1}{2}l - \xi}{\frac{1}{2}l + \xi}\right]^{1/2} d\xi. \tag{36}$$

The moment acting on the foil can be determined similarly from (16), (29), and (32). Neglecting the details of the integration, we get as the final result

$$M = \rho U \int_{-l/2}^{l/2} \gamma(x)\, x\, dx = 2\rho U^2 \int_{-l/2}^{l/2} \frac{d\eta}{d\xi}\left[(\tfrac{1}{2}l)^2 - \xi^2\right]^{1/2} d\xi, \tag{37}$$

and the nondimensional moment coefficient corresponding to (36) is

$$C_M = \frac{M}{\frac{1}{2}\rho U^2 l^2} = \frac{4}{l^2}\int_{-l/2}^{l/2} \frac{d\eta}{d\xi}\left[(\tfrac{1}{2}l)^2 - \xi^2\right]^{1/2} d\xi. \tag{38}$$

The *center of pressure* x_{CP} is the ratio $x_{CP} = M/L$.

5.4 Simple Foil Shapes

To illustrate these results, we consider first the simplest case of a flat plate, or uncambered foil, with an angle of attack $d\eta/dx = \alpha$. Then, it follows from (36) that

$$C_L = 2\pi\alpha, \tag{39}$$

and from (38) that

$$C_M = \tfrac{1}{2}\pi\alpha, \tag{40}$$

Thus, the lift coefficient of an uncambered foil is 2π times the angle of attack, and the center of pressure is at the *quarter-chord* point, $x_{CP} = l/4$. As an indication of the accuracy of the linearized theory, the exact lift coefficient of a flat plate in the absence of separation is

$$C_L = 2\pi \sin \alpha \tag{41}$$

(see problem 5). Thus, the error associated with the linearized theory is insignificant at realistic angles of attack. Similar conclusions follow from the experimental data shown in figure 2.6, where the curve $C_L(\alpha)$ is virtually linear for angles of attack less than 12–15 degrees. For larger angles separation or *stall* occurs, with much greater significance than any errors resulting from the linearization assumption.

The vortex distribution corresponding to the flat plate can be obtained by integrating (32) with $v = -\alpha U$. The result is

$$\gamma(x) = 2\alpha U \left[\frac{\tfrac{1}{2}l + x}{\tfrac{1}{2}l - x}\right]^{1/2}, \tag{42}$$

where (81) and (84) have been used. The function (42) is displayed in figure 5.10.

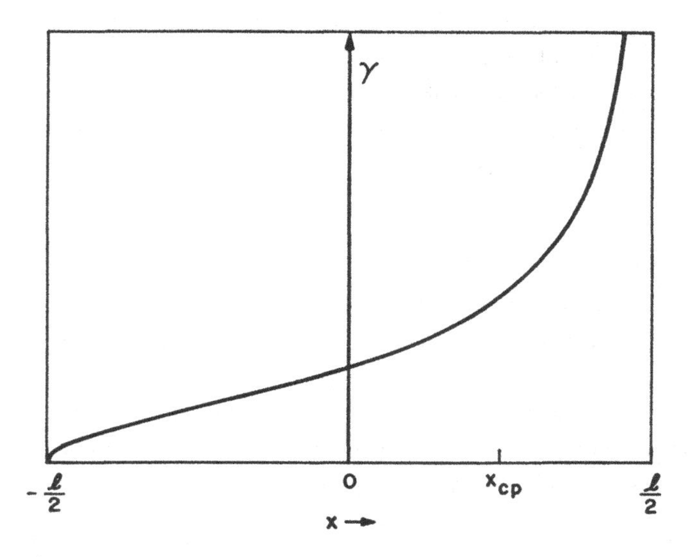

Figure 5.10
The vortex distribution for a fiat plate, given by equation (42).

Since the linearized pressure jump across the foil is proportional to the local vortex strength, figure 5.10 can be regarded as a graph of the pressure loading on the foil. This loading is concentrated near the leading edge and, in fact, becomes infinite as $x \to \frac{1}{2}l$.

The severe pressure gradient near the leading edge will cause separation or cavitation as the angle of attack is increased. Nevertheless, the flat-plate mean-camber line represents the lifting problem for *all* symmetric foils since the thickness distribution does not affect the lift and moment. Thus, these results apply to several cases of practical importance, notably for rudders, keels, and stabilizing fins, where equal performance is required for positive and negative angles of attack.

When the foil has a preferred direction of lift, the mean-camber line can be modified to make the loading more uniform. The simplest possibility is a parabolic-arc mean-camber line, with slope

$$d\eta/dx = \alpha - \beta x. \tag{43}$$

Here α and β are constants; α is the angle of attack of the incoming stream relative to the straight line connecting the leading and trailing edges, and $\beta l^2/8$ is the elevation of the foil at mid chord above this straight line.

Since equations (32, 35, 37) for the vortex strength, lift, and moment are linear functions of the slope $d\eta/dx$, the two terms in (43) can be integrated separately and the results superposed. The angle-of-attack term will give results identical to the flat plate, whereas the effects of the camber will be independent of the angle of attack. This decomposition of the camber and angle of attack applies not only for the parabolic arc, but for any cambered foil. In particular, the lift coefficient $C_L(\alpha)$ can be written in general in the form

$$C_L(\alpha) = C_L(0) + 2\pi\alpha, \tag{44}$$

where the term $C_L(0)$ represents the effects of camber, and the term $2\pi\alpha$ represents the effect of the angle of attack. A similar result applies for the moment.

Proceeding with our analysis of camber for a parabolic arc, we can set the angle of attack equal to zero in (43), in view of the decomposition (44). Evaluating the integrals (32), (35), and (37), we get

$$\gamma(x) = 2U\beta\left[(\tfrac{1}{2}l)^2 - x^2\right]^{1/2}, \tag{45}$$

$$C_L = \tfrac{1}{2}\pi\beta l, \tag{46}$$

$$C_M = 0. \tag{47}$$

For nonzero angles of attack, the corresponding results (42), (39), and (40) can be superposed.

The vortex strength (45) is indeed more uniform than that of the flat plate, and the leading-edge singularity has been removed. This feature is common to all mean-camber lines symmetric about the mid-chord point, since the resulting vortex strength must be symmetric in x and thus will vanish not only at the trailing edge but also at the leading edge. By the same argument, the center of pressure must be at the mid-chord point for such foils, and (47) confirms this for the parabolic arc.

Since a nonzero angle of attack will add the flat-plate load distribution to the cambered foil, the leading-edge singularity will return in this case, and only at the *ideal* angle of attack will there be no leading-edge singularity. The ideal angle of attack corresponds to the situation where the forward stagnation point of the foil coincides with the leading edge, and the slopes of the dividing streamline and the mean-camber line are equal at this point. For mean-camber lines symmetrical about the mid-chord point, the ideal angle of attack is zero, but for more general foil shapes, the ideal angle of attack can be nonzero.

The parabolic arc represents a substantial improvement over the flat plate from the standpoint of uniform loading and pressure distribution, but further improvement is possible if more complicated mean-camber lines are considered. Rather than proceed to analyze alternative foil shapes, however, it is more expedient to approach the optimum loading distribution problem by finding the shape of the mean-camber line such that the loading is a constant. This indirect or *design* problem, with a specified load distribution, is actually simpler to solve, since (30) becomes a definite integral instead of an integral equation with an unknown integrand. In the present case, with γ= constant, and after invoking the boundary condition (22) for the vertical velocity, we have

$$\frac{\partial\eta}{\partial x} = \frac{\gamma}{2\pi U}\,⨍_{-l/2}^{l/2} \frac{d\xi}{\xi - x} = \frac{\gamma}{2\pi U}\log\left|\frac{\tfrac{1}{2}l - x}{\tfrac{1}{2}l + x}\right|. \tag{48}$$

Integrating with respect to x, we obtain the desired mean camber

$$\eta = -\frac{\gamma}{2\pi U}\{(\tfrac{1}{2}l + x)\log(\tfrac{1}{2} + x/l) + (\tfrac{1}{2}l - x)\log(\tfrac{1}{2} - x/l)\}. \tag{49}$$

From (48) it is apparent that requiring uniform load distribution has led to a mean-camber line with infinite slope at both ends. This is not a serious objection at the leading edge where the effects of thickness will alleviate the singularity, but the infinite slope at the trailing edge causes flow separation or cavitation and hence should be avoided. This difficulty might have been anticipated, since the Kutta condition cannot be satisfied unless the loading vanishes at the trailing edge.

A more practical foil shape may be obtained by relaxing the requirement of uniform loading near the trailing edge and assuming a vortex distribution that vanishes there. The simplest family of such foils is the *NACA a* series, with constant loading from the leading edge downstream over a distance al, followed by a linear attenuation as shown in figure 5.11. The corresponding mean-camber lines are shown in figure 5.12, and $a = 0.8$ is a common choice for practical purposes. A detailed account of these and other foil geometries is given by Abbott and von Doenhoff (1959).

5.5 Drag Force on a Two-Dimensional Foil

Thus far in our discussion of two-dimensional steady foils, we have neglected the drag force parallel to the x-axis. One reason for this omission is that lifting surfaces are intended to produce lift forces with a minimum of drag or with a large lift-drag ratio L/D. However, a second reason for

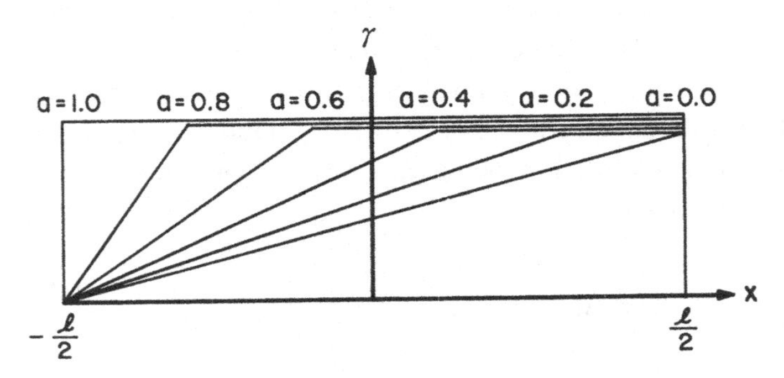

Figure 5.11
Loading of *NACA a* mean-camber lines.

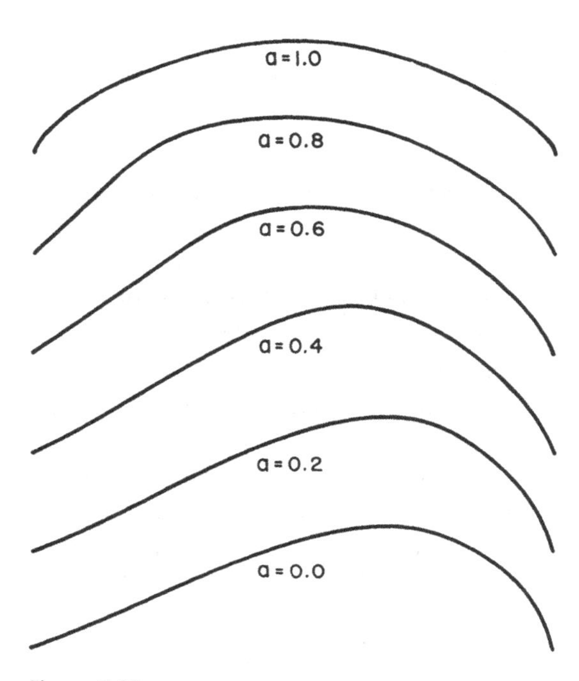

Figure 5.12
Mean-camber lines for *NACA a* series.

omitting drag from the preceding analysis is that *no drag force acts upon a steady two-dimensional foil in an inviscid fluid.* Viscous drag must be anticipated in a real fluid, and experimental results such as those shown in figure 2.7 confirm this, but the magnitude of the total drag force is comparable to the skin-friction drag coefficient for small angles of attack. Subsequently, in dealing with three-dimensional lifting surfaces and with unsteady two-dimensional foils, a nonzero drag force will occur in the inviscid theory, but this cannot be analyzed within the steady two-dimensional theory.

The absence of a drag force, despite the existence of lift, can be verified from the general form of the Kutta-Joukowski theorem (equation 15; see also problem 3). The most general form of this theorem states that the hydrodynamic force vector is equal in magnitude to the product $\rho U\Gamma$, with Γ the total circulation about the foil; it also states that this force is normal to the direction of the streaming flow.

A more elementary proof of zero drag follows from an energy analysis. The velocity field induced by the foil is steady state in a coordinate system moving with the foil and vanishes at large distances from the foil. The kinetic energy of the fluid surrounding the foil will move with the foil, and there can be no net build-up of energy in the fluid downstream. Thus, no work is done in propelling the foil in a steady-state manner, and there can be no drag force.

A curious paradox arises in conjunction with the drag force acting on a flat-plate foil. Given the geometry of this case, it might be expected that the pressure force acts normal to the plate; in that event the drag force must equal the product of the lift force and the tangent of the angle of attack. However, an additional contribution to the hydrodynamic force in this case results from the infinite velocity at the leading edge and the occurrence of a corresponding *leading-edge suction force*. The magnitude of the leading-edge suction force can be calculated from a momentum analysis carried out locally near the leading edge, and the magnitude of the resulting upstream force is such that the drag component of the normal force is precisely canceled. Thus, despite the suggestive geometry, the total hydrodynamic drag force is indeed zero, and the force vector is precisely normal to the direction of the incident stream.

5.6 Two-Dimensional Source and Vortex Distributions

In the previous sections, vortex distributions have been utilized to represent the two-dimensional lifting problem; similarly, source distributions can be used to solve the thickness problem. These two problems are related—both the source and the vortex are conjugate functions, equal to the real and imaginary parts of the complex logarithm. This relationship will be pursued here, not only to derive the corresponding solution of the thickness problem, but also as a necessary first step toward deriving the solution (32) of the integral equation (30).

Complex variables will be used here, as in section 4.7, with $z = x + iy$. The complex potential of a source is given in table 4.1, and that of a vortex by (23). The combination of these two singularities, situated at the point $(\xi, 0)$, is

$$\phi + i\psi = \frac{1}{2\pi}(q - i\gamma)\log(z - \xi). \tag{50}$$

Here q is the source strength and γ is the vortex strength. Differentiation with respect to z gives the corresponding complex velocity

$$u(z) - iv(z) = -\frac{1}{2\pi}(q - i\gamma)\frac{1}{\xi - z}. \tag{51}$$

If (51) is rewritten in polar coordinates, centered at the singular point (ξ, 0), then q is the rate of flux of fluid passing radially outward through a circle of constant radius, and γ is the circulation around the contour.

To represent a two-dimensional hydrofoil, these singularities are distributed continuously along the cut $-l/2 < x < l/2$ of the x-axis, with an arbitrary source strength $q(\xi)$ and vortex strength $\gamma(\xi)$. To simplify the following equations, the coordinates will be nondimensionalized so that $l = 2$. Then the complex velocity of this distribution is

$$u(z) - iv(z) = -\frac{1}{2\pi}\int_{-1}^{1}[q(\xi) - i\gamma(\xi)]\frac{d\xi}{\xi - z}. \tag{52}$$

If the point $z = x + iy$ is close to the boundary $y = 0$, the values of u and v on the foil surface can be calculated. For this purpose we let $z \to x \pm i0$, corresponding to the limiting values on the upper and lower surfaces of the foil. Since there is a pole at the point $z = \xi$. some care is necessary. The point z may approach the real axis from above or below if the contour of integration in (52) is deformed to remain on the original side of the pole, as shown in figure 5.13. Then, from the calculus of residues, the integral is equal to a principal-value integral along the real axis, plus or minus one half of the residue at the singular point $\xi = x$ (see Hildebrand 1976, section 10.15). Thus the limiting values of the velocity components on the upper or lower surfaces of the cut are given by

$$u_{\pm}(x) - iv_{\pm}(x) = -\frac{1}{2\pi}⨍_{-1}^{1}[q(\xi) - i\gamma(\xi)]\frac{d\xi}{\xi - x} \mp \frac{1}{2}i[q(x) - i\gamma(x)], \tag{53}$$

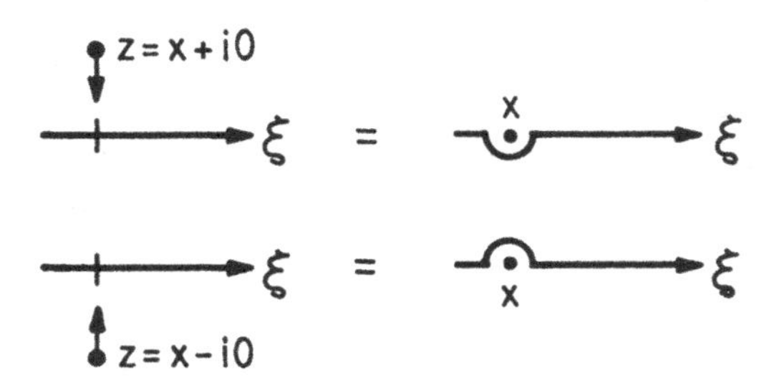

Figure 5.13
Equivalent contours of integration as the point z approaches the real axis.

where $⨍$ denotes the principal-value integral. To emphasize the importance of this equation, let us consider the separate cases of a source distribution ($\gamma = 0$) and a vortex distribution ($q = 0$):

Source Distribution

$$u_\pm(x) - iv_\pm(x) = -\frac{1}{2\pi}⨍_{-1}^{1}\frac{q(\xi)d\xi}{\xi - x} \mp \frac{1}{2}iq(x), \tag{54}$$

Vortex Distribution

$$u_\pm(x) - iv_\pm(x) = -\frac{i}{2\pi}⨍_{-1}^{1}\frac{\gamma(\xi)d\xi}{\xi - x} \mp \frac{1}{2}\gamma(x). \tag{55}$$

For the source distribution, separating the real and imaginary parts of equation (54) gives the resulting velocity components:

$$u_\pm(x) = -\frac{1}{2\pi}⨍_{-1}^{1}\frac{q(\xi)d\xi}{\xi - x}, \tag{56}$$

$$v_\pm(x) = \pm\frac{1}{2}q(x). \tag{57}$$

The horizontal velocity component u is even in y, whereas the vertical component v is odd. From the boundary condition (20) for the thickness problem, it follows that the solution for the flow past a thin symmetrical strut, of thickness $t(x) = y_u(x) - y_l(x)$, can be obtained directly as a distribution of sources with strength

$$q(x) = -Ut'(x). \tag{58}$$

This conclusion is consistent with the result of Green's theorem noted in section 4.11.

A simple physical argument can be used to explain (56–58). From symmetry a source at $(\xi, 0)$ on the real axis will induce a flow radially outward in all directions from this point. Elsewhere on the real axis, the induced velocity will be horizontal. Thus the only contribution to the vertical velocity on or near the real axis will come from the immediate vicinity of the source. Near this point, the net outward flux away from the real axis must equal the source strength, in agreement with (57). On the other hand, the source will induce a horizontal velocity all along the real axis, of magnitude inversely proportional to the distance $x - \xi$ and thus a distribution of sources will give a total horizontal velocity consistent with (56).

For the vortex distribution (55), the velocity components on the cut are given by

$$u_{\pm}(x) = \mp\frac{1}{2}\gamma(x), \tag{59}$$

$$v_{\pm}(x) = -\frac{1}{2\pi}\int_{-1}^{1}\frac{\gamma(\xi)\,d\xi}{\xi - x}. \tag{60}$$

The same pair of equations was derived earlier (29–30) using real variables. The conjugate nature of these results is apparent, relative to (56–57). In particular, the velocity components (u, v) of a source distribution are identical to the conjugate pair $(v, -u)$ of the vortex distribution, except for the change in notation of the singularity strength. Thus, in the lifting problem, the local vortex strength γ can be related to the jump in tangential velocity at the same point on the mean-camber line, but not to the normal velocity.

5.7 Singular Integral Equations

To solve the lifting problem with prescribed foil geometry or a given $v(x)$ on the cut, it is necessary to determine the vortex strength γ from the integral equation (60). Physically, each vortex element along the cut contributes to the net vertical velocity at any other point on the cut, and the problem is to find the appropriate distribution of vortices that induce a prescribed value of $v(x)$. Equation (60) is a *singular* integral equation because the *kernel*, or known part of the integrand, is singular.

The literature on the solution of singular integral equations is extensive; see, for example, the monograph of Muskhelishvili (1953) and a section in the text by Carrier, Krook, and Pearson (1966). Here we shall construct the solution in a manner that exploits our physical understanding of the thickness problem and the conjugate nature of the source and vortex distributions.

Figure 5.14 indicates the linearized boundary-value problem appropriate to each physical situation. The vertical velocity component v_+ denotes the normal velocity on the upper surface of each body, which is assumed known from the boundary conditions (20–21). From the appropriate symmetries, $v(x, y)$ is an odd function of y in the thickness problem, and hence in this case $v(x, 0) = 0$ on the real axis outside the cut. Conversely, for the

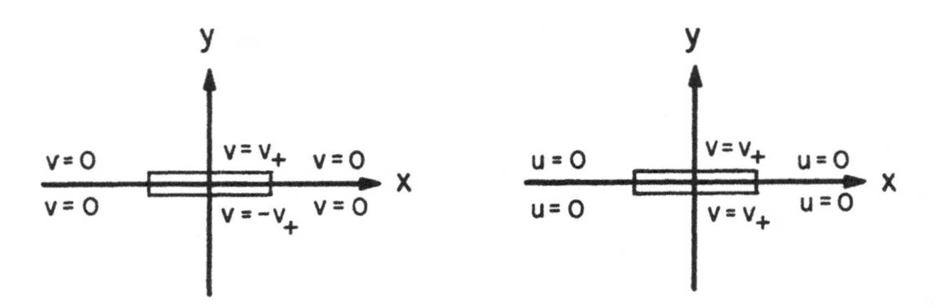

Figure 5.14
Boundary values on the cut real axis in the thickness and lifting problems.

lifting problem, $u(x, y)$ is an odd function of y which must vanish on the real axis outside the cut.

Now let us consider the upper half of the two sketches in figure 5.14. Apparently, the source strength can be found directly in the first case because the vertical velocity is given along the entire upper side of the real axis, from minus infinity to plus infinity, and equation (57) gives an appropriate source distribution that will provide this velocity distribution. On the other hand, the lifting problem is a *mixed* boundary-value problem since the vertical velocity is prescribed only on the cut, with the horizontal velocity prescribed elsewhere on the real axis. However, if the complex velocity $u - iv$ in the lifting problem is multiplied by a suitably chosen analytic function, such as $(1 - z^2)^{1/2}$, which is real on the cut and pure imaginary on the remainder of the real axis, then the product will be a new analytic function whose imaginary part is known all along the real axis. The square-root function $(1 - z^2)^{1/2}$ is clearly not a unique choice, but it is at least a possible one, and the resulting lack of uniqueness must be faced later.

The procedure for converting the lifting problem into an alternative mathematical problem of the same form as the thickness problem is then as follows. A new *pseudo-velocity* function is defined as

$$\tilde{u} - i\tilde{v} = (u - iv)(1 - z^2)^{1/2}. \tag{61}$$

This new function must satisfy the following boundary conditions, on the upper side of the real axis:

$$\tilde{v} = v_+(1 - x^2)^{1/2}, \qquad |x| < 1, \tag{62}$$

$$\tilde{v} = 0, \qquad |x| > 1. \tag{63}$$

Here the second condition (63) results from the fact that $u = 0$ on the real axis outside the cut. Mathematically, this problem is indistinguishable from the upper half of the thickness problem, except for the modification in the vertical velocity due to the square root (62). Thus, the pseudo-velocity function (61) can be generated with a source distribution, of strength given by (57), or by

$$\tilde{q}(x) = 2\tilde{v}(x) = 2v_+(x)(1-x^2)^{1/2}, \qquad |x| < 1, \tag{64}$$

$$\tilde{q}(x) = 0, \qquad |x| > 1. \tag{65}$$

From (52) it follows that in the upper half-plane,

$$\tilde{u}(z) - i\tilde{v}(z) = -\frac{1}{\pi}\int_{-1}^{1}\frac{v_+(\xi)(1-\xi^2)^{1/2}}{\xi - z}d\xi. \tag{66}$$

Thus, from (61), the original velocity function in the upper half-plane must be of the form

$$u - iv = -\frac{1}{\pi}\frac{1}{(1-z^2)^{1/2}}\int_{-1}^{1}\frac{v_+(\xi)(1-\xi^2)^{1/2}}{\xi - z}d\xi. \tag{67}$$

The integral equation (60) for the vortex strength γ can then be solved by using (67) to determine the horizontal velocity component on the upper side of the cut,

$$u(x,0+) = -\frac{1}{\pi}\frac{1}{(1-x^2)^{1/2}}\fint_{-1}^{1}\frac{v_+(\xi)(1-\xi^2)^{1/2}}{\xi - x}d\xi. \tag{68}$$

From (59), the solution for γ is

$$\gamma(x) = \frac{2}{\pi}\frac{1}{(1-x^2)^{1/2}}\int_{-1}^{1}\frac{v_+(\xi)(1-\xi^2)^{1/2}}{\xi - x}d\xi. \tag{69}$$

One deficiency of (69) is the presence of square-root singularities at both the leading and trailing edges. Obviously, it would be coincidental if the Kutta condition at the trailing edge was satisfied at this stage, for we have proceeded so far without regard for the singularities at the two ends of the cut. This problem is related to the nonunique choice of a square-root function used to generate the pseudo-velocity function (61). If we had anticipated the desire to satisfy the Kutta condition at the trailing edge, we might have used the alternative function

$$\tilde{u} - i\tilde{v} = (u - iv)\left(\frac{1-z}{1+z}\right)^{1/2}, \tag{70}$$

instead of (61), and thus obtained the final solution, corresponding to (69), as

$$\gamma(x) = \frac{2}{\pi}\left(\frac{1+x}{1-x}\right)^{1/2} \fint_{-1}^{1} \left(\frac{1-\xi}{1+\xi}\right)^{1/2} \frac{v_+(\xi)}{\xi - x} d\xi. \tag{71}$$

In fact, (71) is the desired solution, singular only at the leading edge. At this point, however, we may feel some concern about the lack of a unique answer to the original integral equation. In essence, (69) and (71) are two *particular* solutions of the integral equation (60), and in general a *homogeneous* solution must be added in a manner analogous to that employed in solving differential equations.

The homogeneous solution of (60) satisfies this integral equation with the left side set equal to zero. To find this function, we could guess, look through a table of integrals, or simply accept that the answer is the inverse square-root function $[1 - x^2]^{-1/2}$. To verify this more systematically, let us find the difference between the two particular solutions (69) and (71). After some algebra, the result is

$$\begin{aligned} &\frac{2}{\pi}\frac{1}{(1-x^2)^{1/2}} \fint_{-1}^{1} \frac{v_+(\xi)}{\xi - x}(1-\xi^2)^{1/2}\left(\frac{\xi - x}{1+\xi}\right) d\xi \\ &= \frac{2}{\pi}\frac{1}{(1-x^2)^{1/2}} \fint_{-1}^{1} v_+(\xi)\left(\frac{1-\xi}{1+\xi}\right)^{1/2} d\xi = \frac{2}{\pi}\frac{C}{(1-x^2)^{1/2}}, \end{aligned} \tag{72}$$

where C is a constant, equal to the last integral. Thus the inverse square-root function on the right side of (72) is a homogeneous solution of the integral equation (60).

To relate the constant C in (72) to the total circulation, we write the general solution for γ in the form

$$\gamma(x) = \frac{2}{\pi}\frac{1}{(1-x^2)^{1/2}} \left\{ \fint_{-1}^{1} \frac{v_+(\xi)(1-\xi^2)^{1/2}}{\xi - x} d\xi + C \right\}, \tag{73}$$

where C is now an arbitrary constant. Integrating (73) between (–1, 1) gives the total circulation,

$$\Gamma = \int_{-1}^{1} \gamma(x)\, dx. \tag{74}$$

The right side of (73) may be integrated as follows:

$$\begin{aligned}\Gamma &= \frac{2}{\pi}\int_{-1}^{1}\frac{dx}{(1-x^2)^{1/2}}\int_{-1}^{1}\frac{v_+(\xi)(1-\xi^2)^{1/2}}{\xi-x}d\xi+\frac{2C}{\pi}\int_{-1}^{1}\frac{dx}{(1-x^2)^{1/2}}\\ &= \frac{2}{\pi}\int_{-1}^{1}d\xi v_+(\xi)(1-\xi^2)^{1/2}\int_{-1}^{1}\frac{dx}{(\xi-x)(1-x^2)^{1/2}}+2C.\end{aligned} \tag{75}$$

Since the inverse square-root is the homogeneous solution of the integral equation (60), the last integral in (75) must be zero, and thus

$$C=\frac{1}{2}\Gamma. \tag{76}$$

Now the Kutta condition can be imposed by choosing the constant $C=\frac{1}{2}\Gamma$ so that the factor in brackets in (73) is zero at $x=-1$, canceling the square-root infinity at this point. Thus,

$$C=\frac{1}{2}\Gamma=-\int_{-1}^{1}\frac{v_+(\xi)(1-\xi^2)^{1/2}}{\xi+1}d\xi=-\int_{-1}^{1}v_+(\xi)\left(\frac{1-\xi}{1+\xi}\right)^{1/2}d\xi, \tag{77}$$

and this is consistent with (72). In this manner the particular solution (71) can be obtained by a systematic analysis from the general solution (73), after imposing the Kutta condition at the trailing edge.

This completes our solution of the integral equation (60) and provides a derivation of (32) and (34). We can go one step further, however, and show how the resulting definite integrals can be evaluated. One type of definite integral required to evaluate the lift and moment coefficients is given by

$$I_n=\int_{-1}^{1}\frac{x^n dx}{(1-x^2)^{1/2}}=\int_{0}^{\pi}(\cos\theta)^n d\theta. \tag{78}$$

Note that (36) and (38) can be expressed as combinations of these integrals, whenever the slope of the mean-camber line is a polynomial in x. The integrals in (78) are elementary integrals; when n is odd they equal zero, and when n is even

$$I_n=\pi\frac{1\cdot3\cdot5\cdot\ldots\cdot(n-1)}{2\cdot4\cdot6\cdot\ldots\cdot(n)}. \tag{79}$$

The corresponding integrals, which must be evaluated to find the vortex strength, are of the form

$$H_n(x)=\int_{-1}^{1}\frac{\xi^n\,d\xi}{(\xi-x)(1-\xi^2)^{1/2}}. \tag{80}$$

These are not elementary, but they can be related to the integrals in (78). First we recall that, since $(1 - \xi^2)^{-1/2}$ is a homogeneous solution of the integral equation (60),

$$H_0(x) = \fint_{-1}^{1} \frac{d\xi}{(\xi - x)(1 - \xi^2)^{1/2}} = 0. \tag{81}$$

For $n \geq 1$ in (80), a recursion formula can be derived as follows:

$$H_n(x) = \fint_{-1}^{1} \frac{[(\xi - x)\xi^{n-1} + x\xi^{n-1}]d\xi}{(\xi - x)(1 - \xi^2)^{1/2}} = I_{n-1} + xH_{n-1}(x). \tag{82}$$

In particular, it follows from (79) and (81) that

$$H_1(x) = \pi, \tag{83}$$

$$H_2(x) = \pi x, \tag{84}$$

and, more generally, $H_n(x)$ is a polynomial of degree $n - 1$.

The integrals defined by (80) are known more generally as *Hilbert transforms;* so far we have considered them only on the cut $-1 < x < 1$. This is the region of principal interest for steady two-dimensional hydrofoils, but in our subsequent treatments of cavity flows and unsteady hydrofoils, where singular effects exist downstream of the foil, it will be necessary to evaluate the resulting Hilbert transforms for values of x exterior to the cut. This generalization is accomplished most easily by replacing x by the complex variable z and using analytic continuation to deduce the values of (80) off the cut. The recursion formula (82) remains valid in this case, but the other results must be modified.

Since (82) is valid, it is necessary to consider only the generalization of (81) which, for complex z, takes the form

$$H_0(z) = \int_{-1}^{1} \frac{d\xi}{(\xi - z)(1 - \xi^2)^{1/2}}. \tag{85}$$

Recalling the discussion following figure 5.13, the limiting behavior of (85) on the cut is given by

$$H_0(z) \to \pm \pi i(1 - x^2)^{-1/2}, \tag{86}$$

for $z \to x \pm i0$, where the real part, $H_0(x)$, vanishes by virtue of (81). Since (85) is an analytic function of z, it follows that

$$H_0(z) = -\pi(z^2 - 1)^{-1/2}, \tag{87}$$

since this is the only analytic function consistent with (86). Here the branch of the square root is prescribed so that $H_0(z) \simeq -\pi/z$ at infinity, a result which can be derived independently from (85) using (79). For $z = x$ outside the cut, the desired result follows in the form

$$H_0(x) = \mp\pi(x^2-1)^{-1/2}, \qquad \begin{cases} x > 1 \\ x < -1 \end{cases} \tag{88}$$

Analogous expressions for $H_n(x)$ can be deduced from (82).

The family of integrals (80) can be expressed in an alternative and particularly useful form, by making the substitutions $x = \cos\chi$. $\xi = \cos\theta$. Since $\cos^n\theta$ can be expressed as a finite Fourier series involving typical terms of the form $\cos m\theta$, $m \le n$, we consider the family

$$G_n(\chi) = \fint_0^\pi \frac{\cos n\theta\, d\theta}{\cos\theta - \cos\chi}; \tag{89}$$

these are known as Glauert integrals. From (81), (83), and (84) it is apparent that the first three members are

$$G_0 = 0, \tag{90}$$

$$G_1 = \pi, \tag{91}$$

$$G_2 = 2\pi\cos\chi. \tag{92}$$

More generally,

$$G_n = \pi\frac{\sin n\chi}{\sin\chi}. \tag{93}$$

This simple result can be confirmed by using the addition formula

$$\cos(n+1)\theta + \cos(n-1)\theta = 2\cos n\theta\cos\theta, \tag{94}$$

to derive from (89) the recursion relation

$$G_{n+1} + G_{n-1} - 2\cos\chi G_n = \int_0^\pi \cos n\theta\, d\theta = 0, \qquad n \ge 1. \tag{95}$$

Using the addition formula

$$\sin(n+1)\chi + \sin(n-1)\chi - 2\cos\chi\sin n\chi = 0, \tag{96}$$

we find that (93) satisfies (95). Since it is consistent with (90–92), it must hold in general.

The last results suggest a Fourier series approach which leads to particularly simple results. In dimensional coordinates, with $x = \frac{1}{2}l\cos\theta$, the vortex strength can be expanded in the following trigonometric series:

$$\gamma = 2U\left\{A_0 \cot\left(\frac{1}{2}\theta\right) + \sum_{n=1}^{\infty} A_n \sin n\theta\right\}. \tag{97}$$

This series has been chosen so that each term satisfies the Kutta condition at the trailing edge $\theta = \pi$. The first term is proportional to the flat-plate load distribution (42), and represents the flat plate if $A_0 = \alpha$. The remaining terms in (97) vanish at the leading edge, and this infinite series is a complete Fourier series representation for any reasonable vortex distribution vanishing at the leading and trailing edges; thus, it follows that (97) can be used to represent practically any realistic vortex distribution, with the leading-edge singularity absorbed in the first term.

Using equation (60), the vertical velocity is given by

$$v = -\frac{1}{2\pi}\int_0^{\pi} \frac{\gamma(\theta_1)\sin\theta_1\, d\theta_1}{\cos\theta_1 - \cos\theta}. \tag{98}$$

Substitute (97) for γ, and use (89), (93) and the trigonometric relations

$$\begin{matrix}\cos\\ \sin\end{matrix}(n-1)\theta - \begin{matrix}\cos\\ \sin\end{matrix}(n+1)\theta = \pm 2\sin\theta \begin{matrix}\sin\\ \cos\end{matrix} n\theta \tag{99}$$

and

$$\cot\left(\frac{1}{2}\theta\right) = \frac{1+\cos\theta}{\sin\theta}. \tag{100}$$

Then it follows from (98) that

$$v = -UA_0 + U\sum_1^{\infty} A_n \cos n\theta. \tag{101}$$

Alternatively, from the boundary condition (22),

$$\frac{d\eta}{dx} = A_0 - \sum_1^{\infty} A_n \cos n\theta. \tag{102}$$

Hence the coefficients A_n can be related to the slope of the mean-camber line, from the usual orthogonality property of Fourier coefficients, by means of the equations

$$A_0 = \frac{1}{\pi}\int_0^{\pi} \frac{d\eta}{dx} d\theta, \tag{103}$$

$$A_n = -\frac{2}{\pi}\int_0^{\pi} \frac{d\eta}{dx} \cos n\theta \, d\theta. \tag{104}$$

Finally, the lift coefficient is given by

$$C_L = \frac{L}{½\rho U^2 l} = \frac{1}{U}\int_0^{\pi} \gamma \sin\theta \, d\theta = \pi(2A_0 + A_1), \tag{105}$$

and the coefficient of the moment about the mid-chord position is

$$C_M = \frac{M}{½\rho U^2 l^2} = \frac{1}{2U}\int_0^{\pi} \gamma \cos\theta \sin\theta \, d\theta = \frac{\pi}{2}\left(A_0 + \frac{1}{2}A_2\right). \tag{106}$$

The results (103–106) are equivalent to (36) and (38).

5.8 Three-Dimensional Vortices

The preceding analysis of two-dimensional lifting surfaces can be extended to three dimensions, using distributions of vortices in the plane $y = 0$. For this purpose the properties of three-dimensional vortices must be developed.

A three-dimensional line vortex, or *vortex filament,* is a singularity situated along an arbitrary curve in space. This singularity has the same local properties as a two-dimensional point vortex—the flow very close to the filament appears locally as a two-dimensional vortex when observed in a plane normal to the filament (see figure 5.15). A smoke ring is a familiar example of a vortex filament that occupies a closed circular contour.

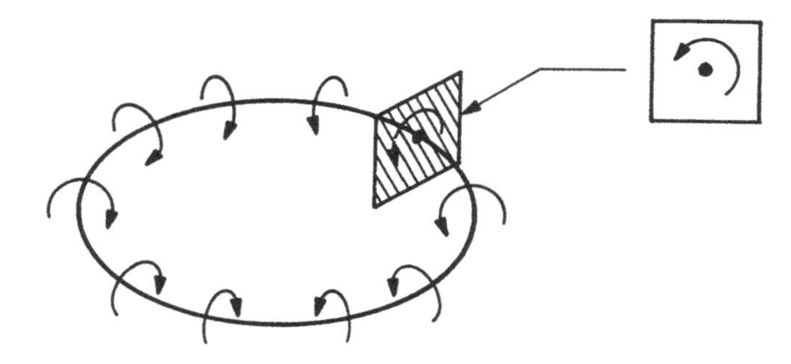

Figure 5.15
Three-dimensional vortex filament.

The flow must be irrotational throughout the fluid region, excluding the line on which the vortex filament is situated. Thus, the circulation around any closed path that does not include the vortex must be zero, while the circulation around a small, closed path about the vortex must be equal to the circulation of the vortex at that point. To see that this circulation is the same for all points along the filament, define a closed contour $C_1 + C_2 + C_3 + C_4$ to form the boundary of a tube surrounding the vortex filament, as shown in figure 5.16. Since $C_1 + C_2 + C_3 + C_4$ is a closed curve, which does not include the vortex, the total circulation around this contour must vanish. The contributions from C_2 and C_4 cancel each other, since these two contours are practically identical except for the change in the positive direction of integration. Thus the contributions from C_1 and C_3 must be equal and opposite, or

$$\oint_{C_1} \mathbf{V} \cdot d\boldsymbol{l} + \oint_{C_3} \mathbf{V} \cdot d\boldsymbol{l} = 0. \tag{107}$$

Figure 5.16 shows that the contours C_1 and C_3 are opposite in their sense of rotation about the vortex filament, and if the first integral in (107) defines the circulation Γ_1. the second integral will be equal to $-\Gamma_3$ and vice versa. Thus $\Gamma_1 = \Gamma_3$, and the circulation of a vortex filament must be constant along its length. As a consequence of this result, the circulation is conserved, and a vortex filament cannot end within the fluid. A vortex filament can terminate at the boundaries of the fluid domain, but otherwise it must form a closed curve, as shown in figure 5.15.

The velocity field induced by a vortex filament can be deduced from the two-dimensional results (24–25). For motion in the x-y plane due to a point vortex at the origin of strength Γ, the velocity components can be written as

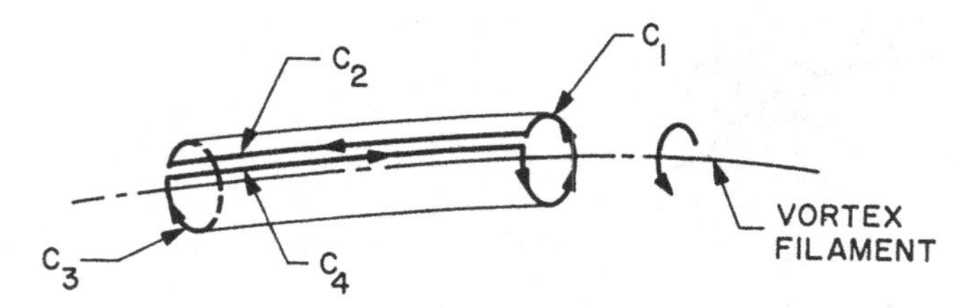

Figure 5.16
Contour used to show constancy of circulation along a vortex filament.

$$u = -\frac{\Gamma y}{2\pi(x^2+y^2)} = -\frac{\Gamma}{2\pi}\frac{\partial}{\partial y}\log r, \tag{108}$$

and

$$v = -\frac{\Gamma x}{2\pi(x^2+y^2)} = -\frac{\Gamma}{2\pi}\frac{\partial}{\partial x}\log r. \tag{109}$$

Thus, the velocity vector in this case can be written as

$$\mathbf{V} = \frac{-\Gamma}{2\pi}\nabla \times \mathbf{k}\log r. \tag{110}$$

Here **k** denotes the unit vector in the z-direction, parallel to the axis of the vortex.

To pass from two to three dimensions, we replace the two-dimensional source potential log r by the potential of a line of three-dimensional sources, of strength m, initially along the z-axis:

$$\begin{aligned}\mathbf{V} &= \nabla \times \frac{m\mathbf{k}}{4\pi}\int_{-\infty}^{\infty}\frac{dz}{(x^2+y^2+z^2)^{1/2}}\\ &= \frac{-m}{4\pi}\int_{-\infty}^{\infty}\frac{(\mathbf{i}y-\mathbf{j}x)\,dz}{(x^2+y^2+z^2)^{3/2}} = \frac{-m(\mathbf{i}y-\mathbf{j}x)}{2\pi(x^2+y^2)}.\end{aligned} \tag{111}$$

The desired two-dimensional velocity components (110) will result from a line distribution of three-dimensional sources normal to the x-y plane, provided the source strength is $m = \Gamma$.

Now we rewrite the last line integral in (111) as

$$\mathbf{V} = -\frac{\Gamma}{4\pi}\int_{-\infty}^{\infty}\frac{\mathrm{R}\times\mathbf{k}\,dz}{R^3}, \tag{112}$$

where $\boldsymbol{R}$ is the position vector $\mathbf{x}\mathbf{i} + y\mathbf{j} + z\mathbf{k}$.

Equation (112) gives a form of the two-dimensional velocity field resulting from a line vortex along the z-axis, which can be generalized to an arbitrary three-dimensional vortex filament of strength Γ. Let the vortex filament be situated on a contour C, with dl the differential element of integration along C as in figure 5.17. Then by analogy with (112) we can define the velocity

$$\mathbf{V} = -\frac{\Gamma}{4\pi}\int_{C}\frac{\mathrm{R}\times d\boldsymbol{l}}{R^3}, \tag{113}$$

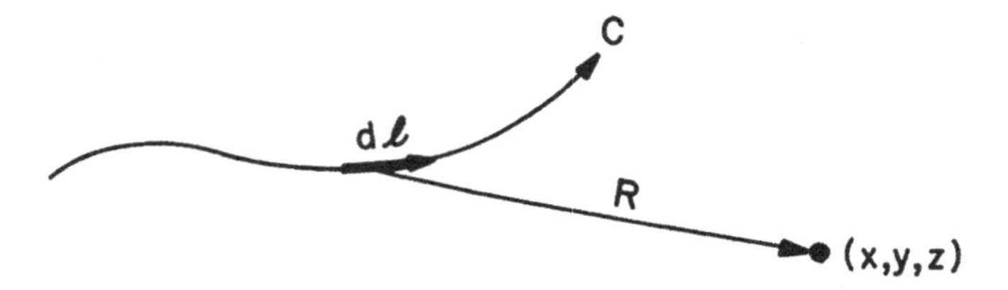

Figure 5.17
Definition sketch for equation (113), showing the position vector **R** and differential line element *dl* along a vortex filament situated on the contour *C*.

where **R** is now the vector to the point (x, y, z) from each point along the curve of integration, as defined in figure 5.17. This more general integral defines a velocity field that is irrotational and divergenceless except on the contour *C*. Near this contour, it possesses the same singularity as the two-dimensional vortex (112), since locally the curve approximates a straight line and near the curve the integral will be dominated by the value of the integrand as $\boldsymbol{R} \to 0$.

Equation (113) is, therefore, the velocity field induced by a three-dimensional vortex filament of strength Γ situated on the curve *C*. This result, well known in electromagnetic theory as the *law of Biot-Savart*, relates the magnetic field to the current in a conducting filament. We have derived it intuitively by guessing and then verifying the final answer. A more systematic derivation, in terms of an arbitrary volume distribution of vorticity reduced ultimately to a line vortex, can be found in Batchelor (1967). An extensive analysis of three-dimensional vortex fields is given by Brard (1972).

5.9 Three-Dimensional Planar Lifting Surfaces

For three-dimensional lifting surfaces and hydrofoils, the powerful method of complex function theory must be abandoned; the real variable z and dummy coordinate ζ are used to denote the coordinate along the span of the foil. However, the thin-wing approximation can be retained if the slope of the foil is sufficiently small over its entire surface (except at the leading edge!). A linearized problem can be formulated in exactly the same manner as for the two-dimensional case in section 5.2. The thickness problem can be solved again in terms of a suitable source distribution, and the lifting problem can be reduced to one involving a mean-camber surface of zero thickness where there is a specified normal velocity

$$v(x,0,z) = -U\frac{\partial}{\partial x}\eta(x,z). \tag{114}$$

This boundary condition must be supplemented by the Kutta condition requiring finite velocity all along the trailing edge.

As in the two-dimensional case, a solution of the boundary condition (114) can be constructed by a suitable distribution of vortices in the plane $y = 0$. In general, such a distribution would be comprised of line vortices of varying strengths and varying orientation angles. In other words, γ in this case is a vector situated in the plane $y = 0$. This vector can be decomposed into a *bound* component, parallel to the z-axis, and a *free* component, parallel to the x-axis. Thus it is necessary to consider only these two specific orientations and the induced velocities corresponding to each. In section 5.8 we showed that the limiting behavior of a line vortex very close to its axis is identical to that of a two-dimensional point vortex of the same strength. Thus the results derived in section 5.2, for a distribution of two-dimensional vortices along the x-axis, will apply locally for the bound vortex distribution in three dimensions. In particular, a bound vortex distribution of density $\gamma_B(x, z)$ will induce a discontinuity in the chordwise component of the fluid velocity, across the plane $y = 0$, in accordance with the relation

$$u(x,\pm 0,z) = \mp\frac{1}{2}\gamma_B(x,z). \tag{115}$$

Similarly, a free vortex distribution of density $\gamma_F(x, z)$ will induce a discontinuity in the spanwise component across $y = 0$ given by

$$w(x,\pm 0,z) = \pm\frac{1}{2}\gamma_F(x,z). \tag{116}$$

Note that the right-hand rule applies here; the circulation of the bound vortices is defined as positive about the z-axis and that of the free vortices positive about the x-axis.

These planar distributions of vortices are known generally as vortex *sheets*, and we may refer separately to the bound vortex sheet and the free vortex sheet. These are not independent, however, since the velocity field adjacent to the sheets must be irrotational. Using (115) and (116) to compute the y-component of the vorticity adjacent to the foil, we get

$$\frac{\partial u}{\partial z} - \frac{\partial w}{\partial x} = 0, \tag{117}$$

or

$$\frac{\partial \gamma_B}{\partial z} - \frac{\partial \gamma_F}{\partial x} = 0. \tag{118}$$

This is essentially the extension to a vortex sheet of the result in section 5.8 for a vortex filament. Each element of bound vorticity cannot change in the z-direction, unless the change in its circulation is balanced by a corresponding change in the free vorticity.

The linearized pressure induced by these vortex sheets can be computed from (14), since the additional velocity component w does not give a linear contribution. Thus, on the upper or lower surface of the vortex sheet,

$$p - p_\infty \simeq \rho u U = \mp \frac{1}{2} \rho U \gamma_B(x,z). \tag{119}$$

Therefore, the pressure jump across the sheet is proportional to γ_B. and the pressure associated with the free vortex sheet is continuous.

The absence of a pressure jump for the free vortex sheet is an important feature of this component of the total vortex distribution, which is responsible for its name. A free vortex sheet, composed of vortex elements parallel to the free stream, can exist in a state of dynamic equilibrium within the free stream, whereas bound vortices with axes perpendicular to the free stream induce a pressure jump that requires an external balancing force, that is, the force acting on the lifting surface.

Now we apply these results to the lifting-surface problem by distributing sheets of bound and trailing vorticity on the plane $y = 0$ throughout the projection of the lifting surface planform. Integration of (119) along the chord, at a section of constant z, gives the sectional lift force

$$L(z) = \rho U \int_{x_T(z)}^{x_L(z)} \gamma_B(x,z)\,dx = \rho U \Gamma(z). \tag{120}$$

Here $\Gamma(z)$ is the total circulation of the section, and the limits of integration denote the x-coordinates of the trailing and leading edges. A second integration in the spanwise direction gives the total lift

$$\bar{L} = \rho U \int_{-s/2}^{s/2} \Gamma(z)\,dz. \tag{121}$$

Ultimately this equation can be used to compute the total lift force, but first the appropriate distributions of bound and trailing vortices must be explored.

Integrating (118) along the chord gives the following equation for continuity of vorticity across the section:

$$\int_{x_T}^{x_L} \left(\frac{\partial \gamma_B}{\partial z} + \frac{\partial \gamma_F}{\partial x} \right) dx = 0, \tag{122}$$

or

$$\Gamma'(z) + [\gamma_F(x,z)]_{x_T}^{x_L} = 0. \tag{123}$$

Thus, any change in the total circulation along the span of the lifting surface must be reflected in a corresponding jump of free vortex density across the section. The circulation Γ clearly must vary with z, especially near the tips $z = \pm\frac{1}{2}s$ where Γ changes from a nonzero value on the lifting surface to zero outside the tips. But where is the resulting free vorticity to go? It too must be conserved and cannot end abruptly at the leading or trailing edges. This apparent dilemma can be resolved by assuming that the lifting surface advances into a uniform and undisturbed fluid, but that it leaves behind a thin wake region of free, or *trailing*, vorticity. The wake extends to infinity downstream, or alternatively to the downstream position of the starting vortex. Since this starting vortex is equal and opposite to $\Gamma(z)$, the system of bound and trailing vortices is rejoined downstream in a closed loop.

The existence of a trailing vortex sheet downstream of the lifting surface is physically permissible since free vortices can exist in a state of dynamic equilibrium with the fluid. With the free vorticity set equal to zero upstream of the leading edge and equal to the trailing vorticity $\gamma_T(z)$ at the trailing edge, (123) determines the value of the trailing vorticity as

$$\gamma_T(z) = \Gamma'(z). \tag{124}$$

We emphasize that γ_T *is* independent of the x-coordinate, by virtue of (118) and because there can be no bound vorticity in the wake.

These vortex distributions are sketched in figure 5.18. Here the lifting surface, (or, strictly, its projection on the plane $y = 0$) contains a distribution

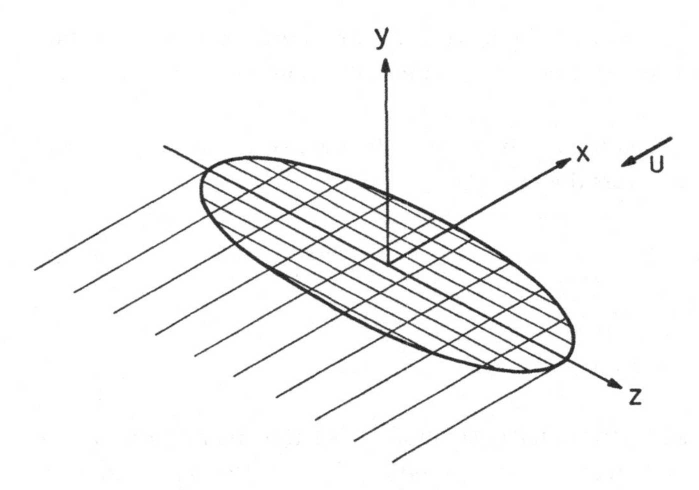

Figure 5.18
Sketch showing examples of the discrete vortices, which make up the continuous vortex sheets on the foil and downstream in the wake.

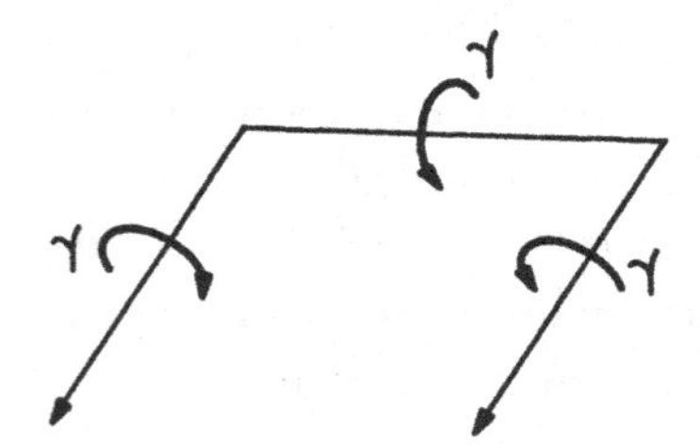

Figure 5.19
Horseshoe vortex of strength γ.

of bound and free vortices; the latter is joined at the trailing edge to a sheet of trailing vortices that extend downstream.

A particularly simple case is the lifting surface of small chord length, or large aspect ratio, and constant sectional loading. The bound vortex distribution is then constant along the span and confined to a small chord, but at the tips, it must be connected to two discrete trailing vortices as shown in figure 5.19. Note that (124) still holds in this case, but the derivative is equal to zero except at the tips,, where it is infinite, and hence γ_T is given by a pair of delta functions.

The line vortex shown in figure 5.19 is a *horseshoe* vortex, which itself satisfies the requirement of conservation of circulation. A system of these may be superposed to represent a more general lifting surface. Horseshoe vortices of equal span may be distributed in the chordwise direction to make a foil of rectangular planform and constant sectional loading γ, as shown in figure 5.20. However, while the loading in this case is constant along the span, the *downwash*, or the velocity $-v\ (x,\ 0,\ z)$ induced on the planform by this vortex distribution, must depend on z, and a pronounced local downwash will exist near the tips induced by the trailing vortices. Thus, from the boundary condition (114), the geometry of the foil in this case is not independent of span wise position and, indeed, must be physically unrealistic near the tips.

Another special case results by distributing horseshoe vortices of varying span along the same chordwise position, as shown in figure 5.21. Here the individual elements combine to form a single bound vortex $\Gamma\ (z)$, and the

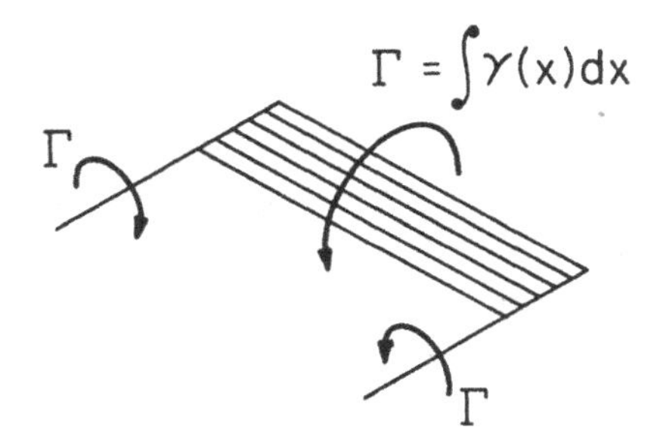

Figure 5.20
Lifting surface with uniform spanwise loading, composed of horseshoe vortices of equal span.

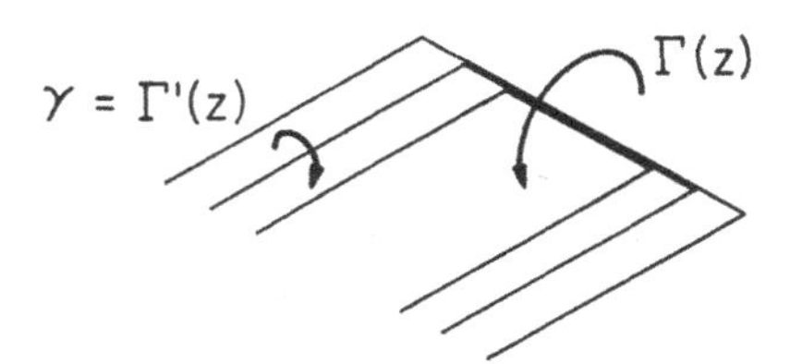

Figure 5.21
Lifting-line, composed of horseshoe vortices with coincident bound elements but different spans.

trailing vorticity is consistent with (124). This case is of practical importance for lifting surfaces of large aspect ratio, where the chord length is small and hence the chordwise position of the vortex elements is unimportant. The resulting approximation, for lifting surfaces of large aspect ratio, is the lifting-line theory that will be treated in section 5.11.

The trailing vortex sheet, which is an inevitable consequence of three-dimensional effects, is detrimental to the performance of a lifting surface. The effects are twofold: the lift is reduced and the drag increased. For the lift, figures 5.19–5.21 show that the sign of the vortices, associated with a positive lift force in the y-direction, induces a downward component of vertical velocity along the foil. This vertical downwash can be superposed with the incident stream velocity vector, resulting in a decreased effective angle of attack, by comparison with the two-dimensional case without downwash. Thus the total lift force $\overline{L}$ is less than would be expected from a two-dimensional analysis at each section.

5.10 Induced Drag

In three dimensions a drag force exists because the trailing vortex sheet increases in length, at a rate proportional to U, with a resulting increase in the total kinetic energy of the fluid at a similar rate. Since this energy must be supplied by work done to overcome the drag force, the *induced* drag D will be equal to the kinetic energy of a slice of fluid far downstream, the slice being of unit width along the x-axis.

Alternatively, from the general force equation (4.90), the drag component is given by the integral

$$D = \frac{1}{2}\rho \iint_{S_c} \left[\left(\frac{\partial \phi}{\partial x}\right)^2 - \left(\frac{\partial \phi}{\partial y}\right)^2 - \left(\frac{\partial \phi}{\partial z}\right)^2 \right] n_x \, dy \, dz. \tag{125}$$

Here the control surface has been taken as a pair of planes x = constant upstream and downstream of the foil, and the remaining surface between these at infinity can be neglected since the perturbation vanishes sufficiently rapidly as $y^2 + z^2 \to \infty$. Indeed, the same must be true upstream but not downstream where the trailing vortex sheet persists. Since the velocity induced by the vortex sheet far downstream is independent of x, it follows that

$$D = \frac{1}{2}\rho \int \int_{\infty}^{-\infty} \left[\left(\frac{\partial \phi}{\partial y} \right)^2 + \left(\frac{\partial \phi}{\partial z} \right)^2 \right] dy\, dz, \tag{126}$$

where the integrand is to be evaluated for $x \to -\infty$. This integral is the two-dimensional kinetic energy in a *Trefftz plane* x = constant, far downstream.

The surface integral in (126) can be replaced by a line integral around the trailing vortex sheet, or the cut $|z| < \frac{1}{2}s$, by means of the divergence theorem. Since $\nabla^2 \phi = 0$, it follows that

$$\begin{aligned} D &= \frac{1}{2}\rho \int\int \nabla\phi \cdot \nabla\phi \, dy\, dz = \frac{1}{2}\rho \int\int \nabla \cdot (\phi \nabla \phi)\, dy\, dz \\ &= \frac{1}{2}\rho \oint \phi \frac{\partial \phi}{\partial n} dl \\ &= -\frac{1}{2}\rho \int_{-s/2}^{s/2} [\phi]_-^+ \frac{\partial \phi}{\partial y} dz. \end{aligned} \tag{127}$$

Note that the contour integral at large radial distance $(y^2 + z^2)^{1/2}$ in the Trefftz plane is zero, since the velocity components v and w vanish at a sufficient rate far from the trailing vortex sheet. Also, in the final step of (127), we have used the fact that $\partial\phi/\partial y = \pm\, \partial\phi/\partial n$ is continuous across the wake, whereas the potential is discontinuous.

The jump in the potential can be related to the bound vorticity by integrating the pressure jump with respect to x and noting that the pressure jump across the wake must vanish. Thus

$$\Gamma(z) = -\int_{x_T}^{x_L} \left(\frac{\partial \phi}{\partial x} \right)_-^+ dx = -\int_{-\infty}^{x_L} \left(\frac{\partial \phi}{\partial x} \right)_-^+ dx = [\phi]_-^+ \big|_{x=-\infty}, \tag{128}$$

since the jump in the potential must vanish at the leading edge. Substituting (128) into (127), we obtain the drag formula

$$D = -\frac{1}{2}\rho \int_{-s/2}^{s/2} \Gamma(z) \frac{\partial \phi}{\partial y} dz, \tag{129}$$

where the integration is taken across the vortex wake far behind the foil.

The vertical velocity component $\partial\phi/\partial y$, far downstream on the trailing vortex sheet, can be computed from the equation (60) for a two-dimensional vortex distribution. With a suitable change in the coordinate system and using (124) to evaluate the trailing vortex density, we get

$$\frac{\partial \phi}{\partial y} = \frac{1}{2\pi} \fint_{-s/2}^{s/2} \frac{d\Gamma(\zeta)}{d\zeta} \frac{d\zeta}{\zeta - z}. \tag{130}$$

(The change of sign from (60) is necessitated by the transformation from the x-y plane to the y-z plane, while defining the vorticity consistently in terms of the right-hand rule.) With this result, (129) may be replaced by a formula involving only the bound circulation:

$$D = \frac{\rho}{4\pi} \int_{-s/2}^{s/2} \Gamma(z)\, dz \fint_{-s/2}^{s/2} \frac{d\Gamma(\zeta)}{d\zeta} \frac{d\zeta}{z - \zeta}. \tag{131}$$

The lift force (121) can be derived by a similar analysis of the flow in the Trefftz plane. Thus, starting with the y-component of (4.90), and including only the linear term, we get the following expression corresponding to (126):

$$\bar{L} = -\rho U \int \int_{-\infty}^{\infty} \partial\phi / \partial y \, dy\, dz = \rho U \int_{-s/2}^{s/2} [\phi]_-^+ dz, \tag{132}$$

and (121) follows from (128).

For a given spanwise distribution of circulation $\Gamma(z)$, equations (121) and (131) can be used to compute the total lift and drag forces. Since the lift is linearly proportional to Γ, it follows from (131) that the induced drag will be proportional to the square of the lift and will depend quadratically on the angle of attack and camber. Thus, the lift-drag ratio L/D is inversely proportional to the angle of attack and camber, and hence $L/D \gg 1$ if the foil satisfies the original assumptions of the linear theory. (At this point it may be a relief to note that the lift-drag ratio is large, since that is presumably the *raison d'être* for the lifting surface!) That the lift-drag ratio is inversely proportional to the angle of attack is another reason for keeping this angle small, in addition to the objectives of avoiding stall, cavitation, and nonlinear effects.

The lift-drag ratio is not a satisfactory figure of merit since this can be made arbitrarily large by reducing the angle of attack and resulting lift force. A more appropriate parameter is the nondimensional ratio

$$\begin{aligned} K &= \frac{\pi A \overline{C_D}}{\bar{C}_L{}^2} = \frac{\pi}{2} \rho U^2 s^2 \frac{\bar{D}}{\bar{L}_2} \\ &= \frac{s^2}{8} \frac{\int \Gamma(z) \int [\Gamma'(\zeta)/(z-\zeta)] d\zeta\, dz}{\left(\int \Gamma(z) dz\right)^2}. \end{aligned} \tag{133}$$

Here A is the aspect ratio, as defined by (1);

$$\bar{C}_L = \frac{\bar{L}}{½\rho S U^2} \tag{134}$$

And

$$\bar{C}_D = \frac{D}{½\rho S U^2} \tag{135}$$

are the three-dimensional lift and drag coefficients based on the planform area S, and the limits of the integrals in (133) are $\pm\frac{1}{2}s$.

With the physically relevant assumptions that the total circulation $\Gamma(z)$ and its derivative $\Gamma'(z)$ are continuous along the span, and that Γ vanishes at the tips $z = \pm\frac{1}{2}s$, this function can be expanded in a Fourier series of the form

$$\Gamma\left(\frac{1}{2}s\cos\theta\right) = 2Us\sum_{n=1}^{\infty} a_n \sin n\theta, \tag{136}$$

where $z = \frac{1}{2}s\cos\theta$. Using (121) and (134), the lift coefficient is given by

$$\bar{C}_L = \frac{2}{US}\int_{-s/2}^{s/2} \Gamma(z)\,dz = \pi A a_1, \tag{137}$$

where the aspect ratio A is defined by (1). The drag coefficient (135) can be computed in a similar fashion from (131) by using the Glauert integrals (89), and it follows that

$$\bar{C}_D = \frac{1}{2\pi U^2 S}\int_{-s/2}^{s/2} \Gamma(z)\,dz \,⨍_{-s/2}^{s/2} \frac{d\Gamma(\zeta)}{d\zeta}\frac{d\zeta}{z-\zeta} = \pi A \sum_{n=1}^{\infty} n a_n{}^2. \tag{138}$$

In deriving (137–138), the orthogonality relation

$$\int_0^{\pi} \sin m\theta \sin n\theta\, d\theta = \begin{cases} \pi/2, & m = n, \\ 0, & m \neq n, \end{cases}$$

has been used, with m and n positive integers.

Since the lift coefficient (137) depends only on the Fourier coefficient a_1. whereas the drag coefficient (138) is a positive-definite sum over all a_n the optimum spanwise distribution is elliptical, with $a_n = 0$ for $n \geq 2$; thus

$$\Gamma(z) \propto (s^2/4 - z^2)^{1/2}. \tag{139}$$

In this case the figure of merit K defined by (133) is equal to 1.0. For any other spanwise loading, K will be greater than unity, and the lifting surface will be less efficient.

5.11 Lifting-Line Theory

The appropriate representation of a three-dimensional lifting surface, in terms of planar vortex sheets, has been outlined in section 5.9. A quantitative solution for the vortex densities requires a surface-integral representation for the downwash velocity on the foil, in terms of these vortex densities, as in the two-dimensional case (30). For a prescribed foil geometry, the boundary condition (114) then yields an integral equation for the unknown vortex density. However, the unknown $\gamma(x, z)$ is now a function of two coordinates, and the corresponding integral equation is a surface integral. Closed-form solutions corresponding to those derived in section 5.7 cannot be derived, and further progress in the three-dimensional case requires additional approximations or a numerical solution. This topic is covered by Robinson and Laurmann (1956) and Thwaites (1960).

Here we shall discuss an approximate technique that adds considerable insight to the understanding of three-dimensional lifting surfaces and permits relatively simple calculations to be made for hydrofoils of large aspect ratio. A complementary approximation for small aspect ratios will be given in chapter 7, as a special case of slender-body theory.

If the aspect ratio is very large, the obvious approximation is a two-dimensional strip theory, with each section treated as a two-dimensional foil with the same geometry and angle of attack. This approach is limited, however, since trailing vorticity must inevitably be shed and will affect both the lift and drag forces on the foil. To account for these effects appropriately for large aspect ratios, Prandtl developed the lifting-line theory. With boundary-layer theory, this must be regarded as an outstanding example of the use of asymptotic approximations to simplify a more complicated problem. The lifting-line approach of Prandtl can be regarded as a second approximation, correcting the strip-theory approach, in the same sense that boundary-layer theory corrects the assumption of inviscid flow.

If the span of the foil is much larger than the chord length, the flow can be described as it would appear to both an observer in the *inner region,* with length scales comparable to the chord, and an observer in the *outer region,* with length scales comparable to the span. The corresponding domains are essentially as sketched in figures 5.2 and 5.21, respectively. Thus, the inner solution appears to be two dimensional, with an appropriate section geometry for the foil and a corresponding inflow velocity and angle of attack. From the standpoint of this inner flow, the three-dimensional effects are slowly varying on the scale of the span, and hence locally constant. This is not to suggest that three-dimensional effects are negligible in the inner region; but being locally constant, these can be allowed for by suitably correcting the inflow velocity vector. Thus, a two-dimensional problem is anticipated for the inner flow, identical to that treated in sections 5.2 to 5.4, but modified in the second approximation by an inflow velocity and direction unknown in advance. Once these are given, the inner solution can be obtained in terms of the two-dimensional foil characteristics, notably the total circulation $\Gamma(z)$ at each section.

Next we consider the corresponding outer solution, as sketched in figure 5.21. Here one is too far away to be concerned about details of the foil geometry, such as its local camber and angle of attack. Instead, the important features are the presence of a bound vortex $\Gamma(z)$ and corresponding trailing vortices with density $\Gamma''(z)$, which govern the outer flow. The trailing vortex sheet, which is absent from the two-dimensional inner solution, induces a significant three-dimensional effect on the inner flow, notably the downwash velocity. To account for this effect, we must calculate the downwash velocity induced by the trailing vortex sheet.

The problem is simplified considerably by restricting our attention to lifting surfaces, as shown in figure 5.21, where there is no sweepback angle. In the outer view of this problem, the foil occupies a straight line or a finite segment $(-\frac{1}{2}s, \frac{1}{2}s)$ of the z-axis. The trailing vortices are situated in the portion of the plane $y = 0$ downstream of this segment, and each individual trailing vortex element coincides with a semi-infinite line $-\infty < x < 0$, $y = 0$, $z = \zeta$. The computation of the induced velocity at points on the z-axis is facilitated by noting from symmetry that a semi-infinite vortex filament, extending from $x = -\infty$ upstream to the plane $x = 0$, will induce in this plane precisely half of the velocity that would result from an infinite line vortex extending from $-\infty$ to $+\infty$ along the same line. The latter problem is two dimensional, and thus the vertical component of the velocity induced by the trailing vortices at $x = 0$ is equal to half of the value (130) in the Trefftz plane at $x = -\infty$, or

$$v_T(0,0,z) = \frac{1}{4\pi} \fint_{-s/2}^{s/2} \frac{d\Gamma}{d\zeta} \frac{d\zeta}{\zeta - z}. \tag{140}$$

Here the subscript T is used to denote the velocity due to the trailing vortices. In general, for $|z| < \frac{1}{2}s$, the integral in (140) is negative and v_T is a *downwash.*

Before using (140) to correct the inflow velocity vector in the inner solution, other possible corrections should be considered, associated with the bound and free vortices within the foil. Since the length of the free vortices is equal to the chord length, which is small compared to the span, the free vortices will have a negligible influence on the downwash compared to the trailing vortices. The bound vortices are more difficult to treat, but from the Biot-Savart integral (113) it can be shown that the correction due to the spanwise variation of γ_B is negligible by comparison with (140).

Returning to the inner solution, the most significant effect of the three-dimensional corrections is to impose a vertical velocity component given by (140), or to change the effective angle of attack by the *induced* angle

$$\alpha_i(z) \equiv \frac{v_T}{U} = \frac{1}{4\pi U} \fint_{-s/2}^{s/2} \frac{d\Gamma}{d\zeta} \frac{d\zeta}{\zeta - z}. \tag{141}$$

The total circulation on a two-dimensional flat plate, and the flat-plate loading portion of an arbitrary cambered foil, is $\Gamma = \pi U l \alpha$. Since (141) is negative on the foil, the total circulation of the lifting line must be decreased, by comparison with its two-dimensional value at each section, by an increment $-\pi U l \alpha_i$. Substituting (141) gives an equation for the total circulation in the form

$$\Gamma(z) = \Gamma_{2D}(z) + \frac{1}{4} l(z) \fint_{-s/2}^{s/2} \frac{d\Gamma(\zeta)}{d\zeta} \frac{d\zeta}{\zeta - z}, \qquad -\frac{1}{2}s < z < \frac{1}{2}s. \tag{142}$$

Here $\Gamma_{2D}(z)$ denotes the two-dimensional circulation of the corresponding section, including the effects of camber and of the geometric angle of attack

α, and the last term in (142) represents the change in the circulation due to the induced angle of attack.

Equation (142) is Prandtl's lifting-line equation. It is an integro-differential equation, since the derivative of the unknown Γ appears in the integrand. In this respect, (142) differs from the integral equation (60) that governs the two-dimensional vortex distribution, and the complication is sufficient that (142) cannot be solved for general planforms without resorting to numerical approximations. Before discussing this matter further, we note that additional end conditions must be imposed to make the solution of (142) unique, as was the case in the two-dimensional situation where the Kutta condition was imposed. The appropriate physical condition requires the circulation to vanish at the tips of the foil, $\Gamma(\pm\frac{1}{2}s)=0$.

Once again, the Fourier sine series (136) may be used. The induced angle of attack (141) takes the form

$$\alpha_i = -\frac{1}{\pi}\sum_{n=1}^{\infty} na_n \int_0^{\pi} \frac{\cos n\theta'\, d\theta'}{\cos\theta' - \cos\theta} = -\sum_{n=1}^{\infty} na_n \frac{\sin n\theta}{\sin\theta}, \tag{143}$$

and the lifting-line equation (142) reduces to a system of linear equations for the coefficients a_n,

$$\sum_{n=1}^{\infty} a_n \sin n\theta = \frac{1}{2Us}\Gamma_{2D}\left(\frac{1}{2}s\cos\theta\right) - \frac{\pi}{2s}l\left(\frac{1}{2}s\cos\theta\right)\sum_{n=1}^{\infty} na_n \frac{\sin n\theta}{\sin\theta}, \quad 0<\theta<\pi. \tag{144}$$

For an elliptical spanwise loading, $a_n = 0$ for $n \geq 2$ and the induced angle (143) is constant along the span.

A specific geometry, where (144) is satisfied by the elliptical loading, is an uncambered lifting surface of elliptical planform,

$$l(z) = l_0 \sin\theta, \tag{145}$$

where l_0 is the chord at mid span. The two-dimensional circulation is given in this case by $\Gamma_{2D} = \pi U\alpha l$, and it follows from (144) that

$$a_1 = \frac{\pi\alpha l_0}{2s}\left(1+\frac{\pi l_0}{2s}\right)^{-1}. \tag{146}$$

For the elliptical planform, the aspect ratio is given by $A = 4s/\pi l_0$, and thus (146) may be replaced by

$$a_1 = \frac{2\alpha}{A+2}. \tag{147}$$

The lift and drag coefficients (137–138) take the simple forms

$$\bar{C}_L = \frac{2\pi A\alpha}{A+2}, \tag{148}$$

$$\bar{C}_D = \frac{4\pi A\alpha^2}{(A+2)^2} = \frac{\bar{C}_L{}^2}{\pi A}. \tag{149}$$

As the aspect ratio tends to infinity, the two-dimensional results are recovered, with zero induced drag and a lift coefficient equal to $2\pi\alpha$.

A systematic approach to lifting-line theory is given by Van Dyke (1975), based on the method of matched asymptotic expansions. This leads to Prandtl's result (142) as well as to higher-order approximations. A notable simplification of this approach is to replace the unknown derivative Γ', to the first approximation, by its two-dimensional value, giving explicit results without the complications of an integral equation.

To illustrate this simplification, let us express the spanwise circulation in the form

$$\Gamma(z) = \Gamma^{(1)}(z) + \Gamma^{(2)}(z) + \Gamma^{(3)}(z) + \cdots, \tag{150}$$

where $\Gamma^{(1)}$ is the first approximation, or the two-dimensional circulation Γ_{2D}, and successive terms represent the corresponding corrections at each stage of the approximation. By assumption $\Gamma^{(2)} \ll \Gamma^{(1)}$, etc., with the understanding that these orders of magnitude are consequences of the underlying assumption $A \gg 1$. Since the integral term in Prandtl's lifting-line equation (142) is proportional to the chord length $l(z)$, this term will be inversely proportional to the aspect ratio, or $O(1/A)$. Thus explicit solutions for the first two terms in (150) can be written as follows:

$$\Gamma^{(1)}(z) = \Gamma_{2D}(z), \tag{151}$$

$$\Gamma^{(2)}(z) = \frac{1}{4} l(z) \fint_{-s/2}^{s/2} \frac{d\Gamma^{(1)}(\zeta)}{d\zeta} \frac{d\zeta}{\zeta - z}. \tag{152}$$

To proceed beyond this point on the basis of (142) would not be consistent with the approximations already made in deriving (142).

For the elliptic planform, the lift coefficient derived from (151–152) is readily obtained, in the form

$$\bar{C}_L^{(1)} = 2\pi\alpha, \tag{153}$$

$$\bar{C}_L^{(1)} + \bar{C}_L^{(2)} = 2\pi\alpha(1 - 2/A). \tag{154}$$

At first glance the second approximation (154) differs from Prandtl's result (148), but if the denominator of (148) is expanded in a Laurent series, in inverse powers of the aspect ratio, it follows that

$$\bar{C}_L = 2\pi\alpha(1 - 2/A) + O(1/A^2). \tag{155}$$

Thus we see that these results are consistent, up to the point where a comparison is justified.

These results are shown in figure 5.22, which includes the conventional lifting-line lift coefficient (148) and the second-order systematic approximation (154), as well as a third-order approximation derived by Van Dyke (1975) and exact lifting-surface calculations. The third-order approximation is not simply a more accurate Laurent series approximation of (148) but a consistent theory including all terms[1] of order $1/A^2$. One example of

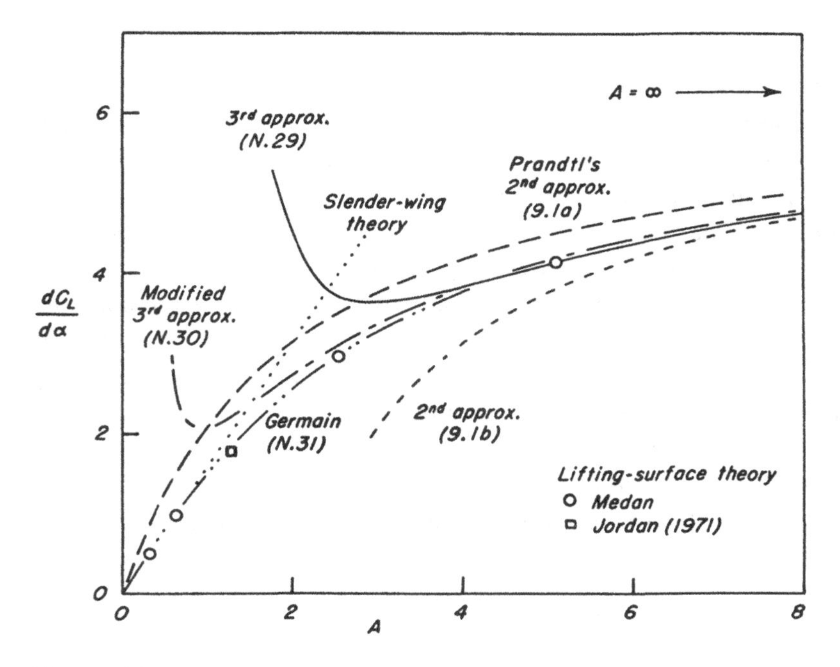

Figure 5.22
Lift coefficient of elliptic wing. (From Van Dyke 1975, fig. 9.4)

the corrections appearing at this order is an effective change not only in the angle of attack but also in the curvature of the streamlines, which represents a change in the camber.

If one attempts to judge the merits of these different approaches from figure 5.22, the conclusions are somewhat mixed. Prandtl's "inconsistent" lifting-line theory is remarkably useful for all aspect ratios from zero to infinity. It overpredicts the lift coefficient, with a relative error of about 5 percent at $A = 8$, increasing to 10 percent at $A = 4$, 20 percent at $A = 2$, and 100 percent at $A = 0$. On the other hand, the consistent results are more accurate for large aspect ratio but become completely invalid as the aspect ratio is reduced. Thus, while Prandtl's approximation is not systematic, the most important higher-order effects seem to have been included through good luck or shrewd insight or a combination of the two. This favorable outcome is not observed often in asymptotic approximations, and contemporary researchers in marine hydrodynamics who lack Prandtl's insight should not expect to be so successful!

5.12 Cavity Flows

Cavitation may have a variety of causes. The most common example is boiling water, where the vapor pressure is increased by raising the water temperature. Generally in marine hydrodynamics applications, cavitation results because the fluid pressure is reduced to the vapor pressure limit. This will occur if the body velocity U is sufficiently large, since the pressure is a decreasing function of fluid velocity. Alternatively, the ambient pressure may be reduced by decreasing the body submergence and corresponding hydrostatic pressure or by evacuating a variable-pressure water tunnel or towing tank. Finally, cavitation may result if the body geometry is modified to reduce the pressure, for instance, if the angle of attack of a hydrofoil is increased.

The relevant nondimensional parameter describing the occurrence of cavitation and the details of the resulting flow is the cavitation number

$$\sigma = \frac{p_0 - p_v}{\frac{1}{2}\rho U^2}. \tag{156}$$

Here p_0 is the ambient pressure of the flow, usually at a large distance from the body, and p_v is the vapor pressure of the fluid.

Large cavitation numbers imply fully wetted flow, without cavitation. As the cavitation number is reduced or the angle of attack increased, local cavitation will occur first near the minimum-pressure position on the foil, initially in the form of small isolated bubbles of vapor. For smaller values of σ, the cavitation will become more widespread, and ultimately the flow will become *supercavitating*, with the entire suction side of the foil contained within the cavity, as shown in figure 5.23.

The lifting-surface theory outlined in the previous sections of this chapter can be used to predict the initial occurrence of cavitation in terms of the minimum pressure on the foil. Subsequently, the cavitation bubbles may become sufficiently large to affect the flow, generally in an unsteady or unstable manner; in the ultimate supercavitating regime, however, the cavity is relatively large and stable, and its effects can be accounted for by a suitable generalization of lifting-surface theory. This procedure will be illustrated in sections 5.13 and 5.14 for steady two-dimensional flow past a symmetric wedge-shaped strut, as well as for the corresponding lifting problem.

Figure 5.23
Supercavitating flow past a two-dimensional foil. Here the lower surface of the foil is a circular arc, at 6° angle of attack, and the cavitation number $\sigma = 0.19$. (From Barker and Ward 1976)

As in the fully wetted case, we assume that the foil and the cavity are thin. The boundary conditions are then linearized, as in section 5.2. On the wetted portion of the body, the kinematic boundary conditions (11–12) are unchanged. On the cavity, these are replaced by an appropriate dynamic boundary condition prescribing the fluid pressure to be equal to the vapor pressure. Using the linearized Bernoulli equation (14) and the cavitation number (156), the appropriate dynamic boundary condition on the cavity surface is given by

$$\rho u U = p_v - p_0 = -\frac{1}{2}\rho U^2 \sigma, \tag{157}$$

or

$$u = -\frac{1}{2} U \sigma. \tag{158}$$

Strictly speaking, the hydrostatic pressure contribution to Bernoulli's equation cannot be neglected in (157) and will give rise to a buoyancy force acting on the cavity, but for sufficiently high speeds this effect is negligible and will be ignored here. The survey articles by Wu (1968, 1972) and Acosta (1973) discuss this and many other aspects of cavitation that are beyond the scope of our present treatment.

5.13 Symmetric Cavity Flows

As our first example we consider the flow past a symmetric strut with a blunt base, or trailing edge, as shown in figure 5.24. Here the coordinates have been nondimensionalized in terms of the chord length of the strut, with the leading edge at $x = 1$, and the nondimensional strut thickness is $2y_0(x)$. Downstream of the trailing edge, at $x = 0$, a cavity of length l exists with linearized pressure prescribed by (158). The resulting boundary conditions are shown in figure 5.25. The problem statement is completed by requiring that Laplace's equation must be satisfied in the fluid, the perturbation velocity and pressure should vanish at infinity, and the velocity at the trailing edge of the foil is finite. The latter condition, like the Kutta condition for the fully wetted lifting foil, is necessary to ensure a smooth transition of the flow from the body surface to the cavity, with continuous slope between these two boundaries.

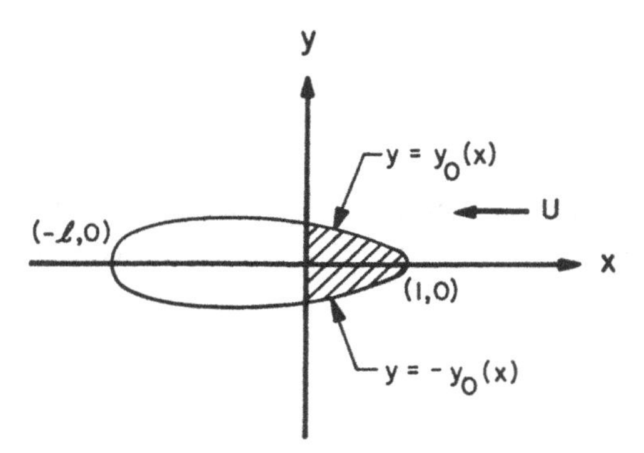

Figure 5.24
Cavity flow past a symmetrical strut.

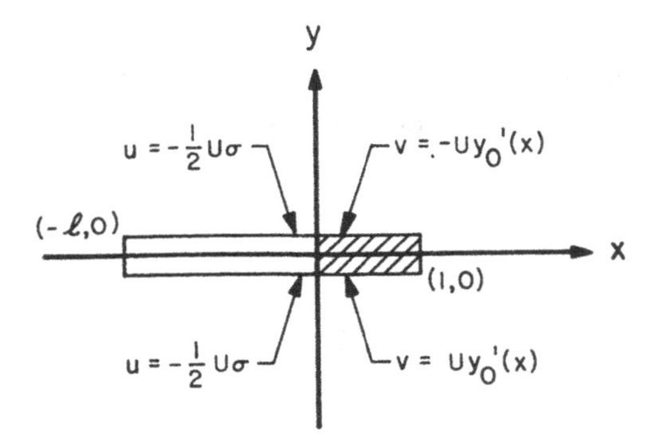

Figure 5.25
Linearized boundary conditions for cavity flow past a symmetric strut.

Since this problem is symmetrical about $y = 0$, it is logical to seek the solution as a source distribution $q(x)$, which represents the thickness effects of both the strut and the cavity. From (54) the resulting velocity components on the boundary surfaces ($-l < x < 1$) are

$$u_{\pm}(x) = -\frac{1}{2\pi}\int_{-1}^{1}\frac{q(\xi)d\xi}{\xi - x}, \tag{159}$$

$$v_{\pm}(x) = \pm\frac{1}{2}q(x). \tag{160}$$

From the boundary condition on the strut, it follows that

$$v_{\pm} = \mp U y_0{}'(x) = \pm \frac{1}{2} q(x), \qquad 0 < x < 1. \tag{161}$$

This determines the source strength on the strut, in a manner identical to the fully wetted thickness problem. Imposing the cavity boundary condition gives the integral equation

$$-\frac{1}{2} U\sigma = -\frac{1}{2\pi} \int_{-1}^{1} \frac{q(\xi)}{\xi - x} d\xi, \qquad -l < x < 0, \tag{162}$$

or using (161),

$$-\frac{1}{2} U\sigma - \frac{1}{\pi} U \int_0^1 \frac{y_0{}'(\xi)}{\xi - x} d\xi = -\frac{1}{2\pi} \int_{-l}^{0} \frac{q(\xi)}{\xi - x} d\xi,. \quad -l < x < 0 \tag{163}$$

This is essentially the same integral equation as (60), which relates the vortex distribution γ of the fully wetted foil to its normal velocity v. The only difference is that the interval (–1, 1) must be replaced by ($-l$, 0). The appropriate solution satisfying the Kutta condition at $x = 0$ is obtained from (73), in the form

$$q(x) = \frac{2}{\pi} \left(\frac{-x}{l + x} \right)^{1/2} \int_{-l}^{0} \left(\frac{l + \xi}{-\xi} \right)^{1/2} \times \left\{ -\frac{1}{2} U\sigma - \frac{1}{\pi} U \int_0^1 \frac{y_0{}'(t)}{t - \xi} dt \right\} \frac{d\xi}{\xi - x}, \qquad -l < x < 0. \tag{164}$$

Here the dummy variable t has been substituted for x in the inner integral.

The integrals with respect to ξ in (164) can be evaluated using the Hilbert transforms derived in section 5.7. The treatment of the first term in braces is straightforward, using (81) and (83); in the second term, however, the order of integration must be inverted, the denominator expanded in partial fractions, and (88) used with (82). The final result is

$$q(x) = -U\sigma \left(\frac{-x}{l + x} \right)^{1/2} - \left[\frac{2}{\pi} U \left(\frac{-x}{l + x} \right)^{1/2} \times \int_0^1 \left(\frac{l + t}{t} \right)^{1/2} \frac{y_0'(t)}{t - x} dt \right], \qquad -l < x < 0. \tag{165}$$

Substitution of the source strength (165) in (159) and (160) gives the velocity and pressure distribution on the strut and the cavity surface. The cavity shape may be found by imposing the kinematic boundary condition (161) on the cavity surface and integration of (165). In particular, requiring the cavity to be closed implies that the integral of the source strength, over the body plus the cavity, must be equal to zero. Integrating (161) and (165) gives the relation

$$0 = \int_{-l}^{1} q(x)dx = -\frac{\pi}{2}U\sigma l - 2U\int_{0}^{1}\left[\left(\frac{l+t}{t}\right)^{1/2} - 1\right]y_0'(t)\,dt - 2U\int_{0}^{1} y_0'(t)dt. \tag{166}$$

Here the last integral comes from the contribution of the sources in the segment (0, 1), and (82), (88) have been used. Reduction of (166) gives an equation between the cavitation number σ and cavity length l,

$$0 = \frac{\pi}{4}\sigma l + \int_{0}^{1}\left(\frac{l+t}{t}\right)^{1/2} y_0'(t)\,dt. \tag{167}$$

Since the pressure acting on the wetted surface of the strut is greater than the vapor pressure p_v, acting on the base, a positive drag force will result, equal to

$$\begin{aligned} D &= -2\int_{0}^{1}(p - p_v)y_0'(x)\,dx \\ &= -2\rho\int_{0}^{1}\left(uU + \frac{1}{2}U^2\sigma\right)y_0'(x)\,dx. \end{aligned} \tag{168}$$

Using equation (159) for the velocity u gives

$$D = \frac{\rho U}{\pi}\int_{0}^{1}\left\{ \fint_{-l}^{1} \frac{q(\xi)}{\xi - x}d\xi - \pi U\sigma \right\} y_0'(x)dx. \tag{169}$$

By substituting equations (161) and (165) for the source strength $q(x)$ on the body and cavity respectively, and (167) for the cavitation number, and by using the Hilbert transforms (80–88), we may express (169) in the form

$$D = \frac{\rho U^2}{\pi}\int_{0}^{1} y_0'(x)\left(\frac{x}{l+x}\right)^{1/2} \times \fint_{0}^{1} y_0'(t)\left(\frac{l+t}{t}\right)^{1/2}\left(\frac{4}{l} - \frac{2}{t-x}\right)dt\,dx. \tag{170}$$

Equation (170) can be recast in a simpler form as follows. First we invert the order of integration. (The more rigorous reader may find justification for this operation in Muskhelishvili 1953, p. 59.) If the dummy variables (x, t) are inverted at the same time, an alternative expression is obtained:

$$D=\frac{\rho U^2}{\pi}\int_0^1 y_0'(x)\left(\frac{l+x}{x}\right)^{1/2}\times\fint_0^1 y_0'(t)\left(\frac{t}{l+t}\right)^{1/2}\left(\frac{4}{l}+\frac{2}{l-x}\right)dt\,dx. \tag{171}$$

Since (170) and (171) are equivalent, the drag can be written as one half of their sum,

$$\begin{aligned}D=\frac{\rho U^2}{\pi}\int_0^1 y_0'(x)\fint_0^1 y_0'(t)\Bigg\{&\frac{2}{l}\left[\left(\frac{x(l+t)}{t(l+x)}\right)^{1/2}+\left(\frac{t(l+x)}{x(l+t)}\right)^{1/2}\right]\\&+\frac{1}{t-x}\left[\left(\frac{t(l+x)}{x(l+t)}\right)^{1/2}-\left(\frac{x(l+t)}{t(l+x)}\right)^{1/2}\right]\Bigg\}dt\,dx.\end{aligned} \tag{172}$$

After algebraic reduction of the integrand it follows that

$$D=\frac{\rho U^2}{\pi l}\int_0^1 y_0'(x)\int_0^1 y'_0(t)\frac{(l+2x)(l+2t)}{[xt(l+x)(l+t)]^{1/2}}dt\,dx, \tag{173}$$

and thus

$$D=\frac{\rho U^2}{\pi l}\left(\int_0^1 y_0'(x)\frac{(l+2x)}{[x(l+x)]^{1/2}}dx\right)^2. \tag{174}$$

The integrals in equations (167) and (174) can be evaluated without difficulty for simple strut shapes. Figures 5.26 and 5.27 show the values of the cavity length and drag force for a simple wedge, as well as two parabolic struts. In all cases the chord length is unity. These figures show that for decreasing values of the cavitation number σ, the cavity length increases while the drag force is reduced. Ultimately, $l \to \infty$ as $\sigma \to 0$, whereas the drag force tends to a finite limiting value dependent upon the body shape. A comparison of the results for these three strut geometries demonstrates the importance of the trailing-edge slope $y_0'(0)$ on the subsequent development of the cavity. In particular, a transition from the convex parabola, where $y_0'(0)=0$, to the wedge and ultimately the cusped case, results in increased cavity length and drag.

The linear theory becomes invalid if the cavity length is substantially shorter than the chord length, or if the cavitation number is too large. This restriction is apparent from the linearized boundary condition (158) on the cavity surface, which implies that $u/U \ll 1$ only if $\sigma \ll 1$. Further discussion of this matter and a comparison of the nonlinear theory with experiments may be found in the paper by Wu, Whitney, and Brennen (1971).

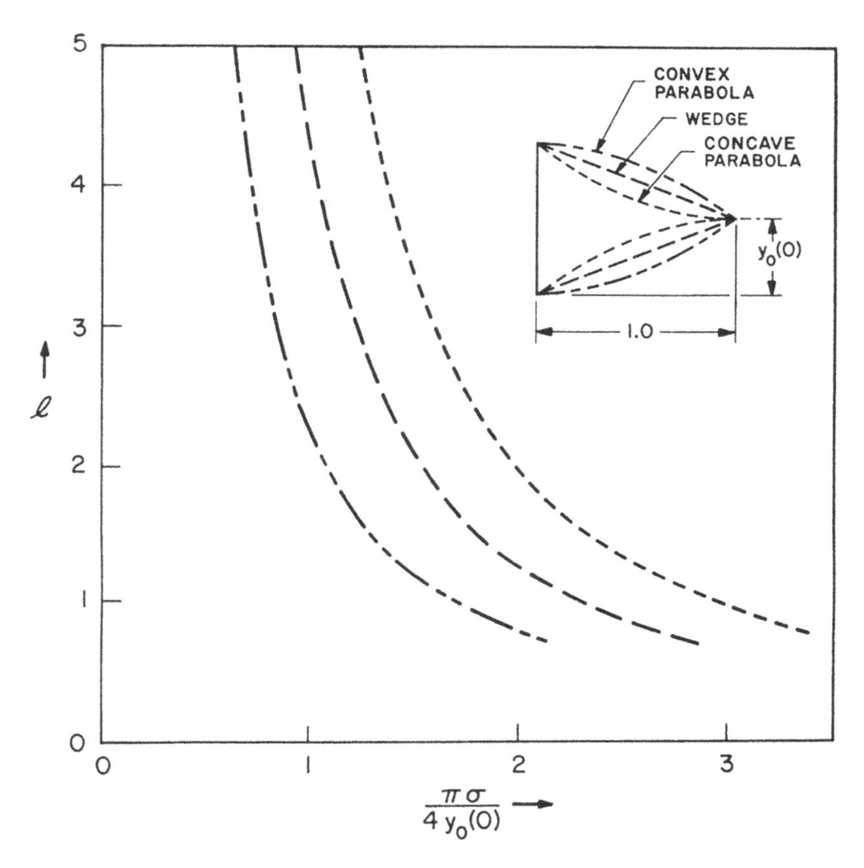

Figure 5.26
Cavity length l for the three symmetric struts shown. Note that $y_0(0)$ and l are nondimensionalized with respect to the unit chord length.

5.14 Supercavitating Lifting Foils

To simplify the problem of a lifting foil in the supercavitating flow regime as much as possible, we will assume that the cavitation number is zero and that the foil geometry is as sketched in figure 5.28. In particular, the leading edge is sharp, and the entire upper side of the foil is situated within the cavity.

Nondimensional coordinates (x, y) are chosen such that the chord length is equal to unity, and the slope of the lower surface of the foil is denoted by the local angle of attack $\alpha(x)$. For reasons of subsequent

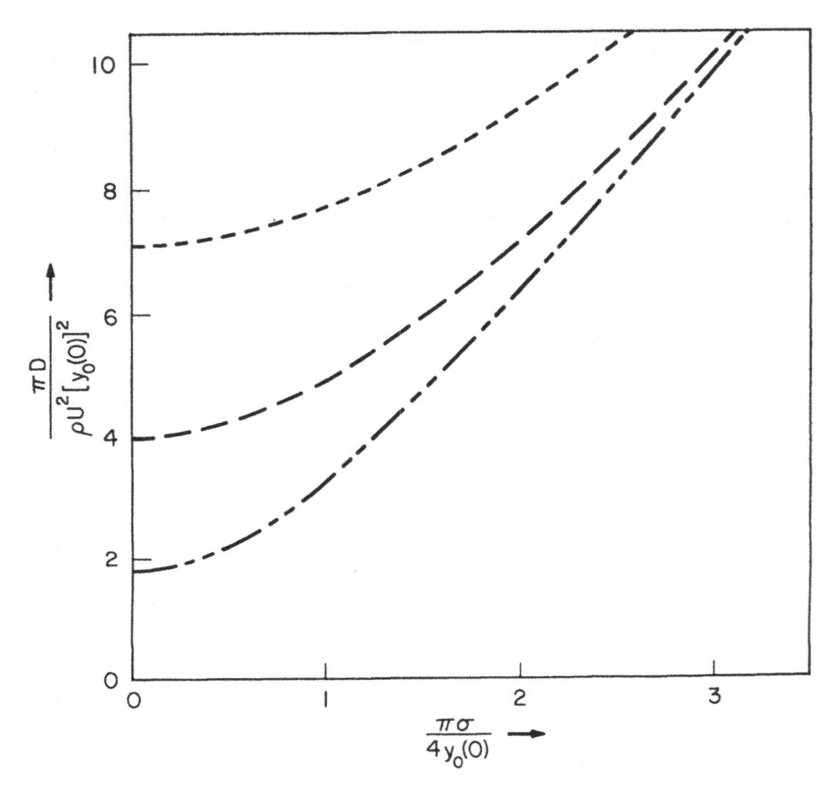

Figure 5.27
Drag coefficient for the three symmetric struts shown in figure 5.26, nondimensionalized in terms of the unit chord length.

Figure 5.28
Supercavitating flow past a lifting foil with sharp leading edge.

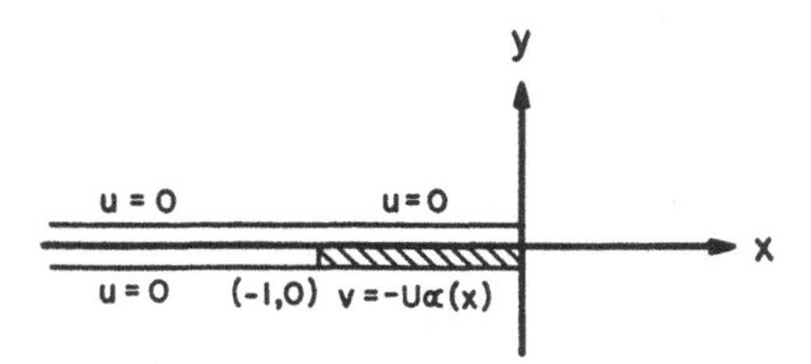

Figure 5.29
Linearized boundary conditions corresponding to the problem shown in figure 5.28.

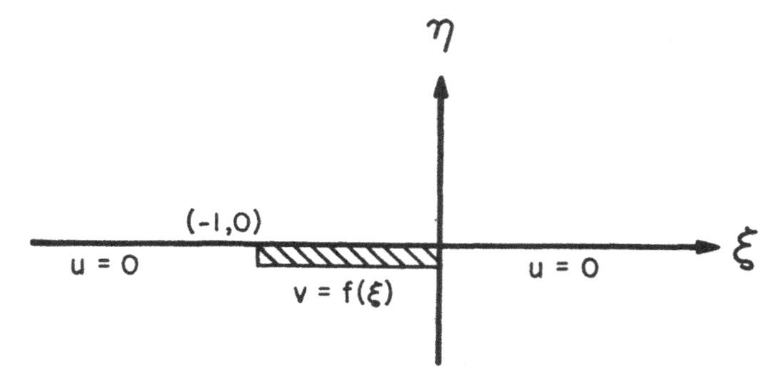

Figure 5.30
Boundary conditions in mapped ζ-plane.

algebraic convenience, the leading edge is situated at the origin; the linearized boundary conditions on the foil and cavity then take the form shown in figure 5.29.

This problem cannot be decomposed readily into symmetric and antisymmetric portions, as in the fully wetted problem, because of the fundamentally different boundary conditions imposed on the upper and lower sides of the segment occupied by the foil in figure 5.29. However, the conformal mapping function $\zeta = -iz^{1/2}$ can be used to map the domain of the fluid into the lower half of the ζ-plane, folding the z-plane about the leading edge of the foil. The complex velocity $u - iv$ is an analytic function of z, and hence also of ζ, which may be mapped from one plane to the other with its values preserved.

The resulting problem in the ζ-plane is as shown in figure 5.30, with $\zeta = \xi + i\eta$. In this mapped plane, the vertical velocity on the lower surface of the foil has been written as

$$v = -U\alpha(x) = -U\alpha(-\xi^2) \equiv f(\xi). \tag{175}$$

Since the fluid domain has been mapped entirely into the lower half of the ζ-plane, the region $\eta > 0$ has no physical significance. Mathematically, it corresponds to the second Riemann sheet of the z-plane, or it could be regarded as the domain of the cavity. In any event, if the boundary conditions are satisfied as shown in figure 5.30, on the lower side of the real axis, the flow in the upper half-plane is immaterial. In accordance with the Schwarz reflection principle, the velocity components u and v can therefore be *reflected* about the ξ-axis, and the resulting flow will be continuous across the real axis, excluding the cut (−1, 0), provided u is an odd function of η. To ensure that $u - iv$ is analytic, it follows from the Cauchy-Riemann equations that v must be an even function of η. Thus, the appropriate reflections into the upper half-plane are given by

$$u(\xi,\eta) = -u(\xi,-\eta), \tag{176}$$

$$v(\xi,\eta) = v(\xi,-\eta). \tag{177}$$

The reflected boundary-value problem shown in figure 5.31 is mathematically identical to the lifting problem of a fully wetted thin foil shown in figure 5.14. In both cases a Kutta condition must be imposed at the trailing edge, and the perturbation velocity should vanish at infinity. *Thus the solution of the cavity-flow problem in the mapped plane is identical to that of the fully wetted lifting problem in the physical plane.*

The solution is given by a line distribution of vortices along the cut (−1, 0), as in equation (55); the appropriate distribution satisfying the Kutta condition at $x = -1$ can be inferred from (71),

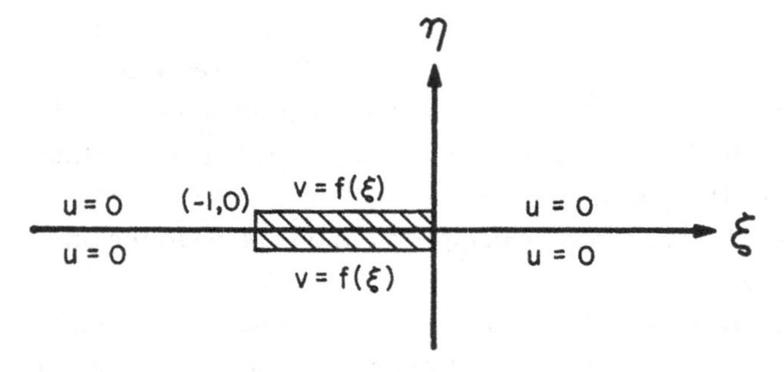

Figure 5.31
Boundary conditions of the reflected complex velocity in the ζ-plane.

$$\frac{1}{2}\gamma(\xi) = \frac{2}{\pi}\left(\frac{1+\xi}{-\xi}\right)^{1/2} \oint_{-1}^{0} \left(\frac{-t}{1+t}\right)^{1/2} \frac{f(t)}{t-\xi} dt, \qquad -1 < \xi < 0. \tag{178}$$

On the lower surface of the foil, the horizontal velocity component $u_{-}(\xi)$ is equal to $\frac{1}{2}\gamma(\xi)$, as in (59), and the lift force can be derived from the linearized Bernoulli equation,

$$L = \int_{-1}^{0} (p - p_v) dx = \rho U \int_{-1}^{0} u_{-}(x) dx. \tag{179}$$

To evaluate the last integral, the variable of integration may be changed from the physical coordinate x to the mapped coordinate ξ. Thus,

$$L = \frac{1}{2}\rho U \int_{-1}^{0} \gamma(\xi) \frac{dx}{d\xi} d\xi = -\rho U \int_{-1}^{0} \gamma(\xi)\xi \, d\xi. \tag{180}$$

This can be recognized as the negative of the moment, about the leading edge, of a fully wetted foil of unit chord length. Thus, with $\alpha(x)$ the slope of the supercavitating foil and $\eta(x)$ the mean-camber line of an equivalent fully wetted foil,

$$\alpha(-x^2)_{\text{cavitating foil}} = (d\eta/dx)_{\text{equivalent fully wetted foil}}. \tag{181}$$

The supercavitating lift coefficient is given by

$$(C_L)_{\text{cavitating}} = -(C_M)_{\text{equivalent}}. \tag{182}$$

Similarly, the moment coefficient for the cavitating foil is equal to the third moment of the equivalent fully wetted foil.

The drag coefficient of the cavitating foil can be related to the lift coefficient of the fully wetted foil, following an analysis similar to that used in deriving (174). The drag force is obtained from (179) after multiplying the integrand by the slope $\alpha(x)$:

$$D = \rho U \int_{-1}^{0} u_{-}(x)\alpha(x) dx = -\rho \int_{-1}^{0} u_{-}(x) v(x) dx. \tag{183}$$

Here the boundary condition (175) has been used to replace the slope by the vertical velocity component. Changing the variable of integration, as in (180), gives

$$D = \rho \int_{-1}^{0} \gamma(\xi) v(\xi) \xi \, d\xi. \tag{184}$$

Using (60) for the vertical velocity gives

$$D = -\frac{\rho}{2\pi} \int_{-1}^{0} \gamma(\xi) \xi \fint_{-1}^{0} \frac{\gamma(t)}{t-\xi} dt \, d\xi. \tag{185}$$

If the order of integration and dummy variables are inverted, as in (171), it follows that

$$\begin{aligned} D &= -\frac{\rho}{4\pi} \int_{-1}^{0} \gamma(\xi) \fint_{-1}^{0} \gamma(t) \left(\frac{-\xi}{t-\xi} + \frac{t}{t-\xi} \right) dt \, d\xi \\ &= \frac{\rho}{4\pi} \int_{-1}^{0} \gamma(\xi) \int_{-1}^{0} \gamma(t) dt \, d\xi \\ &= \frac{\rho}{4\pi} \left(\int_{-1}^{0} \gamma(\xi) d\xi \right)^2 . \end{aligned} \tag{186}$$

The last integral is the total circulation Γ for the equivalent foil, and thus the drag coefficient for the supercavitating foil is given by

$$(C_D)_{\text{cavitating}} = \frac{1}{8\pi} (C_L{}^2)_{\text{equivalent}}. \tag{187}$$

This useful equivalence was first established by Tulin and Burkart in 1955 and has been expounded in several papers (Tulin 1964). For a flat plate of constant angle of attack α, the results from section 5.4 give for the corresponding supercavitating case

$$C_L = \frac{\pi}{2} \alpha, \tag{188}$$

$$C_M = \frac{5\pi\alpha}{32}, \tag{189}$$

$$C_D = \frac{\pi}{2} \alpha^2. \tag{190}$$

Here the center of pressure is shifted from the quarter-chord point to a position 5/16 of the chord downstream from the leading edge. Of greater interest is the lift-drag ratio, which reveals that the total force vector on the foil is precisely normal to the flat plate, inclined at an angle a from the vertical. This situation differs from the fully wetted foil, where the force vector is strictly vertical, due to the leading-edge suction force. In the supercavitating case there is no leading-edge suction force, since the square-root singularity of the equivalent fully wetted flow in the ζ-plane is reduced to a quarter-root singularity in the physical z-plane. Physically, the cavity that originates at the leading edge is of more gradual curvature than the flow around the fully wetted flat plate.

More generally, the lift-drag ratio can be derived from (182) and (187) as

$$(L / D)_{\text{cavitating}} = 8\pi(C_M / C_L{}^2)_{\text{equivalent}}. \tag{191}$$

To maximize this quantity, the ratio C_M/C_L for the equivalent foil must be as large as possible, or the center of pressure must be as far downstream as is practical. Since the pressure on the lower surface must be positive to avoid face cavitation, the optimum lift-drag ratio results when the equivalent center of pressure is at the trailing edge. This situation cannot be attained in practice, but it does imply correctly that efficient supercavitating sections will have a pronounced maximum slope at the trailing edge.

While the equivalence of supercavitating and fully wetted foils is important in predicting the performance and optimizing the design of supercavitating foils, the performance of these two types of foils is *not* identical. As a simple example, the flat-plate lift coefficient is reduced by a factor of four relative to the fully wetted regime, and a nonzero drag force is incurred as well. Thus the supercavitating flow regime is to be avoided if possible, especially given the problems of erosion and noise, which are inevitable consequences of cavitating flows. However, if operation in the supercavitating regime is inevitable, it is preferable to design the foil section to be optimum in this regime, and thus to have a very different form from the fully wetted foil shapes. The latter difference is particularly important at the leading edge where, to avoid both cavitation and separation, a fully wetted foil is designed with a substantial radius of curvature, whereas a supercavitating foil is designed with a sharp leading edge.

5.15 Unsteady Hydrofoil Theory

In general, unsteady motions will be accompanied by corresponding changes in the circulation about the foil or about each section of a three-dimensional lifting surface. We shall confine our attention here to the two-dimensional case, since the situation in three dimensions is qualitatively similar except for the addition of suitable trailing vortices. Applying Kelvin's theorem to a large contour surrounding the foil and its wake during the entire history of its motion, as in figure 5.6, we find that the circulation about this contour must be zero. Thus any change in the circulation $\Gamma(t)$ about the foil must be balanced by an equal and opposite change in the vorticity shed into the wake. The starting vortex shown in figure 5.6 is a particular example where the unsteady effects are restricted to the initial acceleration of the foil, but in the more general case vorticity will be continuously supplied into the wake.

The existence of a vortex distribution throughout the wake is reminiscent of the three-dimensional steady lifting problem of section 5.9. As in that case, the vortices in the wake are responsible both for a drag force acting on the foil and for a downwash which affects the lift. However, unlike the steady lifting-surface problem with its trailing vortices, the present case is purely two-dimensional, and the vortices in the wake must be oriented in the same manner as the bound vorticity on the foil, as shown in figure 5.32.

The downwash induced at the foil by the wake is time-dependent; each individual vortex element induces a contribution to the downwash which diminishes with time as the vortex moves downstream. In this respect, a *memory* is associated with the motion of the foil: the lift force at a given moment will depend on the integrated downwash from the entire wake and thus on the previous time-history of the motion. This memory effect is one of the principal features of the unsteady hydrofoil problem. An analogous situation will be encountered in chapter 6, where the unsteady motions of

Figure 5.32
Unsteady vortex wake behind an oscillating foil. The case shown is for heaving motion of the foil, with a reduced frequency $k = 1.0$. The orientation and length of each arrow denotes the sign and strength of the vorticity at the corresponding position in the wake.

a floating body will generate waves which, in turn, affect the body over a subsequent period of time.

Our discussion of unsteady hydrofoil theory will be limited to the simplest two-dimensional case—linearized flow with constant forward velocity. The linear restriction implies not only that the foil is thin and at a small angle of attack, but that the unsteady motions are small. As in the steady case, the foil can be collapsed onto a cut in the x-axis, and similarly the vortex sheet in the wake is confined to the x-axis. Elsewhere the fluid motion is irrotational, with the perturbation velocity field defined by a velocity potential that vanishes at infinity. Other conditions to be imposed include a kinematic boundary condition on the foil, a Kutta condition at the trailing edge, and a dynamic boundary condition on the vortex wake. Finally, on the assumption that the motion has arisen from an initial state of rest, the total circulation about the foil plus its wake must vanish.

As in sections 5.6 and 5.7, it will be convenient to assume that the coordinates are nondimensionalized on the basis of the half-chord $\frac{1}{2}l$, with the leading and trailing edges of the foil at $x = \pm 1$, respectively, and the wake confined to the line $x < -1$.

Thickness and lifting effects can be treated separately because of linearization. Unless the thickness is variable in time the unsteady effects will be confined to the lifting problem. Thus we consider an unsteady mean-camber line, or a foil of zero thickness with camber and angle of attack that vary with time in some prescribed manner.

If the vertical coordinate of the mean-camber line is denoted by $y = \eta(x, t)$, the kinematic boundary condition on the foil takes the form

$$\begin{aligned} 0 = \frac{D}{Dt}(y-\eta) &= \left(\frac{\partial}{\partial t} - U\frac{\partial}{\partial x} + \nabla\phi\cdot\nabla\right)(y-\eta) \\ &= \frac{\partial\phi}{\partial y} - \frac{\partial\eta}{\partial t} + U\frac{\partial\eta}{\partial x}, \end{aligned} \tag{192}$$

where second-order terms in ϕ and η are neglected. Applying this condition on the cut gives an equation for the vertical velocity component on $y = 0$,

$$v(x,0,t) = \frac{\partial\eta}{\partial t} - U\frac{\partial\eta}{\partial x} \equiv v_0(x,t), \qquad -1 < x < 1. \tag{193}$$

This is the appropriate generalization of the boundary condition (22) for the steady-state lifting problem.

While various techniques can be used to solve this problem, it is natural to seek a generalization of the corresponding steady-state case and thus to represent the foil and its wake by a distribution of vortices. With an obvious extension of (59–60), the velocity components on the upper or lower sides of this vortex sheet will be

$$u_{\pm}(x,t) = \mp\frac{1}{2}\gamma(x,t), \tag{194}$$

and

$$v_{\pm}(x,t) = -\frac{1}{2\pi}\int_{-\infty}^{1}\frac{\gamma(\xi,t)}{\xi - x}d\xi. \tag{195}$$

The condition that the total circulation about the foil plus its wake should equal zero requires that

$$\int_{-\infty}^{1}\gamma(x,t)\,dx = \int_{-\infty}^{-1}\gamma(x,t)dx + \Gamma(t) = 0. \tag{196}$$

The linearized pressure is obtained as in (14), but includes the unsteady contribution from (4.26). Thus

$$p - p_{\infty} = -\rho\left(\frac{\partial\phi}{\partial t} - U\frac{\partial\phi}{\partial x}\right). \tag{197}$$

The dynamic boundary condition on the wake requires that the jump in (197) across the wake should be zero. Differentiating (197) with respect to x and substituting (194), we get

$$\frac{\partial\gamma}{\partial t} - U\frac{\partial\gamma}{\partial x} = 0, \qquad -\infty < x < -1. \tag{198}$$

This is a simple partial differential equation for γ whose general solution is

$$\gamma(x,t) = \gamma(x + Ut), \qquad -\infty < x < -1. \tag{199}$$

Thus, as expected from Kelvin's theorem, the vorticity is convected downstream with velocity U and remains constant in a reference frame moving with the fluid.

If (196) is differentiated with respect to time and (198) utilized to integrate over the wake, it follows that

$$\frac{d\Gamma}{dt} = -U\gamma(-1,t). \tag{200}$$

This equation relates the vorticity shed at the trailing edge to the rate of change of Γ, in precisely the sense that was anticipated initially.

With the vortices in the wake described by (199), we can now construct the complete solution, including an arbitrary vortex distribution on the foil. Separating the integrals over the foil and wake and imposing the kinematic boundary condition (193) on the vertical velocity (195) gives

$$v_0(x,t) + \frac{1}{2\pi}\int_{-\infty}^{-1} \frac{\gamma(\xi+Ut)}{\xi - x} d\xi = -\frac{1}{2\pi} \fint_{-1}^{1} \frac{\gamma(\xi,t)}{\xi - x} d\xi, \qquad -1 < x < 1. \tag{201}$$

This is a singular integral equation for the vortex strength; it is identical to (60) except that the integral[2] on the left side of (201) is unknown. Ignoring this temporarily, the general solution is given by (73–76) in the form

$$\begin{aligned} \gamma(x,t) = \frac{2}{\pi(1-x^2)^{1/2}} \Bigg\{ \fint_{-1}^{1} \frac{(1-\xi^2)^{1/2}}{\xi - x} \Bigg[v_0(\xi,t) \\ + \frac{1}{2\pi} \int_{-\infty}^{-1} \frac{\gamma(\xi'+Ut)}{\xi' - \xi} d\xi' \Bigg] d\xi + \frac{1}{2}\Gamma \Bigg\}. \end{aligned} \tag{202}$$

The double integral in (202) can be reduced in a manner similar to (164), by interchanging the order of integration, using partial fractions, and (81–88). Thus it follows that

$$\begin{aligned} \gamma(x,t) = \frac{2}{\pi(1-x^2)^{1/2}} \Bigg\{ \fint_{-1}^{1} \frac{(1-\xi^2)^{1/2}}{\xi - x} v_0(\xi,t)\, d\xi \\ + \frac{1}{2} \int_{-\infty}^{-1} \frac{(\xi^2 - 1)^{1/2}}{\xi - x} \gamma(\xi + Ut)\, d\xi \Bigg\}. \end{aligned} \tag{203}$$

Here (196) has been used.

If the Kutta condition is invoked as in (77), the terms in braces in (203) must vanish at $x = -1$; this results in an integral equation for the wake vorticity,

$$\int_{-\infty}^{-1} \left(\frac{\xi - 1}{\xi + 1}\right)^{1/2} \gamma(\xi + Ut) d\xi = 2 \int_{-1}^{1} v_0(\xi,t) \left(\frac{1-\xi}{1+\xi}\right)^{1/2} d\xi. \tag{204}$$

If the wake consists of a single starting vortex far downstream with circulation $-\Gamma$, (204) reduces to the steady-state result (34). In general, Laplace transforms can be used to solve the integral equation (204) for the vortex density in the wake, in a manner outlined by Woods (1961).

Before attempting to deal with (204), we shall compute the lift force and moment acting on the foil. Using Bernoulli's equation (197), the lift force is given by

$$L = \rho \int_{-1}^{1} \left(\frac{\partial \phi}{\partial t} - U \frac{\partial \phi}{\partial x} \right)_{-}^{+} dx. \tag{205}$$

From the symmetry of the lifting problem, the velocity potential is odd in y and thus must vanish on the real axis upstream of the leading edge. There is a square-root infinity in the velocity components at $x = 1$, implying a square-root zero of the velocity potential at the leading edge. Thus it follows that

$$\int_{-1}^{1} \frac{\partial \phi}{\partial t} dx = \int_{-1}^{1} \frac{\partial \phi}{\partial t} d(1+x) = -\int_{-1}^{1} (1+x) \frac{\partial u}{\partial t} dx. \tag{206}$$

If this result with (194) is substituted in (205), the lift is given by

$$L = \rho \int_{-1}^{1} \left[(1+x) \frac{\partial \gamma}{\partial t} + U\gamma \right] dx. \tag{207}$$

Similarly, the moment about $x = 0$ is obtained in the form

$$\begin{aligned} M &= \rho \int_{-1}^{1} \left(\frac{\partial \phi}{\partial t} - U \frac{\partial \phi}{\partial x} \right)_{-}^{+} x \, dx \\ &= \rho \int_{-1}^{1} \left[-\frac{1}{2}(1-x^2) \frac{\partial \gamma}{\partial t} + U\gamma x \right] dx. \end{aligned} \tag{208}$$

Substituting the vortex distribution (203), interchanging orders of integration, and using (81–88) in the usual manner gives

$$\begin{aligned} L = &-2\rho \int_{-1}^{1} (1-\xi^2)^{1/2} \frac{\partial v_0(\xi,t)}{\partial t} d\xi + \rho U \Gamma \\ &- \rho \int_{-\infty}^{-1} [1+\xi+(\xi^2-1)^{1/2}] \frac{\partial \gamma(\xi+Ut)}{\partial t} d\xi. \end{aligned} \tag{209}$$

The wake integral in (209) can be integrated by parts after substituting (198) for $\partial\gamma/\partial t$. Using (196), the expression for the lift force reduces to the form

$$L = -2\rho \int_{-1}^{1} (1-\xi^2)^{1/2} \frac{\partial v_0(\xi,t)}{\partial t} d\xi + \rho U \int_{-\infty}^{-1} \gamma(\xi + Ut) \frac{\xi \, d\xi}{(\xi^2 - 1)^{1/2}}. \tag{210}$$

By similar analysis starting with (208), the moment about $x = 0$ is given by

$$\begin{aligned} M &= -\rho \int_{-1}^{1} \xi(1-\xi^2)^{1/2} \frac{\partial v_0(\xi,t)}{\partial t} d\xi \\ &= -2\rho U \int_{-1}^{1} (1-\xi^2)^{1/2} v_0(\xi,t) d\xi \\ &= +\frac{1}{2}\rho U \int_{-\infty}^{-1} \gamma(\xi + Ut) \frac{d\xi}{(\xi^2 - 1)^{1/2}}. \end{aligned} \tag{211}$$

It may be confirmed that (210) and (211) reduce to equations (15) and (37), respectively, if the motion is independent of time and the wake consists only of a starting vortex of circulation $-\Gamma$ at $x = -\infty$.

The first term for the lift in (210) is the added-mass force associated with a generalized vertical acceleration along the plate. Likewise, the first term in (211) is the added moment of inertia, and the second term gives the moment resulting from equation (4.99). The last integrals in (210) and (211) represent the effect of the wake; these introduce memory effects that depend on the past history of the motion.

It is curious that if the moment is transformed to the quarter-chord position, the memory effects associated with the wake vanish. To show this, we use (210) and (211) to compute the moment about the quarter-chord point in the form

$$\begin{aligned} M - \frac{1}{2}L \equiv M|_{x=\frac{1}{2}} &= -2\rho U \int_{-1}^{1} (1-\xi^2)^{1/2} v_0(\xi,t) d\xi \\ &+ \rho \int_{-1}^{1} (1-\xi)(1-\xi^2)^{1/2} \frac{\partial v_0(\xi,t)}{\partial t} d\xi \\ &+ \frac{1}{2}\rho U \int_{-\infty}^{-1} \left(\frac{\xi-1}{\xi+1}\right)^{1/2} \gamma(\xi + Ut) d\xi. \end{aligned} \tag{212}$$

Using (204) to replace the last term, we get

$$\begin{aligned} M|_{x=\frac{1}{2}} &= \rho \int_{-1}^{1} (1-\xi)(1-\xi^2)^{1/2} \frac{\partial v_0(\xi,t)}{\partial t} d\xi \\ &- \rho U \int_{-1}^{1} (1+2\xi) \left(\frac{1-\xi}{1+\xi}\right)^{1/2} v_0(\xi,t) d\xi. \end{aligned} \tag{213}$$

From (213) it is clear that the moment about the quarter-chord point is not affected by memory-dependent downwash from the wake and can be expressed in terms of the instantaneous normal velocity and acceleration of the plate. This is not to say that the moment is independent of the vortices in the wake, which give a contribution represented by the last term of (212). Since (213) can be calculated directly for any prescribed motion of the foil, it is only necessary to consider the lift force, which can be combined with (213) to give the moment about any point along the foil.

5.16 Oscillatory Time Dependence

The simplest case of unsteady motion is sinusoidal time dependence, with period $2\pi/\omega$. In this case the normal velocity of the foil can be expressed in the form

$$v_0(x,t) = \mathrm{Re}[v_0(x)e^{i\omega t}], \tag{214}$$

and the wake vorticity must be of a similar form,

$$\gamma(x+Ut) = \mathrm{Re}[\gamma_0 e^{i(\omega t + kx)}]. \tag{215}$$

Here Re denotes the real part, and $k = \omega/U$ is the *reduced* frequency. This parameter can be interpreted physically as π times the ratio of the chord length to the wavelength of the wake vortices.

Substitution of (214–215) in (203) gives an equation for the unknown constant γ_0 in the form

$$\gamma_0 = 2\int_{-1}^{1}\left(\frac{1-\xi}{1+\xi}\right)^{1/2} v_0(\xi)d\xi \Big/ \int_{-\infty}^{-1}\left(\frac{\xi-1}{\xi+1}\right)^{1/2} e^{ik\xi}\, d\xi. \tag{216}$$

Postponing consideration of the numerator, which will depend on the prescribed motion of the foil, we consider the integral in the denominator which depends only on the reduced frequency k. Strictly, this integral is not convergent because the oscillatory motion assumed in (214–215) violates the assumption of an initial state of rest. One technique for avoiding this difficulty is to make ω and k slightly complex, with negative imaginary parts, so that (214) and (215) vanish for $t \to -\infty$. An equivalent but simpler approach is to assume that the motion starts abruptly from a state of rest at some time $t_0 \ll t$. The vorticity (215) is replaced by zero if $x < (t_0 - t)U$, and the infinite lower limit of the integral in the denominator of (216) is replaced by $(t_0 - t)U$.

With the latter interpretation, the denominator of (216) becomes

$$\int_{(t_0-t)U}^{-1} \left(\frac{\xi-1}{\xi+1}\right)^{1/2} e^{ik\xi}\, d\xi = (1+i\, d/dk) \int_{(t_0-t)U}^{-1} \frac{d\xi}{(\xi^2-1)^{1/2}} e^{ik\xi}. \tag{217}$$

In the limit $t_0 \to -\infty$, the integral on the right-hand side of (217) is proportional to the Hankel function $H_0^{(2)}(k)$, where $H_n^{(2)}(k) = J_n(k) - iY_n(k)$ defines the Hankel function in terms of the Bessel functions of the first and second kinds. The necessary properties of these functions are given by Abramowitz and Stegun (1964). Thus it follows that

$$\begin{aligned} \lim_{t_0\to\infty} \int_{(t_0-t)U}^{-1} \left(\frac{\xi-1}{\xi+1}\right)^{1/2} e^{ik\xi}\, d\xi &= -\frac{\pi}{2} i(1+i\, d/dk) H_0^{(2)}(k) \\ &= -\frac{\pi}{2}[iH_0^{(2)}(k) + H_1^{(2)}(k)]. \end{aligned} \tag{218}$$

If this result is substituted in (216), the parameter γ_0 is determined as

$$\gamma_0 = -\frac{4}{\pi}\int_{-1}^{1}\left(\frac{1-\xi}{1+\xi}\right)^{1/2} v_0(\xi)\, d\xi / \{iH_0^{(2)}(k) + H_1^{(2)}(k)\}. \tag{219}$$

The wake integral in (210) can be evaluated in a similar manner, with (215) and (219) used for $\gamma(\xi + Ut)$. The lift force is given by the expression

$$\begin{aligned} L = -2\rho U\, e^{i\omega t} &\left\{ \left(\frac{H_1^{(2)}(k)}{H_1^{(2)}(k) + iH_0^{(2)}(k)} \right) \int_{-1}^{1} \left(\frac{1-\xi}{1+\xi}\right)^{1/2} v_0(\xi)\, d\xi \right. \\ &\left. + ik \int_{-1}^{1} (1-\xi^2)^{1/2} v_0(\xi)\, d\xi \right\}. \end{aligned} \tag{220}$$

Thus, for sinusoidal time dependence, the memory effects of the wake can be expressed as frequency-dependent force coefficients proportional to the ratio of the Hankel functions:

$$C(k) = \frac{H_1^{(2)}(k)}{H_1^{(2)}(k) + iH_0^{(2)}(k)}. \tag{221}$$

This ratio is called the *Theodorsen function* and is plotted in an Argand diagram in figure 5.33. For low frequencies, the limiting value of the

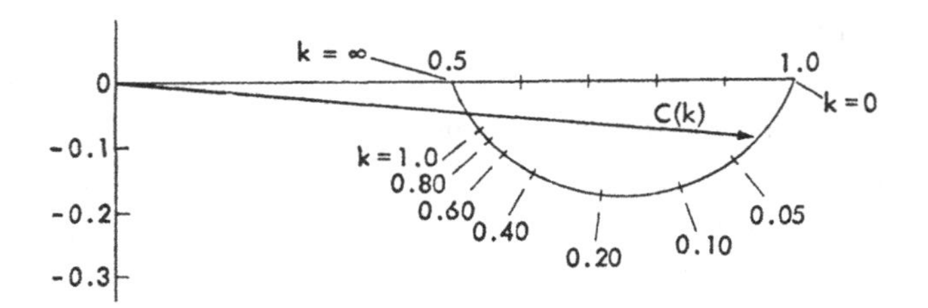

Figure 5.33
The Theodorsen function (221) representing the reduction in amplitude and shift in phase of the lift force due to a vertical oscillation h. (From Bisplinghoff, Ashley, and Halfman 1965)

Theodorsen function is 1.0, and the lift and moment reduce to their quasi-steady-state values.

The cases of greatest practical interest are oscillatory rigid-body motions in the vertical direction (heave) and about the origin (pitch). These displacements of the foil will be denoted by

$$h(t) = \mathrm{Re}\, h_0\, e^{i\omega t}, \tag{222}$$

and

$$\alpha(t) = \mathrm{Re}\, \alpha_0\, e^{i\omega t}, \tag{223}$$

respectively, where h_0 and α_0 are complex constants. The resulting oscillatory mean-camber line is $\eta = h + \alpha x$, and the normal velocity follows from (193),

$$v_0(x,t) = \dot{h} + \dot{\alpha} x - U\alpha. \tag{224}$$

Here a dot denotes the time derivative. Substituting these results with (214) in (220) and evaluating the integrals using (78–79), we find that the lift force is given by

$$L = -2\pi\rho U^2\, \mathrm{Re}\left\{e^{i\omega t} C(k)\left[ikh_0 - \left(1 + \frac{1}{2}ik\right)\alpha_0\right]\right\} - \pi\rho(\ddot{h} - U\dot{\alpha}). \tag{225}$$

The last term in (225) has been left in a general time-dependent form to emphasize its origin in the added-mass sense. This contribution to the lift force can be derived from equation (4.115), after noting that the velocity of the foil in a body-fixed reference frame is given by

$$(U\cos\alpha + \dot{h}\sin\alpha, \dot{h}\cos\alpha - U\sin\alpha) \simeq (U, \dot{h} - U\alpha). \tag{226}$$

No consideration has been given to the drag force, which results in part from the normal component on the foil and in part from the leading-edge suction force. This component is particularly important for flapping propulsion and has been studied extensively in the context of bird and fish propulsion. A particularly simple result follows by considering the heaving motion of a flat plate; here there is no normal component of the drag force along the plate, and the leading-edge suction force is by its nature a negative drag, or positive propulsion force. This interesting area of unsteady hydrofoil theory has been studied by Wu (1971) and by other authors cited in that reference. More recent contributions can be found in Wu, Brokaw, and Brennen (1975).

5.17 The Sinusoidal Gust Problem

If the velocity of the hydrofoil is steady, but the surrounding fluid is in a state of nonuniform motion, the resulting interaction between the foil and the fluid will be unsteady. Examples of this are the motion of hydrofoils in waves and of propeller blades in a spatially nonuniform wake. This type of unsteady lifting-surface problem, which is known as a *gust* problem, differs from those problems treated in sections 5.15 and 5.16 where the foil is unsteady and the undisturbed velocity field of the surrounding fluid is uniform.

The boundary condition on the foil, in the presence of a gust, can be derived from (192–193) by adding the gust potential ϕ_g to the perturbation potential ϕ associated with the foil. The boundary condition (193) takes the modified form

$$v(x,0,t) = \frac{\partial n}{\partial t} - U\frac{\partial n}{\partial x} - v_g(x,0,t), \tag{227}$$

where v_g denotes the upwash of the gust on the foil. Thus, if the foil is fixed, the results obtained in sections 5.15 and 5.16. can be utilized for the gust problem simply by setting $v_0 = -v_g$.

We shall restrict our attention here to gusts that are sinusoidal functions of $x + Ut$, and thus steady with respect to the free stream. If α_0 is the amplitude of the resulting change in angle of attack, v_g can be written in the form

$$v_g(x,0,t) = \mathrm{Re}\{U\alpha_0 \, e^{ikx+i\omega t}\}. \tag{228}$$

Substituting this quantity for $-v_0$ in (220) and evaluating the resulting integrals in terms of the Bessel functions J_0 and J_1 gives the lift force in the form

$$L = -2\rho U^2 \,\mathrm{Re}\left\{\alpha_0 \, e^{i\omega t}\left[\frac{2i/\pi k}{H_1^{(2)}(k) + iH_0^{(2)}(k)}\right]\right\}. \tag{229}$$

The quantity in brackets in (229) is the *Sears function*, which is plotted in figure 5.34. For large argument k, the Hankel functions in the denominator oscillate proportionally to e^{-ik}, and thus the phase of (229) varies rapidly, as indicated in figure 5.34. This can be avoided if the phase of the sinusoidal gust is measured with reference to the leading edge rather than the mid-chord position, and thus (229) is multiplied by the quantity e^{-ik}. The result is slowly varying, as shown in figure 5.34. Physically, this change suggests that the foil responds to the gust most significantly when the gust is at the leading edge. For this reason the leading edges of propeller blades are often raked, so that the phase of the unsteady load along the blade will be distributed more uniformly as the blade enters a nonuniform region.

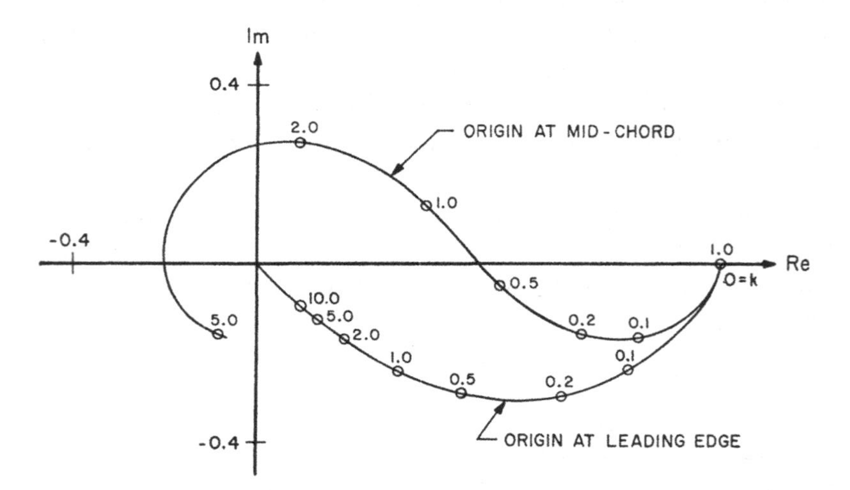

Figure 5.34
The Sears function for a sinusoidal gust, corresponding to the factor in brackets in (229). The phase of the gust is measured relative to mid chord or leading edge in the respective cases shown. Figures marked on the curves are the values of the reduced frequency k.

5.18 Transient Problems

As noted in section 5.15, the general case of arbitrary time-dependence can be treated directly by Laplace transform techniques. Equivalently, the frequency ω in the oscillatory time-dependent results of section 5.16 can be made imaginary, and the superposition of these solutions used to generate an arbitrary transient disturbance. We shall avoid performing this analysis here, but two simple results will be noted—where the vertical velocity of the foil is changed suddenly in a step-function manner, and where the foil moves into a similar gust.

Since any constant angle of attack may be added without affecting the unsteady problem, we assume that initially the foil moves with zero angle of attack, and at time $t = 0$ this is changed by imposing a small vertical velocity component. The resulting lift is given proportional to *Wagner function* shown in figure 5.35. This lift force immediately achieves a nonzero value precisely half of the steady-state value and subsequently increases monotonically to that limit. To develop 90 percent of the steady lift, the foil must travel about six chord lengths.

This problem is identical to that of a foil accelerating instantaneously from rest at $t = 0$ to a constant velocity U, with a constant angle of attack, since in both cases the perturbation vanishes for $t < 0$. Thus the Wagner function in figure 5.35 predicts the initial development of lift during the rapid acceleration of a foil with constant α.

An inverse situation will develop if the foil stops suddenly. This case can be related to the Wagner function by subtracting a constant angle of attack equal to the value before stopping. On this basis it follows that the lift force upon stopping is reduced immediately to half of its steady value, and subsequently a more gradual attenuation to zero persists over a relatively long period.

The corresponding gust problem is that of a *sharp-edged* gust. The resulting lift force, known as the *Kussner function*, is shown in figure 5.35. In this case the lift force rises from an initial value of zero. Half of the ultimate steady-state lift develops before the gust passes the trailing edge, and 90 percent is developed when the foil moves a distance of about eight chord lengths beyond the gust. The details of this and other transient motions are given by Woods (1961).

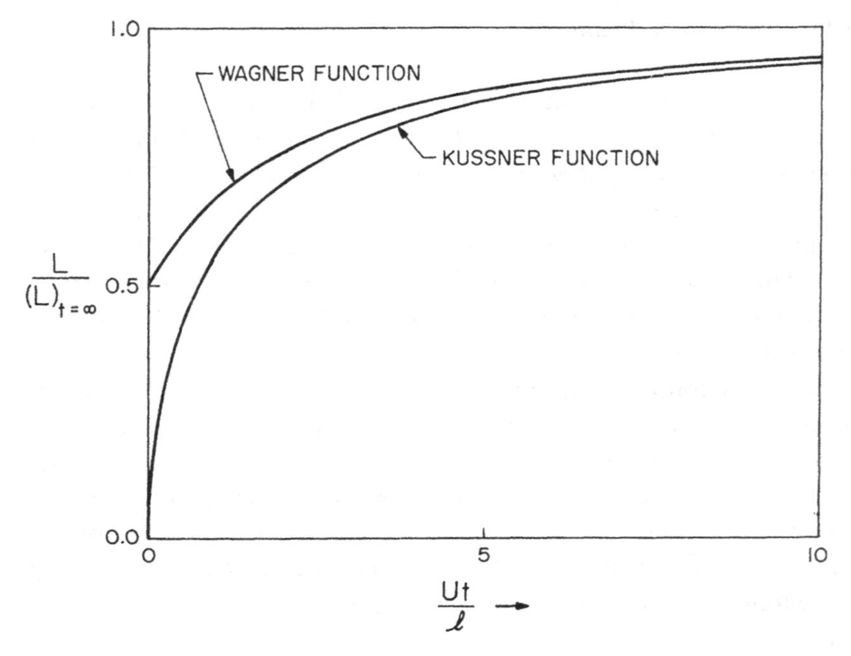

Figure 5.35
Transient lift force due to a sudden change in vertical velocity (Wagner's function) and due to passage through a sharp-edged gust (Kussner's function). The ordinate denotes the number of chord lengths traveled after increasing the angle of attack, or after encountering the gust at the leading edge.

Problems

1. Integrate the normal and tangential components of the complex velocity (51) around a circular contour centered at the singular point, and confirm that q and γ are the total flux rate and circulation, respectively.

2. For a two-dimensional point vortex at $r = 0$, show that $\nabla \times \mathbf{V} = 0$ and $\nabla \cdot \mathbf{V} = 0$ if $r > 0$.

3. Derive the Kutta-Joukowski theorem (15) for steady two-dimensional flow, without linearizing the Bernoulli equation. Start with (4.90) and assume that at large distances from the foil the velocity potential consists of the combination of a free stream of velocity U and a point vortex of circulation Γ.

4. Show that the complex velocity function $u - iv = -(i\Gamma/2\pi)\,(z^2 - 1)^{-1/2}$ satisfies the condition of zero normal velocity on a flat plate situated on the cut (−1, 1), and behaves like a vortex of circulation Γ at large distances from the origin. Why is this not the correct solution for a flat-plate hydrofoil?

5. Derive equation (41) for the nonlinear lift coefficient of a flat plate by superposition of a flow tangential to the plate with velocity $U \cos a$, a flow of the form (4.58) normal to the plate, and a suitable amount of circulation to satisfy the Kutta condition. The complex velocity for the latter component is given in problem 4.

6. What is the ideal angle of attack for a flat plate?

7. If the lift coefficient of a cambered two-dimensional hydrofoil is 0.8, at $\alpha = 5$ degrees, what is C_L at $\alpha = 10$ degrees?

8. According to the Weissinger "quarter-three-quarter-chord" approximation, the bound vortex distribution of a lifting surface can be approximated by a single point vortex situated at the quarter-chord position $x = l/4$ with circulation chosen to satisfy the boundary condition at the three-quarter-chord position $x = -l/4$. Show from the Kutta-Joukowski theorem that the resulting lift coefficient is exact for the two-dimensional parabolic mean-camber line (43) and that the quarter-and three-quarter chord positions form the only combination that gives the correct result in this case.

9. From the approximation described in problem 8 and the method of images, show that a two-dimensional hydrofoil moving above a horizontal plane will experience an increased lift force due to the "ground effect" of the plane surface, whereas the interference between two foils with the same angle of attack in a "biplane" configuration will decrease the lift force on each.

10. Show that a uniform distribution of two-dimensional dipoles, oriented in the vertical direction and distributed along the negative x-axis, is identical to a point vortex at the origin.

11. Find the lift and drag forces on a sailboat keel, assuming that the keel profile is one half of an ellipse with high aspect-ratio and that the bottom of the hull is a large flat plane coincident with the minor axis of the ellipse.

12. Write the complex potential of a pair of vortices, with opposite strengths, separated by a distance $2a$. A free vortex in a fluid is generally assumed to be convected with the velocity of the fluid at the point where

the vortex is situated. On this basis, what is the free motion of the vortex pair discussed above? What is the free motion if the two vortices have the same circulation?

13. A two-dimensional uncambered hydrofoil is pivoted about a point at distance d downstream from the leading edge. For what values of d is the hydrofoil statically stable at zero angle of attack? Discuss qualitatively the situation that would result if the pivot is further downstream.

14. An uncambered three-dimensional lifting surface with aspect ratio $A = 5$ experiences a lift coefficient of 0.4 at an angle of attack $\alpha = 5$ degrees. What is the minimum induced drag coefficient in this condition and at $\alpha = 10$ degrees?

15. A two-dimensional hydrofoil consists of a parabolic arc situated at an angle of attack, as in equation (43). Compare the lift and drag forces of this foil in the fully wetted and supercavitating conditions.

16. A hydrofoil set at an angle of attack of 7 degrees relative to the zero-lift angle moves through a wave system such that the unsteady component of the angle of attack is ±4 degrees. Compare the magnitudes of the mean and unsteady lift forces, assuming the hydrofoil travels a distance of five chord lengths in each period of oscillation.

References

Abbott, I. H., and A. E. von Doenhoff. 1959. *Theory of wing sections*. New York: Dover.

Abramowitz, M., and I. A. Stegun, eds. 1964. *Handbook of mathematical functions*. Washington: U.S. Government Printing Office.

Acosta, A. J. 1973. Hydrofoils and hydrofoil craft. *Annual Review of Fluid Mechanics* 5:161–184.

Barker, S. J., and T. M. Ward. 1976. *Experiments on two- and three-dimensional cavitating hydrofoil models with and without flaps*. California Institute of Technology, Guggenheim Aeronautical Laboratory report HSWT-1130.

Batchelor, G. K. 1967. *An introduction to fluid dynamics*. Cambridge: Cambridge University Press.

Bisplinghoff, R. L., H. Ashley, and R. L. Halfman. 1965. *Aeroelasticity*. Reading, Mass.: Addison-Wesley.

Brard, R. 1972. Vortex theory for bodies moving in water. In *Ninth symposium on naval hydrodynamics*, eds. R. Brard and A. Castera, pp. 1187–1284. Washington: U.S. Government Printing Office.

Carrier, G. F., M. Krook, and C. E. Pearson. 1966. *Functions of a complex variable*. New York: McGraw-Hill.

Cox, G. G., and W. B. Morgan. 1972. The use of theory in propeller design. *Marine Technology* 9:419–429.

Hildebrand, F. B. 1976. *Advanced calculus for applications*. 2nd ed. Englewood Cliffs, N. J.: Prentice-Hall.

Kerwin, J. E. 1973. Computer techniques for propeller blade section design. *International Shipbuilding Progress* 20:227–251.

Muskhelishvili, N. I. 1953. *Singular integral equations*. Groningen, Netherlands: Nordhoff.

Robinson, A., and J. A. Laurmann. 1956. *Wing theory*. Cambridge: Cambridge University Press.

Thwaites, B., ed. 1960. *Incompressible aerodynamics*. Oxford: Oxford University Press.

Tulin, M. P. 1964. Supercavitating flows—small perturbation theory. *J. Ship Res.* 7 (3): 16–37.

Van Dyke, M. 1975. *Perturbation methods in fluid mechanics*. 2nd ed. Stanford: Parabolic Press.

Woods, L. C. 1961. *The theory of subsonic plane flow*. Cambridge: Cambridge University Press.

Wu, T. Y. 1968. Inviscid cavity and wake flows. *Basic Developments in Fluid Dynamics* 2:1–116.

Wu, T. Y. 1971. Hydromechanics of swimming propulsion, Part 1. *Journal of Fluid Mechanics* 46:337–355.

Wu, T. Y. 1972. Cavity and wake flows. *Annual Review of Fluid Mechanics* 4:243–284.

Wu, T. Y., A. K. Whitney, and C. Brennen. 1971. Cavity-flow wall effects and correction rules. *Journal of Fluid Mechanics* 49:223–256.

Wu, T. Y., C. J. Brokaw, and C. Brennen eds. 1975. *Proceedings of the symposium on swimming and flying in nature*, vol. 2. New York: Plenum.

6 Waves and Wave Effects

A unique aspect of marine hydrodynamics, relative to other branches of fluid mechanics, is the importance of wave effects on the free surface. The complexity of this topic is apparent from observations of the waves generated by a storm at sea, by a moving ship in calm water, or simply by throwing a pebble into a pond. These are diverse examples of *surface waves*; for ocean engineers and naval architects these are the most common and important wave phenomena in the ocean.

Other wave phenomena that also exist in the ocean may be of comparable significance in special circumstances. *Internal waves*, like surface waves, result from the balance of kinetic and potential energy, but they are found in internal regions of density stratification beneath the sharp interface with the atmosphere. Because of the small density differences involved, internal waves occur at much lower frequencies, with periods on the order of several minutes. As a result, the influence of internal waves on bodies in the ocean is generally negligible unless the body has an unusual low-frequency resonance.

Waves of even lower frequencies exist, including *inertial waves* associated with the Coriolis acceleration due to the earth's rotation, and tides generated by changes in the potential energy of heavenly bodies. At the high-frequency end of the spectrum are capillary waves and ripples, which may be observed on the ocean surface and in small wave tanks, but these do not affect large vessels or structures. A description of these different types of wave motion from the oceanographic standpoint is given by Phillips (1966) and by Neumann and Pierson (1966).

In marine hydrodynamics we are concerned with the effects of the wave environment on floating or fixed structures. Of particular interest are the wave loads on fixed structures and the oscillatory motions of vessels free to respond to the waves. Also of engineering importance are the waves generated by the motion of bodies in otherwise calm water, including the prediction of wave resistance for a moving ship.

These problems are amenable to a theoretical description based on the assumptions that the fluid is ideal and that the wave motions are sufficiently small to linearize. Nonlinear effects are important in special circumstances, such as the breaking of waves in shallow water or locally near a ship's bow. Viscous stresses are significant for the wave forces on small bodies, as noted in section 2.13, and in the generation of waves by wind. Nevertheless, we shall find here, as in chapter 5, that useful results can be obtained by neglecting nonlinear and viscous effects; in fact, these assumptions are practically essential if the complexity is to be kept in reasonable bounds.

6.1 Linearized Free-Surface Condition

The fluid velocity **V** is expressed by the gradient of a velocity potential ϕ, and the effects of the free surface must be expressed in terms of appropriate boundary conditions on this surface. The physical nature of a free surface requires both a kinematic and a dynamic boundary condition—that is, the normal velocities of the fluid and of the boundary surface must be equal, and the pressure on the free surface must be atmospheric. In both boundary conditions the simplifications resulting from linearization are significant. Similar assumptions were used in chapter 5 to derive the kinematic boundary condition on the wetted surface of a hydrofoil and the dynamic boundary conditions on both trailing vortex sheets and cavity surfaces.

We shall begin by stating the "exact" boundary conditions, assuming only that the fluid is ideal and that surface tension at the free surface is negligible (see section 2.3). The linearized results will follow by neglecting second- and higher-order terms in the wave amplitude and associated fluid motions.

A Cartesian coordinate system (x, y, z) is adopted, with $y = 0$ the plane of the undisturbed free surface and the y-axis positive upward. The vertical elevation of any point on the free surface may be defined by a function $y = \eta(x, z, t)$. In the special case of two-dimensional fluid motion,[1] parallel to the x-y plane, the dependence on z will be deleted.

The exact kinematic boundary condition can be derived most readily by requiring that the substantial derivative of the quantity $y - \eta$ vanish on the free surface. The result of this condition is that, on $y = \eta$,

$$0 = \frac{D}{Dt}(y - \eta) = \frac{\partial \phi}{\partial y} - \frac{\partial \eta}{\partial t} - \frac{\partial \phi}{\partial x}\frac{\partial \eta}{\partial x} - \frac{\partial \phi}{\partial z}\frac{\partial \eta}{\partial z}. \tag{1}$$

If the wave elevation η is sufficiently small, the slopes $\partial\eta/\partial x$ and $\partial\eta/\partial z$ will be small quantities compared to one and of the same order of magnitude as $\partial\eta/\partial t$. Similarly, if the fluid velocity is a small quantity proportional to the wave motion, the derivatives of the velocity potential in (1) will be small first-order quantities. Thus, the last two terms in (1) are of second order and may be neglected in the linearized kinematic boundary condition, which is therefore given by

$$\frac{\partial \eta}{\partial t} = \frac{\partial \phi}{\partial y}. \tag{2}$$

This approximate boundary condition simply states that the vertical velocities of the free surface and fluid particles are equal, ignoring the small departures of that surface from the horizontal orientation. (A similar boundary condition is given by equation (5.193) for the unsteady hydrofoil, but with an additional contribution due to the free-stream velocity.)

The dynamic boundary condition is obtained from Bernoulli's equation (4.26). Assuming the atmospheric pressure p_a is independent of position on the free surface and choosing the constant of integration $C(t)$ suitably, we find that the exact condition to be satisfied on the free surface is

$$-\frac{1}{\rho}(p - p_a) = \frac{\partial \phi}{\partial t} + \frac{1}{2}\nabla\phi \cdot \nabla\phi + gy = 0. \tag{3}$$

Substituting the free-surface elevation η for y and solving for η,

$$\eta = -\frac{1}{g}\left(\frac{\partial \phi}{\partial t} + \frac{1}{2}\nabla\phi \cdot \nabla\phi\right). \tag{4}$$

If we neglect the last term, which is a second-order quantity in the fluid velocity, the linearized equation for the free-surface elevation is

$$\eta = -\frac{1}{g}\frac{\partial \phi}{\partial t}. \tag{5}$$

Strictly speaking, these boundary conditions should be imposed on the exact free surface $y = \eta$. However, it is consistent with the linearizations already carried out to impose the first-order boundary conditions (2) and (5) on the undisturbed plane of the free surface, $y = 0$. This additional simplification of the linearization procedure can be justified formally by expansion of the velocity potential and its derivatives in Taylor series about the plane $y = 0$, as will be shown in section 6.4.

On the plane $y = 0$, the dynamic boundary condition (5) can be differentiated with respect to time and combined with the kinematic condition (2). This gives a single boundary condition for the velocity potential,

$$\frac{\partial^2 \phi}{\partial t^2} + g\frac{\partial \phi}{\partial y} = 0, \quad \text{on } y = 0. \tag{6}$$

6.2 Plane Progressive Waves

The simplest solution of the free-surface condition (6), which nevertheless has great practical significance, is the plane progressive wave system. This motion is two dimensional, sinusoidal in time with radian frequency ω, and propagates with phase velocity V_p such that, to an observer moving with this velocity, the wave appears steady-state. Thus, the free-surface elevation must be of the general form

$$\eta(x,t) = A\cos(kx - \omega t + \varepsilon), \tag{7}$$

where the positive x-axis has been chosen to coincide with the direction of wave propagation. Here A is the wave amplitude, and ε is an arbitrary phase angle which can be set equal to zero by suitable choice of the origin $x = 0$. Hereafter this will be assumed, with $\varepsilon = 0$. The parameter

$$k = \omega / V_p \tag{8}$$

is the *wavenumber*, the number of waves per unit distance along the x-axis. Clearly,

$$k = 2\pi / \lambda, \tag{9}$$

where the wavelength λ is the distance between successive points on the wave with the same phase, as shown in figure 2.1.

The solution of this problem is expressed in terms of a two-dimensional velocity potential $\phi(x, y, t)$, which must satisfy Laplace's equation, the free-surface condition (6), an appropriate boundary condition on the bottom of the fluid, and must yield the wave elevation (7) from (5). Clearly this potential must be sinusoidal in the same sense as (7); therefore we seek a solution of the form

$$\phi(x,y,t) = \text{Re}[Y(y)e^{-ikx+i\omega t}]. \tag{10}$$

From Laplace's equation, Y must satisfy the ordinary differential equation

$$\frac{d^2Y}{dy^2} - k^2Y = 0, \tag{11}$$

throughout the domain of the fluid.

The most general solution of (11) is given in terms of exponential functions, in the form

$$Y = Ce^{ky} + De^{-ky}. \tag{12}$$

Here C and D are constants to be determined from the boundary conditions on the free surface and bottom.

For now we assume that the fluid depth is infinite; hence (11) must hold for $-\infty < y < 0$. To avoid an unbounded motion deep beneath the free surface, the constant D in (12) must vanish; thus

$$Y = Ce^{ky}. \tag{13}$$

If this function is substituted in (10), the velocity potential is given by

$$\phi = \text{Re}[Ce^{ky-ikx+i\omega t}]. \tag{14}$$

Use (5) to compute the linearized wave elevation, with $y = 0$, and compare the result with the prescribed plane wave (7); then the constant C must take the value

$$C = igA / \omega. \tag{15}$$

The potential (14) can be rewritten in the form

$$\phi = \frac{gA}{\omega} e^{ky} \sin(kx - \omega t). \tag{16}$$

At this point in our derivation, the solution appears complete; yet the free-surface condition (6) has not been imposed. Therefore (16) is too general, and if this potential is substituted in (6) an additional condition is obtained, relating the wavenumber and the frequency in the form

$$k = \omega^2 / g. \tag{17}$$

Thus the frequency and wavenumber are mutually dependent parameters related by the *dispersion* relation (17). Of course, the frequency ω can be replaced by the wave period $T = 2\pi/\omega$, just as the wavenumber k can be replaced by the wavelength $\lambda = 2\pi/k$.

The phase velocity V_p can be determined from (8), and with (17) it follows that

$$V_p = \omega / k = g / \omega = (g / k)^{1/2} = (g\lambda / 2\pi)^{1/2}. \tag{18}$$

The last of these equivalent relations is consistent with the conclusion, based on dimensional analysis in section 2.3, that the phase velocity is proportional to the square root of the wavelength.

While the wave moves with the phase velocity V_p, the fluid itself moves with a much smaller velocity given by the gradient of the potential (16). The velocity components (u, v) of the fluid are

$$u = \frac{\partial \phi}{\partial x} = \omega A e^{ky} \cos(kx - \omega t), \tag{19}$$

$$u = \frac{\partial \phi}{\partial y} = \omega A e^{ky} \sin(kx - \omega t). \tag{20}$$

Since the horizontal and vertical velocity components have the same magnitude but a phase difference of a quarter period, the fluid particles move through circular orbits of radius Ae^{ky}. On the linearized free surface $y = 0$, the amplitude of this motion is equal to the wave amplitude A, in accordance with the kinematic boundary condition (2). Comparison of (19–20) with the wave elevation (7) shows that the horizontal velocity component is a maximum beneath the crest and trough. Beneath the crest the velocity is positive, in the same direction as the wave propagation, and beneath the trough the flow is in the opposite direction. The vertical velocity component is a maximum beneath the nodes $\eta = 0$, rising or falling with the free surface. This velocity field is shown in figure 6.1.

Within the accuracy of the linear theory, the fluid particles move in small circular orbits proportional to the wave amplitude; they remain in the same mean position as the wave propagates through the fluid with a phase velocity independent of the wave amplitude. Some nonlinear effects that modify this situation will be discussed in sections 6.4 and 6.5, but

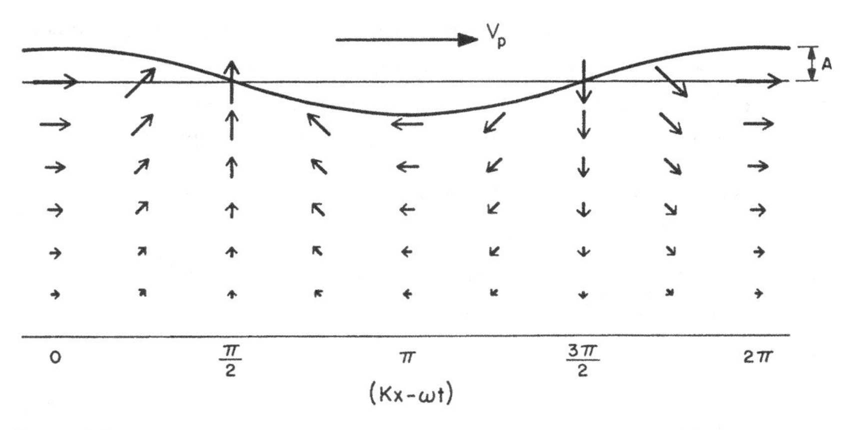

Figure 6.1
Velocity field of a plane progressive wave in deep water. The phase velocity and fluid velocity vectors are to the same scale. This example corresponds to the case $A/\lambda = 1/20$.

for most practical purposes the linear results described here are extremely accurate.

As the depth of submergence beneath the free surface increases, the fluid velocity (19–20) is attenuated exponentially. For a submergence of half a wavelength, $ky = -\pi$ and the exponential factor is reduced to 0.04. Thus waves in deep water are confined to a relatively shallow layer near the free surface, with negligible motion beneath a depth of about $\frac{1}{2}\lambda$. On this basis one can anticipate that if the fluid depth is finite, but greater than half a wavelength, the effects of the bottom will be negligible. This conclusion will be confirmed in section 6.3.

6.3 Finite-Depth Effects

For a fluid of constant depth h with the bottom a rigid impermeable plane, the boundary condition

$$\partial\phi / \partial y = 0, \quad y = -h, \tag{21}$$

must be imposed. Returning to the general solution (12), both exponential functions must be retained, with the constants C and D suitably chosen to satisfy (21).

The necessary algebra can be reduced by noting that the two independent solutions $e^{\pm ky}$ may be replaced by linear combinations of the hyperbolic sine and cosine function. The only combination that satisfies (21) is proportional to $\cosh[k(y + h)]$. Thus, for a fluid of finite depth,

$$Y(y) = C\frac{\cosh k(y+h)}{\cosh kh}, \tag{22}$$

where the constant C has been chosen such that (15) remains valid. Substituting (22) in (10) and using (15) gives the velocity potential for finite depth in the form

$$\phi = \frac{gA}{\omega}\frac{\cosh k(y+h)}{\cosh kh}\sin(kx-\omega t). \tag{23}$$

Finally, for (23) to satisfy the free-surface condition (6), it follows that

$$k\tanh kh = \omega^2 / g. \tag{24}$$

The dispersion relation (24) is an implicit equation which determines the wavenumber k and wavelength $\lambda = 2\pi/k$, for given values of the depth h and frequency ω. Conversely, for given values of the wavelength and depth, (24) can be used to calculate the frequency.

As the depth h tends to infinity, equations (23–24) reduce to the infinite-depth results (16–17) respectively. At a depth of $\frac{1}{2}\lambda$, the discrepancy between (17) and (24) is less than half a percent, confirming the estimates at the end of section 6.2.

The fluid velocity components can be computed as in (19–20), and for finite depth it follows that

$$u = \frac{\partial\phi}{\partial x} = \frac{gAk}{\omega}\frac{\cosh k(y+h)}{\cosh kh}\cos(kx-\omega t), \tag{25}$$

$$v = \frac{\partial\phi}{\partial y} = \frac{gAk}{\omega}\frac{\sinh k(y+h)}{\cosh kh}\sin(kx-\omega t). \tag{26}$$

Once again there is a phase difference of a quarter period between these two velocity components, but because the magnitudes are unequal, the trajectories of fluid particles will be elliptical. The major axis of these ellipses is horizontal, and the velocity distribution is as shown in figure 6.2. Figure 6.7(a) is a time-exposure photograph that shows these trajectories vividly.

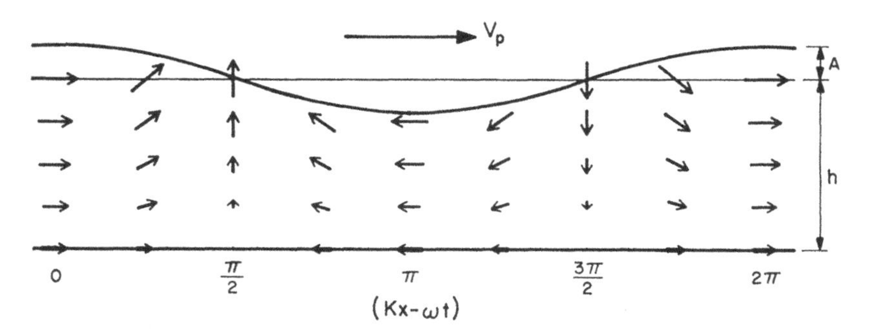

Figure 6.2
Velocity field of a plane progressive wave in finite depth. The phase velocity and fluid velocity vectors are to the same scale. This example corresponds to the case $A/\lambda = 1/20$ and $A/h = 1/5$.

From (8) and (24) the phase velocity for finite depth can be expressed in the form

$$V_p = \omega / k = [(g/k)\tanh kh]^{1/2}. \quad (27)$$

This tends to the deep-water limit (18) for $kh \gg 1$. The opposite limit, $kh \ll 1$, is the regime of *shallow-water waves*. In this case (27) can be approximated using the Taylor series for the hyperbolic tangent, and the leading-order approximation for the phase velocity is

$$V_p \simeq (gh)^{1/2}, \quad kh \ll 1. \quad (28)$$

The ratio of the phase velocity (27) to the infinite-depth limit (18) is plotted in figure 6.3, together with the depth-wavelength ratio h/λ. For a fixed value of ω, it follows from (8) and (9) that $V_p/V_{p\infty} = \lambda/\lambda_\infty$. Both ratios increase monotonically from zero to one, with increasing depth. If $h > \frac{1}{2}\lambda$, the phase velocity and wavelength are essentially equal to their infinite-depth values. Dimensional plots of the phase velocity, period, and wavelength are given by Wiegel (1964) and Le Méhauté (1976).

In general, the phase velocity (27) depends on the wavenumber and the wavelength. Thus water waves are *dispersive*, as noted in section 2.3, with long waves traveling faster than shorter waves. The shallow-water limit (28) is an exception, where the phase velocity depends only on the depth, and the resulting wave motion is *nondispersive*.

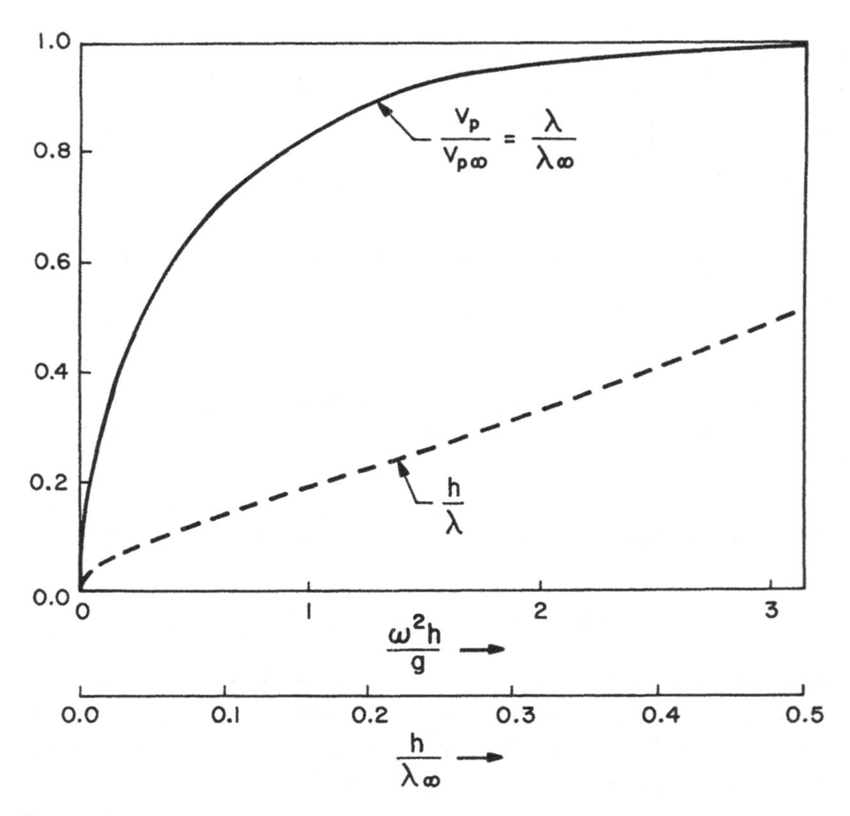

Figure 6.3
Phase velocity, wavelength, and depth ratios of a plane progressive wave. The deep-water limiting values $V_{p\infty}$ and λ_∞ can be determined from equation (18).

For a fluid of nonuniform depth, these results can be applied locally, provided the change of depth is small over distances comparable to a wavelength. If waves of constant frequency propagate from deep to shallow water, as at a beach, it follows from figure 6.3 that the wavelength and phase velocity will decrease. Figure 6.4 illustrates this phenomenon in a small wave tank.

6.4 Nonlinear Effects

In this section we shall consider briefly some of the simpler effects neglected in the linearized theory. First the nonlinear free-surface boundary condition

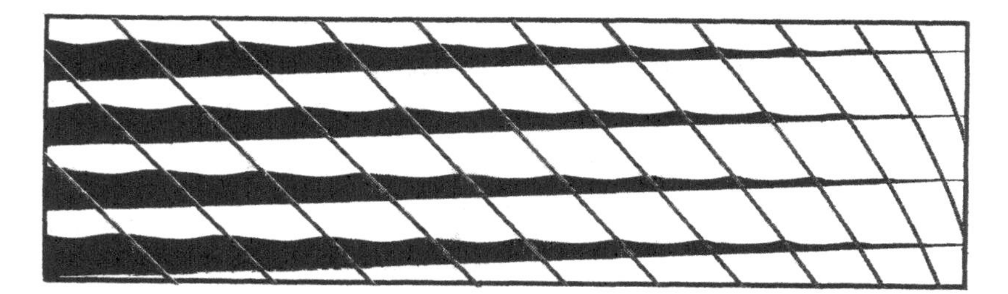

Figure 6.4
Sequence of photographs, taken through the transparent side of a wave tank, showing the propagation of plane progressive waves into shallow water. The water is darkened with dye, and the waves are generated by an oscillating vertical wedge at the left side of the tank. Each wave crest is connected in successive photographs by diagonal lines, which advance in time with the phase velocity. Both the phase velocity and wavelength are reduced as the waves propagate toward the shallow end of the tank. The interval between successive photographs is 0.25s, and the wave period is 0.4s. The water depth is 0.11m at the left end, decreasing to zero at the right. Nonlinear distortion occurs when the depth is comparable to the wave height.

must be derived to replace (6). This can be done, starting with the exact kinematic and dynamic boundary conditions (1) and (4), but the necessary derivatives of (4) are difficult to evaluate since this equation holds only on the curvilinear surface $\eta(x, z, t)$.

A more expedient approach is to replace the kinematic condition (1) by the statement that the substantial derivative of the pressure is zero on the free surface. This is a rather pragmatic mixture of the dynamic and kinematic boundary conditions, since the statement that $Dp/Dt = 0$ on the free surface implies that this is precisely the appropriate moving surface on which the pressure is constant. Substituting (3) for the pressure, we obtain the desired boundary condition in the form

$$0 = \left(\frac{\partial}{\partial t} + \nabla\phi \cdot \nabla\right)\left(\frac{\partial \phi}{\partial t} + \frac{1}{2}\nabla\phi \cdot \nabla\phi + gy\right). \tag{29}$$

Working out the indicated derivative gives

$$\frac{\partial^2 \phi}{\partial t^2} + g\frac{\partial \phi}{\partial y} + 2\nabla\phi \cdot \nabla\frac{\partial \phi}{\partial t} + \frac{1}{2}\nabla\phi \cdot \nabla(\nabla\phi \cdot \nabla\phi) = 0. \tag{30}$$

The free-surface boundary condition (30) is exact and explicit, except that it must be applied on the unknown surface $y = \eta$ defined by (4). If only the linear terms are retained in (30), this equation reduces to (6).

A systematic procedure can be used to expand the boundary condition (30), as well as (4), from the exact free surface to the plane $y = 0$, using Taylor series expansions of the potential and its derivatives. A typical example is

$$\phi(x,\eta,z,t) = \phi(x,0,z,t) + \eta\left(\frac{\partial\phi}{\partial y}\right)_{y=0} + \frac{1}{2}\eta^2\left(\frac{\partial^2\phi}{\partial y^2}\right)_{y=0} + \cdots. \tag{31}$$

Using this scheme for each of the derivatives in (4) and (30), we can obtain a sequence of boundary conditions valid on the known surface $y = 0$. The first three members are

$$\frac{\partial^2\phi}{\partial t^2} + g\frac{\partial\phi}{\partial y} = 0 + O(\phi^2), \tag{32}$$

$$\frac{\partial^2\phi}{\partial t^2} + g\frac{\partial\phi}{\partial y} + 2\nabla\phi\cdot\nabla\frac{\partial\phi}{\partial t} - \frac{1}{g}\frac{\partial\phi}{\partial t}\frac{\partial}{\partial y}\left(\frac{\partial^2\phi}{\partial t^2} + g\frac{\partial\phi}{\partial y}\right) = 0 + O(\phi^3), \tag{33}$$

$$\begin{aligned}
&\frac{\partial^2\phi}{\partial t^2} + g\frac{\partial\phi}{\partial y} + 2\nabla\phi\cdot\nabla\frac{\partial\phi}{\partial t} + \frac{1}{2}\nabla\phi\cdot\nabla(\nabla\phi\cdot\nabla\phi) \\
&\quad - \frac{1}{g}\frac{\partial\phi}{\partial t}\frac{\partial}{\partial y}\left(\frac{\partial^2\phi}{\partial t^2} + g\frac{\partial\phi}{\partial y} + 2\nabla\phi\cdot\nabla\frac{\partial\phi}{\partial t}\right) \\
&\quad - \frac{1}{g}\left(-\frac{1}{g}\frac{\partial\phi}{\partial t}\frac{\partial^2\phi}{\partial y\partial t} + \frac{1}{2}\nabla\phi\cdot\nabla\phi\right)\frac{\partial}{\partial y}\left(\frac{\partial^2\phi}{\partial t^2} + g\frac{\partial\phi}{\partial y}\right) \\
&\quad + \frac{1}{2g^2}\left(\frac{\partial\phi}{\partial t}\right)^2\frac{\partial^2}{\partial y^2}\left(\frac{\partial^2\phi}{\partial t^2} + g\frac{\partial\phi}{\partial y}\right) = 0 + O(\phi^4).
\end{aligned} \tag{34}$$

The symbol $O()$ indicates the order of magnitude of the neglected terms.

Equation (34) can be simplified by invoking Laplace's equation and (32) to eliminate the last group of terms. Nevertheless, the complexity of this systematic procedure increases rapidly with the order of accuracy of the boundary condition.

Plane progressive waves in deep water are of particular interest and importance, and equations (32–34) can be utilized to obtain the third-order expansion for plane waves.

Direct substitution of the first-order plane-wave potential (16) in the second-order boundary condition (33) reveals that the second-order terms in (33) vanish. Therefore the first-order potential is a solution of the second-order boundary value problem, and we can state that

$$\phi = \frac{gA}{\omega} e^{ky} \sin(kx - \omega t) + O(A^3). \tag{35}$$

On the other hand, the free-surface elevation must be corrected for second-order effects, using the systematic expansion of (4) in the form

$$\begin{aligned}\eta &= -\frac{1}{g}\left(\frac{\partial \phi}{\partial t} + \frac{1}{2}\nabla\phi \cdot \nabla\phi\right)_{y=\eta} \\ &= -\frac{1}{g}\left(\frac{\partial \phi}{\partial t} + \frac{1}{2}\nabla\phi \cdot \nabla\phi\right)_{=0} \\ &\quad + \eta \frac{\partial}{\partial y}\left\{-\frac{1}{g}\left(\frac{\partial \phi}{\partial t} + \frac{1}{2}\nabla\phi \cdot \nabla\phi\right)\right\}_{y=0} + \cdots \\ &= -\frac{1}{g}\left(\frac{\partial \phi}{\partial t} + \frac{1}{2}\nabla\phi \cdot \nabla\phi - \frac{1}{g}\frac{\partial \phi}{\partial t}\frac{\partial^2 \phi}{\partial y \partial t}\right)_{y=0} + O(\phi^3).\end{aligned} \tag{36}$$

Combining equations (35) and (36) yields the following second-order results:

$$\begin{aligned}\eta &= A\cos(kx - \omega t) - \tfrac{1}{2}kA^2 + kA^2\cos^2(kx - \omega t) \\ &= A\cos(kx - \omega t) + \tfrac{1}{2}kA^2\cos 2(kx - \omega t).\end{aligned} \tag{37}$$

The last term in (37), which represents the second-order correction to the free-surface profile, is positive both at the crests $(kx - \omega t) = 0, 2\pi, 4\pi, \ldots$ and at the troughs $(kx - \omega t) = \pi, 3\pi, 5\pi, \ldots$. The crests will be steeper, and the troughs flatter, as a result of this nonlinear correction.

For the third-order free-surface condition (34), substituting the plane-wave potential (35) in the nonlinear terms in (34) eliminates all but one term; thus the boundary condition for the third-order plane-wave solution is given by

$$\frac{\partial^2 \phi}{\partial t^2} + g\frac{\partial \phi}{\partial y} + \frac{1}{2}\nabla\phi \cdot \nabla(\nabla\phi \cdot \nabla\phi) = 0 + O(\phi^4). \tag{38}$$

The first-order solution (16) will satisfy this third-order boundary condition provided the dispersion relation (17) is corrected for a second-order effect of the form

$$\omega^2 = gk(1 + k^2A^2) + O(k^3A^3). \tag{39}$$

The relation (8) can be utilized with (39) to obtain the corresponding correction for the phase velocity,

$$\begin{aligned} V_p = \omega / k &= (g / k)^{1/2}(1 + k^2A^2)^{1/2} + O(k^3A^3) \\ &= (g / k)^{1/2}(1 + \tfrac{1}{2}k^2A^2) + O(k^3A^3). \end{aligned} \tag{40}$$

Thus the phase velocity depends weakly on the wave amplitude, and waves of large height travel faster than small waves. By analogy with the more significant dependence of the phase velocity on wavelength, this dependence on amplitude is known as *amplitude dispersion.*

If (39) is used to replace the linear dispersion relation (17), the linearized velocity potential (16) and (35) is accurate up to and including terms of order A^3—that is, the error in this simple velocity potential is $O(A^4)$. The absence of second- and third-order terms in the potential holds only for deep water, however. The nonlinear corrections for finite depth, including the effects of surface tension, are given by Wehausen and Laitone (1960).

Nonlinear solutions for plane waves based on systematic power series in the wave amplitude are known as *Stokes' expansions*. The most extensive of these is due to Schwartz (1974), who utilized a digital computer to carry out the infinite-depth expansion to order A^{117} and the finite-depth results to a more modest point. Ultimately one might be concerned with the convergence of these power series. In 1918–1924 it was established by Levi-Civita and Struik that these power series converged, at least for sufficiently small wave amplitudes. Almost fifty years later it has been shown that nonlinear plane waves are unstable if they are contaminated by side-band waves of slightly different frequencies.

Thus, in practice, it is impossible to propagate a plane wave system in a long wave tank, and waves that appear stable near the wavemaker will deteriorate completely after traveling several hundred wavelengths. Further details of this and related topics are contained in Lighthill (1967).

A complementary approach to nonlinear waves has been the study of the *highest* wave. It can be shown that the limiting form for a plane wave system has a sharp crest, with a local stagnation point at the crest, and with the free surface inclined 30 degrees from the horizontal on each side of the crest. In deep water this will occur when the ratio of wave height to wavelength is approximately 0.14, or about 1/7, as shown in figure 6.5. Details of this analysis are summarized by Wehausen and Laitone (1960).

A description of nonlinear progressive waves was first derived in 1802 by Gerstner, who showed that the free-surface boundary conditions could be

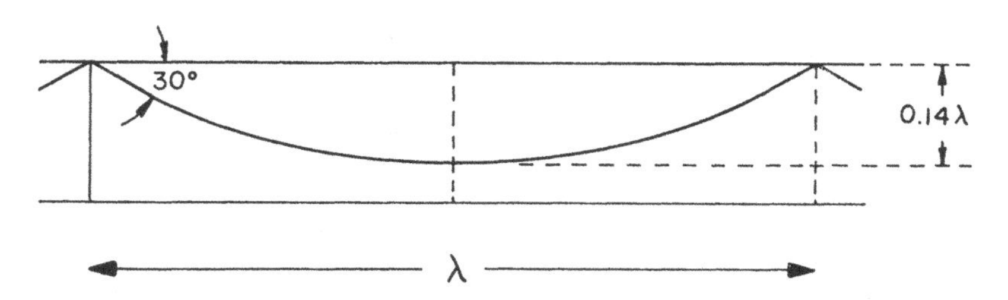

Figure 6.5
The "steepest wave" profile.

satisfied exactly by a trochoidal[2] wave profile. This trochoidal wave system has been used extensively in the field of naval architecture, but it is scientifically deficient since the velocity field is not irrotational. The practical appeal of the trochoidal wave results from the similarity of its appearance to actual waves, but the same is true of the second-order Stokes wave profile (37), which actually agrees with the second-order approximation of a trochoid! Descriptions of the trochoidal wave system may be found in Lamb (1932), as well as in Wehausen and Laitone (1960). Wehausen and Laitone note that while the occurrence of vorticity in ocean waves cannot be ruled out, due to viscous effects, the *sign* of vorticity in Gerstner's wave is opposite that expected from the action of shear stresses due to wind blowing over the free surface. In any event, the mechanism of wave generation by wind involves more complex processes than a simple shear stress on the interface, as discussed by Phillips (1966).

6.5 Mass Transport

One of the more interesting nonlinear features of plane progressive waves is the occurrence of a second-order mean drift of the fluid particles, in the same direction as the wave propagation. This effect can be calculated most easily for the infinite-depth case, since the velocity potential (16) is exact to second order in the wave amplitude. The existence of a net flux follows because the horizontal velocity component (19) is equal in magnitude and opposite in sign beneath the crest and trough, at points of equal depth y below the *mean* free surface. Since u is positive beneath the crest, where the total elevation of the fluid is greater, the total horizontal flux beneath the crest will exceed that beneath the trough, and on the average a net mass transport will occur.

The orbital motion of a particular fluid particle can be computed in terms of the *Lagrangian* coordinates $[x_0(t), y_0(t)]$ which define the position of a particle. These must satisfy the relation s

$$\frac{dx_0}{dt} = u(x_0, y_0, t), \tag{41}$$

$$\frac{dy_0}{dt} = v(x_0, y_0, t). \tag{42}$$

If x_0 and y_0 differ by a small amount, of order A, from the fixed position (x, y), Taylor series can be used to expand (41–42) in the form

$$\frac{dx_0}{dt} = u(x,y,t) + (x_0 - x)\frac{\partial u}{\partial x} + (y_0 - y)\frac{\partial u}{\partial x} + O(A^3), \tag{43}$$

$$\frac{dy_0}{dt} = v(x,y,t) + (x_0 - x)\frac{\partial v}{\partial x} + (y_0 - y)\frac{\partial v}{\partial y} + O(A^3). \tag{44}$$

Integrating (19) and (20) with respect to time gives the first-order trajectories

$$(x_0 - x) = \int u \, dt = -Ae^{ky}\sin(kx - \omega t) + O(A^2), \tag{45}$$

$$(y_0 - y) = \int v \, dt = Ae^{ky}\cos(kx - \omega t) + O(A^2). \tag{46}$$

If these expressions are substituted with (19–20) in (43–44), it follows that

$$\frac{dx_0}{dt} = \omega Ae^{ky}\cos(kx - \omega t) + \omega kA^2 e^{2ky} + O(A^3), \tag{47}$$

$$\frac{dy_0}{dt} = \omega Ae^{ky}\sin(kx - \omega t) + O(A^3). \tag{48}$$

The second-order vertical motion of a given particle of fluid is strictly periodic, with the velocity given by (48), but the horizontal velocity (47) contains a steady *Stokes' drift*. Integrating (47) vertically with respect to all particles of fluid $-\infty < y_0 < 0$ yields the total mean flux

$$Q = \tfrac{1}{2}\omega A^2 = \tfrac{1}{2}kA^2 V_p. \tag{49}$$

The same result could be derived directly from a computation of the average flux across a vertical column of fluid $-\infty < y < \eta$ (see problem 3).

The presence of a mean drift is obvious from the observation of small vessels floating in waves, although here the mean motion of the vessel may be affected more by wind forces and other nonlinear effects such as surfing. Mass transport can result in the piling up of water at a beach, with an associated increase in the local mean depth. In some cases a long-shore current is associated with this phenomenon. This effect is discussed by Wiegel (1964). In shallow water, viscous effects significantly modify the mean drift flow, in a manner described by Longuet-Higgins (1953) and by Ünlüata and Mei (1970).

6.6 Superposition of Plane Waves

The plane progressive wave described in sections 6.2 and 6.3 is a single, discrete wave system, with a prescribed monochromatic component of frequency ω and wavenumber k, moving in the positive x-direction. This wave system can be generalized in several respects. First, the wave profile (7) rewritten in the form

$$\eta = A\cos(kx + \omega t), \tag{50}$$

corresponds to a plane progressive wave moving in the *negative* x-direction. The subsequent relations derived in sections 6.2 and 6.3 are unchanged, except that the phase function $(kx - \omega t)$ is replaced by $(kx + \omega t)$ throughout, and the signs of some expressions must be reversed.

More generally, in three dimensions, a plane wave moving in an arbitrary direction, at a polar angle θ in the horizontal x-z plane, can be analyzed by rotating the coordinate system appropriately. As shown in figure 6.6, the coordinate x' may be defined to coincide with the direction of wave propagation. Since $x' = x\cos\theta + z\sin\theta$, the appropriate generalization of (7) is given by

$$\eta = A\cos(kx\cos\theta + kz\sin\theta - \omega t), \tag{51}$$

with similar expressions for the velocity potentials (16) and (23). Note that for $\theta = \pi$, (51) reduces to (50).

In the linearized theory, solutions may be superposed without violating the boundary conditions or Laplace's equation. Thus considerable scope

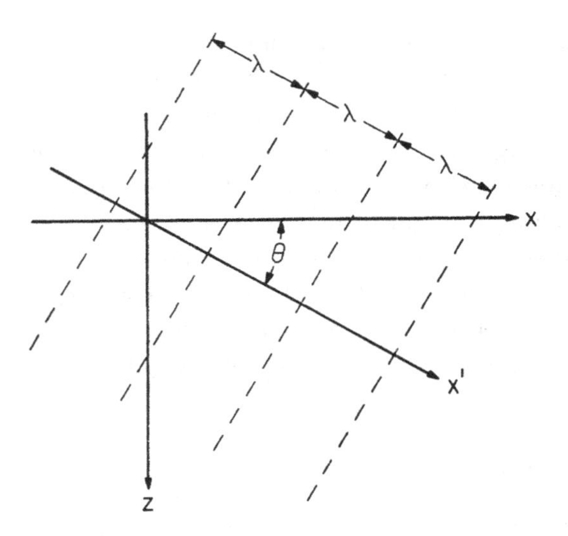

Figure 6.6
View from above of a plane progressive wave system, moving parallel to the x'-axis at an angle θ relative to x. The wave crests are shown by dashed lines.

for further generalizations is provided by the superposition of plane-wave solutions. The simplest example is a *standing* wave formed by adding two identical plane waves moving in opposite directions. The sum of these two wave systems is

$$\begin{aligned}\eta &= A\cos(kx-\omega t)+A\cos(kx+\omega t)\\ &= 2A\cos kx\cos\omega t,\end{aligned} \tag{52}$$

and the velocity potential for deep water is

$$\phi = -\frac{2gA}{\omega}e^{ky}\cos kx\sin\omega t. \tag{53}$$

In finite depth the corresponding result is obtained without difficulty from (23).

The standing wave (52) is sinusoidal in time, for fixed position x, and vice versa. This wave motion is therefore oscillatory but not progressive, as shown in figure 6.7(c); it is typical of the "sloshing" motions in closed containers such as swimming pools, tanks, and the wells of some drilling or research ships. If the container is rectangular of width w, with vertical walls at $x = 0$ and $x = w$, the boundary conditions of zero horizontal velocity on

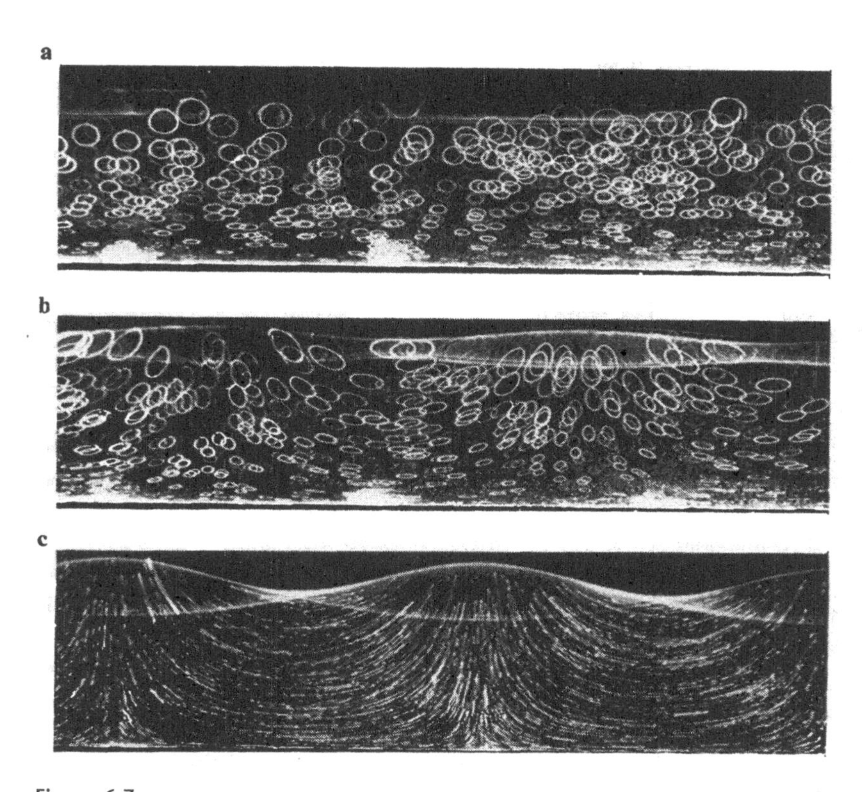

Figure 6.7
Particle trajectories in a plane progressive wave (a), a partial reflected wave (b), and a standing wave (c). These correspond respectively to a reflection coefficient of 0, 0.38, and 1.0 in equation (56). Note that the reflection coefficient can be measured from the maximum and minimum of the envelope, using (56). These photographs are based on time exposures, and are reproduced from a more extensive series of observations made by Ruellan and Wallet (1950).

these walls will be satisfied by (53) if $kw = n\pi$, with n any integer. Therefore a discrete spectrum of eigenfrequencies exist where standing waves are resonant in such a container.

Standing waves are also of physical relevance if the plane waves are incident upon a perfectly reflecting vertical wall, say at $x = 0$, and the solution (52–53) corresponds to the two-dimensional case where the wave crests are parallel to the wall. If the fluid domain is $x < 0$, the first term in (52) is the *incident* wave and the second term is the *reflected* wave. For oblique waves, the appropriate generalization follows from (51) in the form

$$\begin{aligned}\eta &= A\cos(kx\cos\theta + kz\sin\theta - \omega t) + A\cos(kx\cos\theta - kz\sin\theta + \omega t)\\ &= 2A\cos(kx\cos\theta)\cos(kz\sin\theta - \omega t).\end{aligned} \tag{54}$$

Note that the corresponding potential will satisfy the boundary condition of zero normal velocity on the wall $x = 0$. The reflected wave that has been added in (54) is consistent with Snell's law and with the method of images.

If plane waves are incident upon a beach, which can absorb a portion of the incident wave energy, or upon a vessel, which reflects part of the wave energy and transmits the remainder, the amplitude of the reflected wave will be reduced. In this case the reflected and incident waves in (52) or (54) will differ, and their ratio defines a reflection coefficient R. In the two-dimensional case, with the incident wave (7) traveling from $-\infty$, (52) is replaced by

$$\eta = A\,\mathrm{Re}[e^{-ikx+i\omega t} + Re^{ikx+i\omega t}]. \tag{55}$$

Here a complex notation has been used since the phase of the reflected wave may be nonzero, with the corresponding coefficient R complex. For a beach, the magnitude of the reflection coefficient is generally small.

The presence of small reflected waves in a wave tank is generally apparent in terms of the amplitude modulation of the wave system. Thus, (55) can be rewritten in the form

$$\eta = A\,\mathrm{Re}[e^{-ikx+i\omega t}(1 + Re^{2ikx})]. \tag{56}$$

The factor in parenthesis is the amplitude variation, which varies in x with twice the wavenumber or half the wavelength. This structure usually can be noted, by observing the change in amplitude of a particular wave crest. Figure 6.7(b) shows an exaggerated example. The reflection coefficient can be measured by recording the wave amplitude at two or more positions and analyzing the measured data in conjunction with (56).

More general wave motions, which are no longer monochromatic, can be obtained by superposing plane waves of different wavenumbers or frequencies. In the two-dimensional case, we begin by forming a discrete sum of waves of the form

$$\eta = \sum_{n=1}^{N} \mathrm{Re}[A_n \exp(-ik_n x + i\omega_n t)], \tag{57}$$

with a corresponding representation of the velocity potential (16) or (23) Here k_n denotes the wave number of the nth wave component, and ω_n its frequency. These two parameters must satisfy the appropriate dispersion relation (17) or (24) for each value of n. The amplitude A_n of each component is arbitrary and may be chosen to suit the desired solution, including the possibility that each A_n is complex with a different phase.

If the total number of discrete waves N tends to infinity, while the difference between adjacent frequencies and wave numbers reduces to zero, the sum (57) will tend to an integral over the continuous spectrum of frequencies,

$$\eta(x,t) = \text{Re}\int_{-\infty}^{\infty} A(\omega)\exp[-ik(\omega)x + i\omega t]\,d\omega. \tag{58}$$

Here $k(\omega) > 0$ must satisfy the appropriate dispersion relation (17) or (24); negative values of the frequency have been admitted in this representation, to allow for the possibility of plane waves moving in the negative x-direction, without explicitly writing out a second term of the form (50). Equation (58) can alternatively be expressed in an equivalent form as an integral over all possible wave numbers $-\infty < k < \infty$, with $\omega(k) > 0$ determined by the appropriate dispersion formula.

These distributions of wave systems can be extended from two to three dimensions by introducing the oblique wave (51) and summing or integrating over all possible wave directions θ. In the most general case of a continuous distribution, a two-dimensional integral representation follows of the form

$$\eta(x,z,t) = \text{Re}\int_{0}^{\infty} d\omega \int_{0}^{2\pi} d\theta\, A(\omega,\theta)\exp[-ik(\omega)(x\cos\theta + z\sin\theta) + i\omega t]. \tag{59}$$

This expression will be used to analyze ship waves in section 6.10. It also can be utilized, with $A(\omega, \theta)$ a random variable, to represent a spectrum of ocean waves. The latter application will be discussed in section 6.20.

The integral in (58) is recognized as a Fourier integral, and (59) is a two-dimensional integral. Thus a large class of physically relevant wave systems can be represented, and the inversion formulas from Fourier theory can be used to find the appropriate amplitude functions $A(\omega)$ or $A(\omega, \theta)$ required to generate a prescribed wave elevation η. This procedure can be utilized to solve the *Cauchy-Poisson* problem for the wave motion resulting from a transient initial disturbance such as would result by throwing a pebble into a pond. The solution of this interesting problem is derived by Lamb (1932), Stoker (1957), and Wehausen and Laitone (1960).

6.7 Group Velocity

Distributions of wave systems, such as those represented by the sum (57) or the integrals (58–59), are difficult to describe in simple terms. Since the component waves each travel with a different phase velocity, a continuously changing wave pattern results. Nevertheless, if we consider a narrow band of the component waves, with nearly equal wavelength and direction, a characteristic of the resulting distribution is that the waves travel in a group. The *group velocity* V_g can be derived from a dynamic analysis of energy flux, but a simpler approach to this subject follows from a purely kinematic study of the group of waves formed by two nearly equal plane waves.

Two adjacent components of the discrete spectrum (57) can be written in the form

$$\begin{aligned}\eta &= \mathrm{Re}\{A_1 e^{-ik_1x+i\omega_1t} + A_2 e^{-ik_2x+i\omega_2t}\}\\ &= \mathrm{Re}\left\{A_1 e^{-ik_1x+i\omega_1t}\left[1+\frac{A_2}{A_1}e^{-i\delta kx+i\delta\omega t}\right]\right\},\end{aligned} \tag{60}$$

where $\delta k = k_2 - k_1$ and $\delta\omega = \omega_2 - \omega_1$. Here, as with the partial reflected wave in (56), the factor in brackets represents an amplitude modulation; but unlike (56), this factor will be slowly varying in space *and* time. The wave form (60) is similar to the beat-frequency effect in electromagnetic wave motions, where one refers to the first exponential factor in (60) as the *carrier* wave.

This type of wave motion is illustrated in figure 6.8, which shows a group of carrier waves enclosed by a slowly-varying envelope. The wavelength and period of the carrier are the usual parameters $\lambda = 2\pi/k_1$ and $T = 2\pi/\omega_1$. By analogy, the wavelength of the group is $2\pi/\delta k$, and its period in time is $2\pi/\delta\omega$. Of particular interest is the group velocity V_g, given by the ratio

$$V_g = \frac{\delta\omega}{\delta k}. \tag{61}$$

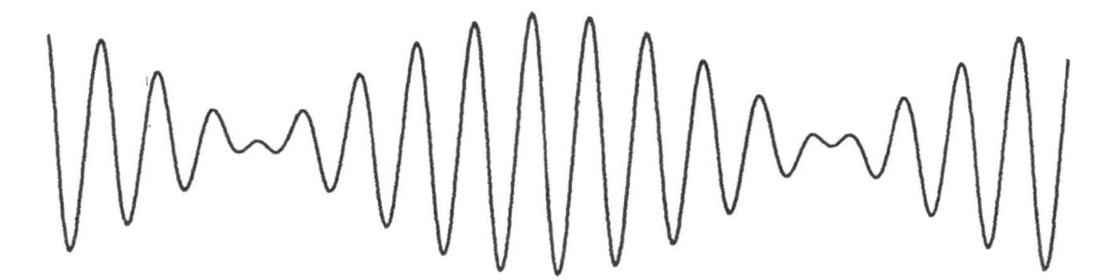

Figure 6.8
Wave group resulting from the superposition of two nearly equal plane waves as in (60). The individual waves travel with the phase velocity, while the envelope travels with the group velocity (61).

In the limiting case where $\delta\omega \to 0$, $\delta k \to 0$, (60) will tend to a monochromatic plane wave, but if x and t are large, the amplitude modulation in (60) will persist so long as the products $x\delta k$ and $t\delta\omega$ are finite. Under this circumstance the group velocity (61) will approach the finite limit

$$V_g = \frac{d\omega}{dk}, \tag{62}$$

in accordance with the classical definition of a derivative. Thus, a group of waves of nearly equal frequency and wavenumber will propagate with the velocity (62), as opposed to the phase velocity $V_p = \omega/k$ of the individual wave components in the group.

In general, the phase and group velocities differ, unless $\omega/k = d\omega/dk$, which will occur only if the frequency and wavenumber are directly proportional and the phase velocity is constant. This exception occurs in the shallow-water limit $kh \ll 1$.

In deep water, the group velocity can be evaluated by solving the dispersion relation (17) for $\omega(k)$ and substituting this in (62). Thus

$$V_g = \tfrac{1}{2}(g/k)^{1/2} = \tfrac{1}{2}\omega/k = \tfrac{1}{2}V_p, \tag{63}$$

and the group velocity is precisely half of the phase velocity. In finite depth, the dispersion relation (24) can be combined with (62) to give

$$V_g = \left(\frac{1}{2} + \frac{kh}{\sinh 2kh}\right)V_p. \tag{64}$$

This expression reduces to (63) if kh is large, whereas in the shallow-water limit the factor in parenthesis in (64) reduces to unity. The value of the group velocity is plotted in figure 6.9, as a ratio of the infinite-depth phase velocity. This figure shows that the group velocity is less than the phase velocity; their ratio decreases from one to one-half as the water depth increases. The curve in figure 6.9 also shows that for fixed frequency and increasing depth, the group velocity attains a maximum value somewhat greater than its infinite-depth limit, at an intermediate depth where h/λ is about 0.2.

That individual waves move faster than the group implies that in a wave system with a *front*, propagating into otherwise calm water, the individual waves will overtake the front and vanish. This phenomenon can be observed in a wave tank by starting the wavemaker from rest at a

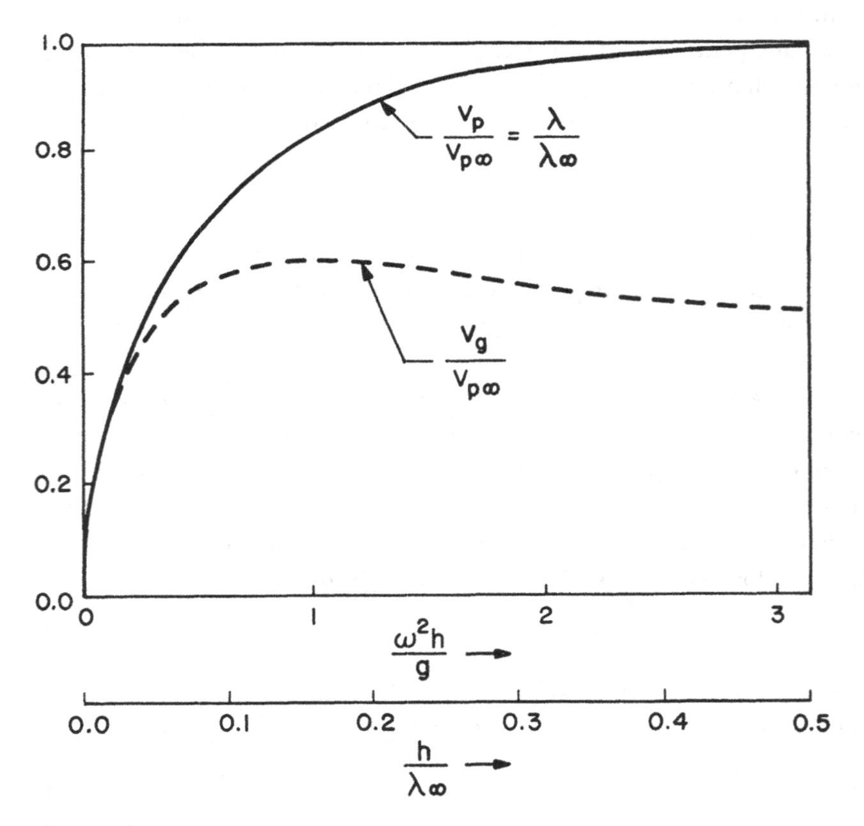

Figure 6.9
Phase velocity and group velocity ratios for a plane progressive wave as a function of the depth.

given frequency so that a group of waves will propagate down the tank as shown in figure 6.10. The front of the group moves at a slower velocity than the individual wave crests behind it, and these crests appear to vanish as they approach the front. After the wavemaker is turned off, the "back" of the wave system moves with the group velocity; thus "new" waves that move with the phase velocity must arise from calm water at the back. At first glance it appears that the energy within the wave system is destroyed in some sense, contrary to the principal of conservation of energy. In fact, there is no contradiction, since the wave energy moves not with the phase velocity but with the group velocity. To confirm this statement requires an analysis of the wave energy and an alternative approach to the group velocity based upon the rate of energy flux in a plane progressive wave system.

6.8 Wave Energy

Water waves are the result of a balance between kinetic and potential energy in the fluid, and the total energy is the sum of these two components. Thus, in a prescribed volume $\mathcal{V}$, the total energy is given by the integral

$$KE + PE = \rho \iiint_{\mathcal{V}} \left(\frac{1}{2} V^2 + gy \right) d\mathcal{V}. \tag{65}$$

First, we will focus attention on a vertical column, extending throughout the depth of the fluid and bounded above by the free surface; the energy density E, per unit area of the mean free surface above this column, is given by

$$E = \rho \int_{-h}^{\eta} \left(\frac{1}{2} V^2 + gy \right) dy = \frac{1}{2} \rho \int_{-h}^{\eta} V^2 dy + \frac{1}{2} \rho g (\eta^2 - h^2). \tag{66}$$

The potential energy $-\frac{1}{2}\rho g h^2$ of the fluid below the equilibrium plane $y = 0$ is of no interest, since this is a constant unrelated to the wave motion. (Moreover, this term is a source of embarrassment in the limit $h \to \infty$!) Therefore this contribution to (66) is omitted from the subsequent equations.

For plane progressive waves in infinite depth, the energy density can be computed using the results of section 6.2. Substituting (19–20) in the last integral in (66) and carrying out this integration over the depth yields

Figure 6.10
Sequence of photographs showing a plane progressive wave system advancing into calm water. The water is darkened with dye, and the lower half of the water depth is not shown. The wave energy is contained within the heavy diagonal lines, and propagates with the group velocity. (The boundaries of the wave group diffuse slowly with time, due to dispersion.) The position of one wave crest is connected in successive photographs by the light line, which advances with the phase velocity. Each wave crest moves with the phase velocity, equal to twice the group velocity of the boundaries. Thus each wave crest vanishes at the front end and, after the wavemaker is turned off, arises from calm water at the back. The interval between successive photographs is 0.25s and the wave period is 0.36s. The wavelength is 0.23m and the water depth is 0.11m.

$$E = \frac{\rho\omega^2 A^2}{4k} e^{2k\eta} + \frac{1}{2}\rho g \eta^2. \tag{67}$$

For small-amplitude waves, the energy density (67) is proportional to the square of the wave amplitude A, and to this order the exponential factor in (67) may be set equal to one. Equivalently, (66) shows that the kinetic energy is $O(A^2)$, and the contribution to the last integral between $y = 0$ and $y = \eta$ will give a higher-order contribution proportional to A^3. Using the dispersion relation (17) in the first term of (67) and substituting (7) for the free-surface elevation we obtain

$$E = \tfrac{1}{4}\rho g A^2 + \tfrac{1}{2}\rho g A^2 \cos^2(kx - \omega t). \tag{68}$$

The first and second terms on the right side of (68) represent the kinetic and potential energies, respectively. If these are averaged with respect to time over one cycle of the wave motion, their contributions are identical, and the total mean energy density is given by the expression

$$\bar{E} = \tfrac{1}{2}\rho g A^2, \tag{69}$$

where a bar denotes the time average.

For finite depth the corresponding relations can be derived with somewhat more effort. The expression for the potential energy in (68) is unchanged, and the mean value of the kinetic energy is identical to the first term in (68). Thus (69) holds in general, along with the conclusion that the mean energy is divided equally between kinetic and potential contributions (see problem 4).

To analyze the rate of energy flux, we consider the rate of change with respect to time of the total energy (65). Allowing the boundary surface S of the volume $\mathcal{V}$ to move with the velocity U_n and using the transport theorem (3.11) gives

$$\begin{aligned} \frac{dE}{dt} &= \rho \frac{d}{dt} \iiint_{\mathcal{V}} \left(\frac{1}{2}V^2 + gy\right) d\mathcal{V} = \rho \iiint_{\mathcal{V}} \frac{\partial}{\partial t}\left(\frac{1}{2}V^2 + gy\right) d\mathcal{V} \\ &\quad + \rho \iint_S \left(\frac{1}{2}V^2 + gy\right) U_n \, dS. \end{aligned} \tag{70}$$

Since the vertical coordinate y is independent of time, the only contribution to the integrand of the last volume integral is from the kinetic energy term, which takes the form

$$\frac{\partial}{\partial t}\left(\frac{1}{2}\nabla\phi\cdot\nabla\phi\right)=\nabla\phi\cdot\nabla\frac{\partial\phi}{\partial t}=\nabla\cdot\left(\frac{\partial\phi}{\partial t}\nabla\phi\right)-\frac{\partial\phi}{\partial t}\nabla^2\phi. \tag{71}$$

The last term vanishes by Laplace's equation, and the remaining contribution to the volume integral can be evaluated by the divergence theorem. Thus

$$\iiint_{\mathcal{V}}\frac{\partial}{\partial t}\left(\frac{1}{2}V^2\right)d\mathcal{V}=\iint_{S}\frac{\partial\phi}{\partial t}\frac{\partial\phi}{\partial n}dS. \tag{72}$$

Substituting this result in (70) gives the time rate of change of energy in the form

$$\frac{dE}{dt}=\rho\iint_{S}\left\{\frac{\partial\phi}{\partial t}\frac{\partial\phi}{\partial n}+\left(\frac{1}{2}V^2+gy\right)U_n\right\}dS. \tag{73}$$

Substituting Bernoulli's equation (3), we can recast (73) in the alternative form

$$\frac{dE}{dt}=\rho\iint_{S}\left\{\frac{\partial\phi}{\partial t}\frac{\partial\phi}{\partial n}-\left(\frac{p-p_a}{\rho}+\frac{\partial\phi}{\partial t}\right)U_n\right\}dS. \tag{74}$$

Once again we restrict our attention to a volume of fluid, with vertical sides, bounded above by the free surface and below by the bottom. Physically no contribution to (74) from the last two surfaces is anticipated, and this can be confirmed by noting that $\partial\phi/\partial n = U_n = 0$ on the bottom, whereas on the free surface $\partial\phi/\partial n = U_n$ and $p = p_a$. Thus the energy flux occurs across the vertical control surface bounding the fluid column.

Equation (74) holds with U_n an arbitrary normal velocity of the vertical surface of the column. If the column moves with constant velocity U, in the direction of the wave propagation, energy must enter one side of the column at the same mean rate as it leaves on the other, as shown in figure 6.11. Thus the mean rate of energy flux across any vertical control surface x = constant, per unit width in the z-direction, is given by

$$\overline{\frac{dE}{dt}}=\overline{\rho\int_{-h}^{\eta}\left\{\frac{\partial\phi}{\partial t}\frac{\partial\phi}{\partial x}+\left(\frac{1}{2}V^2+gy\right)U\right\}dy}, \tag{75}$$

where U is the velocity of this control surface in the + x-direction.

If (75) is set equal to zero, U must be equal to the mean velocity of the energy flux in the fluid. On this basis it follows that

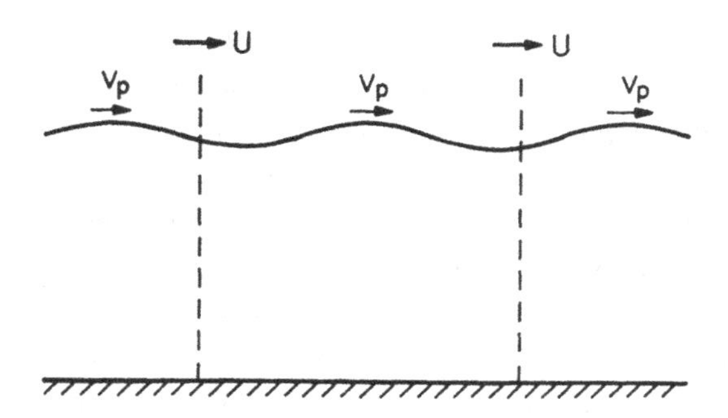

Figure 6.11
Vertical control surfaces used to measure the rate of energy flux. Since the fluid domain bounded by these control surfaces is of constant width, the average energy within this domain must be constant and the rate at which energy crosses each vertical surface must be equal to (75).

$$U = -\rho \overline{\int_{-h}^{\eta} \frac{\partial \phi}{\partial t} \frac{\partial \phi}{\partial x} dy} \Big/ \rho \overline{\int_{-h}^{\eta} \left(\frac{1}{2} V^2 + gy \right) dy}. \tag{76}$$

The denominator is the energy density (66), which can be evaluated from (69). As in (66–68), the contribution of the interval $0 < y < \eta$ to the integral in the numerator of (76) is of third order in A, and the leading-order contribution of order A^2 can be computed with the upper limit set equal to zero. On this basis, using the infinite-depth potential (16) gives

$$\begin{aligned} U &= -\rho \int_{-\infty}^{0} \overline{\frac{\partial \phi}{\partial t} \frac{\partial \phi}{\partial x}} dy \Big/ \frac{1}{2} \rho g A^2 = \frac{2gk}{\omega} \int_{-\infty}^{0} e^{2ky} \overline{\cos^2(kx - \omega t)} dy \\ &= \tfrac{1}{2} g / \omega = \tfrac{1}{2} V_p, \end{aligned} \tag{77}$$

where (18) has been used to substitute the phase velocity V_p. Comparing this result with (63), we see that the energy's velocity U is equal to the group velocity V_g. The same conclusion holds in the general case of finite depth h, as can be confirmed by substituting (23) for the potential in the numerator of (77).

The mean rate of energy flux across a *fixed* control surface is the product of the energy density and the group velocity,

$$\overline{\frac{dE}{dt}} = V_g \overline{E}. \tag{78}$$

This result can be calculated directly from (75) with $U = 0$.

Equation (78) can be used in conjunction with the energy-density (69) to predict the change in height of a sinusoidal wave as it propagates through a region of gradually changing depth. Provided the change in depth is small over distances on the order of a wavelength, these relations can be applied locally at each depth, and it can be assumed moreover that the reflected energy is negligible. It then follows that the energy-flux (78) is constant for all depths, and since the frequency ω remains constant, the local wave height must be inversely proportional to the square root of the group velocity. Referring to figure 6.9, the *shoaling coefficient* can be deduced as the inverse square root of the group velocity shown by the dashed line. Simultaneously, the wavelength and phase velocity will decrease with depth (see the solid line in figure 6.9). Taking both effects into account, the steepness of the waves will increase as the waves near the beach, leading ultimately to wave breaking when the wave height is comparable to the local depth.

This description of two-dimensional waves incident upon a beach is based on the assumption of no energy reflection, which is valid provided the bottom slope is small. Figure 6.4 illustrates this situation on a small scale, with the bottom slope of the wave tank inclined at an angle of 2.5 degrees. The increasing steepness of the waves is apparent, but the change in wave amplitude is not obvious. This can be explained from figure 6.9 by noting that the shoaling coefficient exceeds 1.0 only if the local depth is less than 0.06 times the deep-water wavelength. Thus, for virtually the entire length of the wave tank in figure 6.4, the shoaling coefficient is less than one and the wave amplitude is attenuated.

This dynamic analysis of energy flux provides an alternative interpretation of the group velocity, complementary to that based on the kinematic argument in section 6.7. From the physical viewpoint the equivalence of these results is not surprising, particularly since each group of waves shown in figure 6.8 is bounded by a region of relatively small wave amplitudes and correspondingly small amounts of energy. If no energy is transferred between adjacent groups, then the energy must travel with the velocity of the group. A similar argument holds for the propagation of the front in figure 6.10, between a region of a calm water and a system of plane waves; in this situation, no wave energy can be transferred across the front.

Since the group velocity is less than the phase velocity, the energy in a plane progressive wave system will be propagated at a slower speed than the individual wave crests and will be "left behind" in the fluid. This seems physically reasonable and explains why ship waves are observed downstream of the disturbance that generates them. There are exceptions to this situation, however. For very short ripples governed by surface tension, the group velocity *exceeds* the phase velocity and the resulting waves occur *upstream* of the disturbance. This situation can be observed by slowly moving a suitable obstacle in a basin of water or by observing the waves generated by a calm stream flowing past a fish line. For the latter reason, this is known as the *fish-line problem*; details may be found in Lamb (1932). Small capillary waves will occur upstream of large bodies as well, but usually their presence is obscured and of no practical significance.

Alternative derivations and interpretations of the group velocity exist. A particularly elegant approach is given by Wehausen and Laitone (1960), based on the propagation of a packet of wave energy in a narrow-band spectrum. An interpretation based on the kinematic analysis of a general slowly-varying wave motion is emphasized by Whitham (1974). In section 6.11, we shall find that the group velocity emerges in yet another, but related sense, in conjunction with Kelvin's method of stationary phase for approximating the far-field asymptotic form of wave motions.

6.9 Two-Dimensional Ship Waves

Having developed the analysis of wave energy radiation for plane progressive waves, we can now consider the problem of wave generation by moving vessels and the associated wave resistance. We begin with the simplest case, the two-dimensional motion that would be generated by moving a long cylindrical obstacle normal to its axis.

Since our approach will be based on energy conservation, we can ignore the details of the flow near the body and focus attention on the wave system far downstream. If the body moves with constant velocity U in otherwise calm water and if the motion has reached a steady state, the only waves that can exist downstream move with a phase velocity $V_p = U$. Any other waves would either overtake the body, or drop further behind, in an unsteady manner. Since the phase velocity is fixed, the wavelength and the wave number must take prescribed values; for the deep-water case they are given by $\lambda = 2\pi U^2/g$ and $k = g/U^2$.

The waves generated by the body contain energy that must be imparted to the fluid as work done by the body on the fluid. Thus, the body will experience a drag force D due to its wave resistance; the purpose of our analysis is to relate D to the wave amplitude A far downstream.

Figure 6.12 shows a sketch of this situation in a fixed frame of reference, as well as in a reference frame moving with the body. The fixed system shown in (a) pertains to our analysis of the energy density and flux. The values of these quantities are indicated in the figure in terms of the wave amplitude A. The body and waves move with velocity U in the + x-direction, and thus there is a positive flux of energy across the fixed control surface downstream.

Since the control surface is fixed and the body is moving, the length of the fluid region between the two will increase with velocity U, and the total energy in this region will increase at a rate equal to the product of U and the

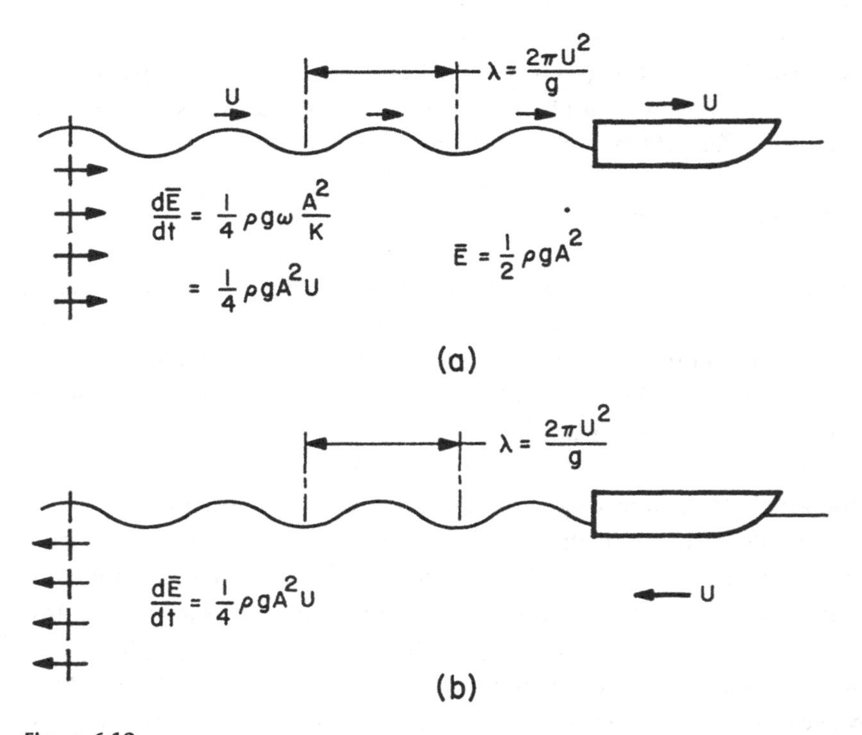

Figure 6.12
Two-dimensional ship waves as viewed by a fixed observer (a), and in a reference frame moving with the ship (b).

energy density $\frac{1}{2}\rho g A^2$. The energy input necessary to balance this increase results in part from the work done at a rate DU by the body, in opposition to the wave drag D. In accordance with (78), energy also enters the fluid region across the control surface downstream, at a rate equal to the product of the energy density and group velocity $\frac{1}{2}U$. From energy conservation it follows that

$$\tfrac{1}{2}\rho g A^2 U = DU + \tfrac{1}{4}\rho g A^2 U. \tag{79}$$

Solving this equation for the drag D yields

$$D = \tfrac{1}{4}\rho g A^2. \tag{80}$$

The same conclusion follows from the alternative analysis based on the moving reference frame of figure 6.12(b). Here the position of the control surface is fixed relative to the body, and the total energy of the intermediate fluid region is constant. The rate of work done by the body on the fluid is unchanged by the transformation to a moving coordinate system, but across the downstream control surface the energy now moves at the relative velocity $V_g - U = -\frac{1}{2}U$. If the latter is multiplied by the energy density, the energy balance takes the form

$$DU - \tfrac{1}{4}\rho g A^2 U = 0, \tag{81}$$

which is equivalent to (79) and (80).

Thus, if a two-dimensional body generates waves of amplitude A because of its steady motion on the free surface, the associated wave drag is $\frac{1}{4}\rho g A^2$. This result by itself does not provide a means for predicting the wave resistance, however, since we do not have a method for calculating the wave amplitude A generated by the body. Physically, the latter quantity will depend on the body geometry, as well as the velocity U. One might expect that increasing the body size or the velocity will increase A and thus the wave resistance. However, the latter conclusion overlooks the possibilities of destructive interference, which may be emphasized with a simple example.

Let us consider a single small disturbance, which generates waves of amplitude a. The resulting free-surface profile, in a coordinate system moving with the disturbance, will have the form

$$\eta = a\cos(kx + \varepsilon), \tag{82}$$

where $k = g/U^2$ and the phase angle ε is unimportant. From (80), the wave resistance of this disturbance by itself is given by $\frac{1}{4}\rho g a^2$.

Next we consider the result of superposing a second disturbance, of equal but opposite magnitude, situated at a distance l downstream of the first. The total free-surface elevation resulting from this superposition of two disturbances is

$$\begin{aligned}\eta &= a\cos(kx+\varepsilon) - a\cos(kx+\varepsilon+kl)\\ &= \mathrm{Re}\{ae^{i(kx+\varepsilon)}(1-e^{ikl})\}.\end{aligned} \tag{83}$$

The total wave amplitude downstream of *both* disturbances is

$$A = a|1-e^{ikl}| = 2a|\sin(\tfrac{1}{2}kl)|, \tag{84}$$

and the associated wave resistance is given by

$$D = \rho g a^2 \sin^2(\tfrac{1}{2}kl). \tag{85}$$

The amplitude (84) and drag (85) are oscillatory functions of the velocity and length; the drag varies between zero and four times the corresponding value for a single disturbance.

If the Froude number $F = U/(gl)^{1/2}$ is introduced and (85) is plotted as shown in figure 6.13, the importance of interference effects is obvious, especially for the lower speeds. For spacing l equal to an integer number of wavelengths, the waves and resulting wave resistance are zero, whereas for spacings of an integer plus a half times the wavelength, the disturbance is a maximum.

This simplified example can be related in practice to the bow and stern wave systems of a more realistic vessel, particularly if the bow flow is represented by a point source and the stern by a point sink, as in the Rankine bodies of chapter 4. To the extent that this analogy is valid it can be presumed that the length l is the distance between the source and sink, which will be less than the overall length of the body. Similar interference effects between the bow and stern exist in the three-dimensional case, but with reduced magnitude, as we shall see in sections 6.12 and 6.13.

Analogous results follow if two positive disturbances are superposed. Thus, if a pair of two-dimensional vessels moves in tandem, with one in the other's wake, the total wave resistance of the pair will fluctuate as in figure 6.13 but with a shift of $\frac{1}{2}\lambda$ in the definition of l. However, the drag in question is the sum of the individual components acting on each vessel,

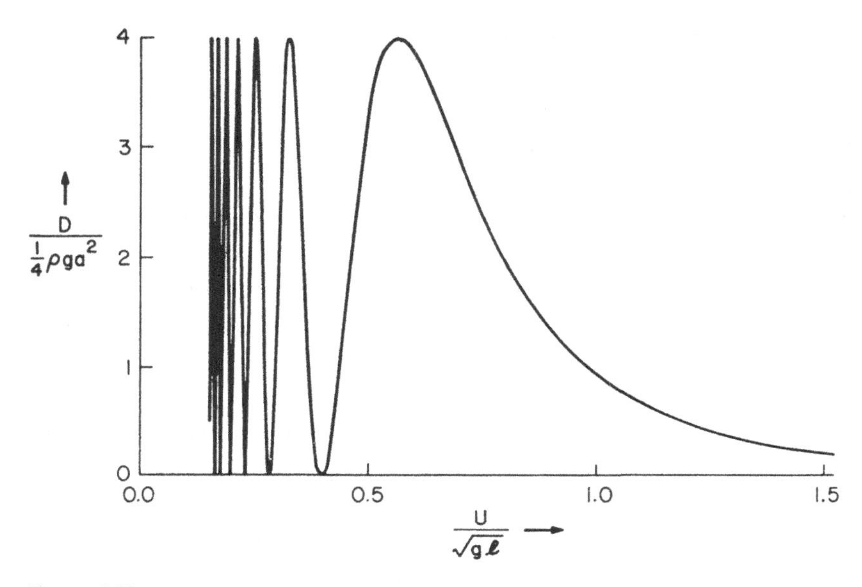

Figure 6.13
Wave drag of a pair of two-dimensional disturbances. The strength of these is equal and opposite, and their separation distance is *l*.

and vanishing of the total drag does not imply that each vessel experiences no drag. What in fact occurs is that the upstream vessel experiences a drag, and the downstream vessel an equal and opposite thrust force, much like a surfboard riding on a wave system generated in this case by the upstream vessel. Similar considerations apply to convoys of ships in three dimensions, but with reduced interference as noted above.

These simplified results are independent of the body geometry that produces the assumed waves and apply equally whether the body is floating on the free surface or submerged. The simplest specific example, where the wave amplitude A and wave resistance are related explicitly to the body shape and speed, is a submerged circular cylinder; an approximate theory for this case is outlined by Lamb (1932). The case of a planing body is of particular interest; in many respects it resembles the flow beneath a lifting surface, but with the added complications of the free surface. More extensive results for a wide variety of floating and submerged bodies are surveyed by Wehausen (1973) and in other references listed in section 6.13.

6.10 Three-Dimensional Ship Waves

Two-dimensional results can be generalized to more realistic three-dimensional motions if the single plane wave system in the wake, shown in figure 6.12, is replaced by a suitable distribution of waves moving at all possible oblique angles θ. The most general wave distribution in three dimensions is given by the double integral (59). If this is transformed to a reference system moving with the ship in the positive x-direction with velocity U, the appropriate expression with x replaced by $x + Ut$ is given by

$$\eta(x,z,t) = \mathrm{Re}\int_0^{\infty} d\omega \int_0^{2\pi} d\theta\, A(\omega,\theta) \exp[-ik(x\cos\theta + z\sin\theta) + i(\omega - kU\cos\theta)t]. \quad (86)$$

Here $k(\omega)$ is the wavenumber corresponding to a given frequency ω in accordance with the dispersion relation (17) for infinite depth or (24) for finite depth. As in section 6.9, we shall treat only the former possibility.

If the motion is steady state in the reference system which moves with the ship, the expression (86) must be independent of time. As in the corresponding two-dimensional case, this implies a restriction on the phase velocity of the waves and thus on the wavenumber. In the present situation, (86) will be independent of time provided

$$kU\cos\theta - \omega = 0; \quad (87)$$

this is equivalent to stipulating that the phase velocity of each admissible wave component is given by

$$V_p = \frac{\omega}{k} = U\cos\theta. \quad (88)$$

In this sense (see figure 6.6), a system of plane progressive waves moving at an oblique angle θ with respect to the x-axis will appear steady state to an observer moving along the x-axis with velocity $V_p \sec\theta$.

The restriction (87) can be used to eliminate one of the variables of integration in (86). Retaining the wave angle θ, and noting that (87–88) require that $\cos\theta > 0$, we can replace (86) by the single integral

$$\eta(x,z) = \mathrm{Re}\int_{-\pi/2}^{\pi/2} d\theta\, A(\theta) \exp[-ik(\theta)(x\cos\theta + z\sin\theta)]; \quad (89)$$

for the deep-water case, from (17) and (87),

$$k(\theta) = g / U^2 \cos^2 \theta. \tag{90}$$

The corresponding expression for finite depth follows from (24), in combination with (87). In either case, the velocity potential corresponding to (89) is of the same general form and can be inferred from (16) or (23).

The expression (89) is known as the *free-wave distribution* of a given ship. This is characterized by the amplitude and phase of the complex function $A(\theta)$; here $A(\theta)$ takes a role analogous to that of the constant A in section 6.9. Thus energy arguments can be used to express the wave resistance in terms of the function $A(\theta)$, which depends on the shape of the ship hull and its forward velocity. This reduction will be deferred to section 6.12.

If the distance downstream from the position of the ship is very large, the integral (89) can be simplified and the classical ship-wave pattern will be obtained, as derived by Kelvin in 1887. Strictly speaking this asymptotic approximation should be based upon the stationary-phase approximation of section 6.11. However, the Kelvin wave pattern is of sufficient interest that a heuristic argument will be used in the interim, based on the concept of the group velocity.

We recall from section 6.7 that in two dimensions wave groups travel with the group velocity $V_g = d\omega/dk$. Since the free-wave distribution (89) appears locally as a slowly varying two-dimensional wave system, the same result should hold here. Thus if x' is a local coordinate, normal to the wave crests as in figure 6.6 and defined with respect to a fixed reference frame, the significant waves will travel in groups such that $x'/t = V_g$ is determined by

$$\frac{d}{dk}(kx' - \omega t) = 0. \tag{91}$$

If this expression for the derivative of the phase of the local waves is transformed into the frame of reference of (89), the significant waves will satisfy the equation

$$\frac{d}{dk}[k(\theta)(x\cos\theta + z\sin\theta)] = 0. \tag{92}$$

Since k and θ are related by (90), the vanishing of (92) is equivalent to the condition

$$\frac{d}{d\theta}\left(\frac{x\cos\theta + z\sin\theta}{\cos^2\theta}\right) = 0, \tag{93}$$

except for the isolated points $\theta = 0$, $\pm\pi/2$ where $d\theta/dk$ is infinite or zero, respectively. Ignoring these singular points and evaluating the derivatives in (93) yields

$$x\sec^2\theta\sin\theta + z\sec^3\theta(1+\sin^2\theta) = 0. \tag{94}$$

Thus, the significant waves, moving in a direction θ, will be situated along the radial line

$$\frac{z}{x} = -\frac{\cos\theta\sin\theta}{1+\sin^2\theta}. \tag{95}$$

A plot of (95) is shown in figure 6.14, and it is apparent that the maximum value of the ratio z/x is defined by

$$z\,/\,x = 2^{-3/2} \simeq \tan(19°28'), \tag{96}$$

this maximum occurs when the wave angle is

$$\theta = \pm\sin^{-1}(1/\sqrt{3}) = \pm 35°16'. \tag{97}$$

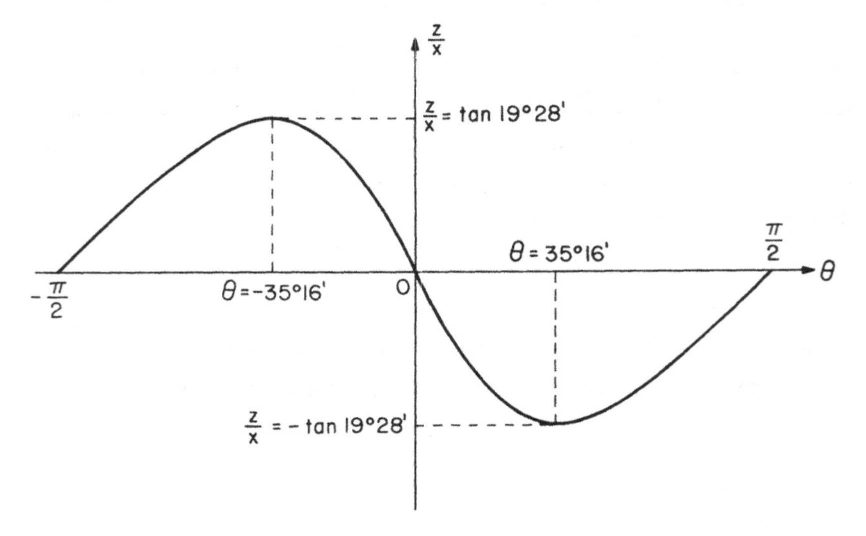

Figure 6.14
Plot of equation (95).

Thus, the waves are confined to a sector symmetrical about the negative x-axis, with included semiangles of $19\frac{1}{2}°$. Within this sector, for a given point (x, z) and corresponding ratio z/x, two solutions of (95) exist; hence two waves with distinct angles θ move in different directions. By requiring that the phase function in parentheses in (93) be a constant, we may calculate the loci of the crests as shown in figure 6.15.

The waves on the negative x-axis move in the same direction as the ship, with $\theta = 0$ as in the two-dimensional case. As the lateral coordinate z is increased, the angle of these *transverse* waves changes, in accordance with (95) and figure 6.14, reaching a value of ±35° on the boundaries. Larger values of θ correspond to the *diverging* waves; these have shorter wavelength and converge toward the origin as shown in figure 6.15. On the boundaries where these two wave systems meet with a common angle of ±35°, they differ in phase by a quarter wavelength. This particular detail will be explained from the method of stationary phase.

The wave systems generated by real ships are shown in two aerial photographs, figures 6.16 and 6.17. In both cases, the qualitative features of the Kelvin wave pattern are confirmed. Figure 6.17, taken directly above the wake of the ship, permits a quantitative confirmation of the two angles noted, although in this case the transverse waves are not obvious. Figure 6.17 also shows a commonly observed feature, namely that the apex of the sectors containing the Kelvin waves is displaced upstream from the ship's bow by an amount typically as large as one ship length.

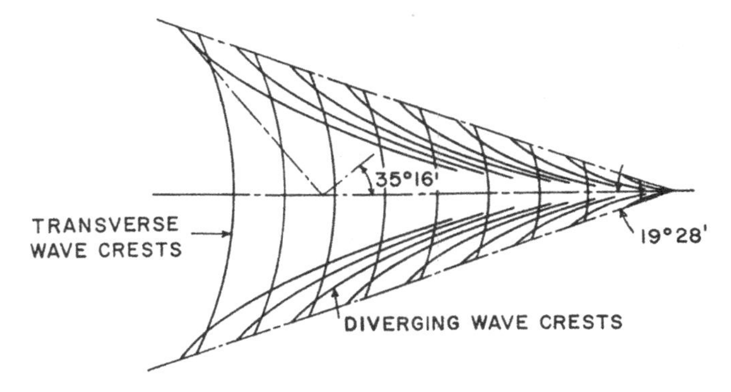

Figure 6.15
The Kelvin ship-wave pattern.

Figure 6.16
Kelvin ship waves as observed in an aerial photograph. The transverse waves can be seen in this photograph, although the steeper diverging waves appear to be dominant. Note the nodal lines which divide the diverging waves along two or three radial lines, with 180° phase shifts across these lines. (Courtesy of U. S. Navy)

A departure from the linearized potential theory that appears in figures 6.16 and 6.17 is a pronounced breaking wave, originating at the bow and trailing aft on either side of the ship into the wake. The effect of this breaking wave on the resistance of the ship, a particularly interesting topic in contemporary naval architecture, will be discussed in section 6.14.

6.11 The Method of Stationary Phase

A more mathematical approach to the Kelvin ship-wave system can be based upon the method of stationary phase, developed originally by Kelvin

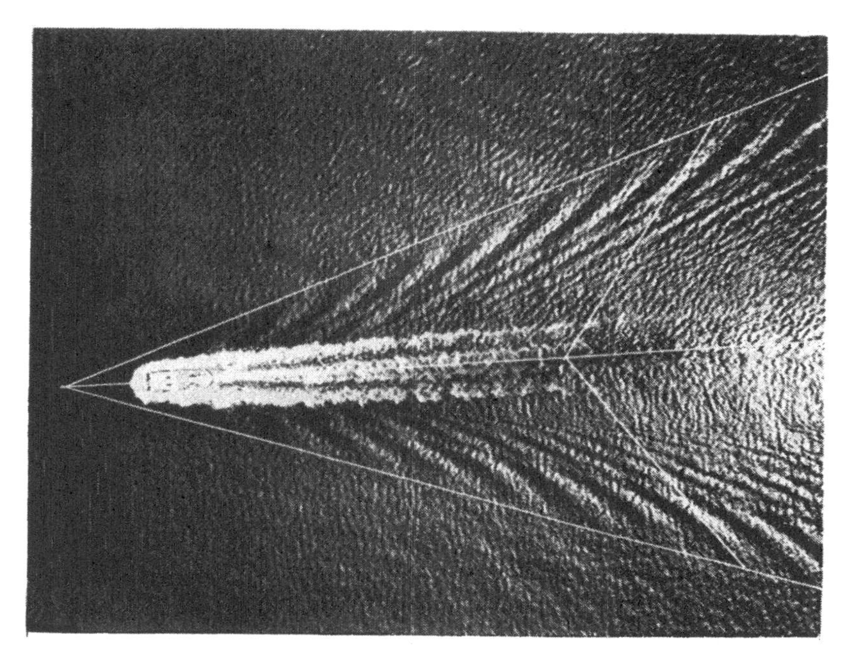

Figure 6.17
Aerial photograph of ship waves with the angles $19\frac{1}{2}°$ and 35° superposed. A single nodal line can be detected on each side by the 180° phase shift as in figure 6.16. (From Newman 1970)

for the same purpose. This will furnish an alternative viewpoint to the heuristic argument based on the group velocity and will provide the necessary information regarding the wave amplitude for subsequent use in analyzing the energy flux and wave resistance.

An asymptotic approximation is desired for the integral (89), based on the assumption that the distance from the origin is very large compared to the wavelength. To derive the stationary-phase approximation in a more general context, let us consider the integral

$$I = \int F(\theta) e^{iRG(\theta)}\, d\theta, \tag{98}$$

where $F(\theta)$ and $G(\theta)$ are arbitrary regular functions and R is a large parameter such as the polar radius. The function F may be complex, but G will be assumed real. For large values of R, the phase RG will vary rapidly, and the integrand of (98) will be a highly-oscillatory function. The resulting oscillations of the integrand will tend to cancel out over the range of integration, except locally at points of *stationary phase* where the derivative $G'(\theta)$ vanishes.

Thus the principal contribution(s) to the integral (98) will come from the point(s) of stationary phase, assuming that such points exist in the range of integration. To evaluate the contribution to (98) from the vicinity of such a point, say at $\theta = \theta_0$, the function $G(\theta)$ may be expanded in a Taylor series of the form

$$G(\theta) = G(\theta_0) + \tfrac{1}{2}(\theta - \theta_0)^2 G''(\theta_0) + \cdots. \tag{99}$$

Here G'' is the second derivative, and the term involving the first derivative has been deleted since, by definition,

$$G'(\theta_0) = 0. \tag{100}$$

Substituting (99) in the integral (98), the contribution from the vicinity of this point will be given by

$$I \simeq \int_{0_0-\varepsilon}^{\theta_0+\varepsilon} d\theta\, F(\theta) \exp\{iR[G(\theta_0) + \tfrac{1}{2}(\theta - \theta_0)^2 G''(\theta_0)]\}. \tag{101}$$

Here ε is a small quantity whose value depends on the variation of F and G but not on R.

The integral (101) may be simplified by substituting the new variable of integration

$$\xi = (\theta - \theta_0)(\tfrac{1}{2}|RG''(\theta_0)|)^{1/2}. \tag{102}$$

It follows that

$$I \simeq \frac{F(\theta_0)}{(\tfrac{1}{2}|RG''|)^{1/2}} \int_{-\varepsilon(\frac{1}{2}|RG''|)^{1/2}}^{\varepsilon(\frac{1}{2}|RG''|)^{1/2}} d\xi \exp[i(RG \pm \xi^2)], \tag{103}$$

where the function F has been approximated by its value at θ_0, and the $\pm$ sign is the same as the sign of the second derivative $G''(\theta_0)$. In the limit $R \to \infty$, the last integral can be evaluated using the definite integrals

$$\int_{-\infty}^{\infty} \begin{matrix}\sin\\ \cos\end{matrix}\, \xi^2 d\xi = (\pi / 2)^{1/2}. \tag{104}$$

Substituting (104) in (103) gives the desired asymptotic approximation of (99) in the form

$$I \simeq F(\theta_0)\left(\frac{2\pi}{|RG''|}\right)^{1/2} \exp[i(RG \pm \pi/4)]. \tag{105}$$

This expression is valid, for sufficiently large values of R, unless $G'' = 0$.

If there are multiple points of stationary phase in the domain of integration, the contribution from each may be treated separately, according to (105), and added together to give the total contribution from all relevant points. The singular case $G'' = 0$ occurs if two points of stationary phase coalesce, and here it is necessary to retrace the steps from the Taylor series (99) with the addition of a term involving the third derivative. Further details are given by Lamb (1932) and Stoker (1957).

Having derived the stationary-phase approximation (105), we return to the ship-wave integral (89). Cartesian coordinates (x, z) are retained, with the understanding that the results are valid for sufficiently large values of the polar radius $R = (x^2 + z^2)^{1/2}$. The relevant phase function is

$$RG(\theta) = (g/U^2)(x\sec\theta + z\sec^2\theta\sin\theta), \tag{106}$$

and the points of stationary phase are given by (93). Thus the method of stationary phase confirms the analysis of the ship-wave pattern that was based on the group velocity in section 6.10.

We can now obtain more information regarding the asymptotic form of the free-wave distribution (89), including the amplitude of the waves; from (105) the waves take the form

$$\eta = \operatorname{Re} A(\theta)\left(\frac{2\pi}{|RG''|}\right)^{1/2} \exp\{i[RG(\theta) \pm \pi/4]\}. \tag{107}$$

This expression is to be summed over both points of stationary phase defined by (93) and figure 6.14. The second derivative G'' can be computed, and its sign is opposite that of the slope of the curve plotted in figure 6.14. Thus, the plus or minus sign in the phase of (107) corresponds respectively to the transverse and diverging wave systems; a net phase shift of 90° occurs between these as shown in figure 6.15.

The amplitude of the waves in (107) is proportional to $R^{-1/2}$. This rate of attenuation could be anticipated since in three dimensions the energy flux (73–74) spreads out radially in the horizontal plane.

These results can be extended to ship waves in finite depth, starting with (89) and the appropriate relation from (23) for $k(\theta)$. For subcritical Froude numbers, $U/(gh)^{1/2} < 1$, the wave pattern is similar to the deepwater situation, but with a larger included angle which ultimately approaches 90° as $U/(gh)^{1/2} \to 1$. At supercritical Froude numbers, transverse waves are ruled out, since these are unable to move with sufficient phase velocity to maintain a steady position. The diverging waves that remain are confined within an included angle similar to the Mach cone in supersonic aerodynamics. The details are worked out in an early paper by Havelock (1963); some corrections appear in Kinoshita (1943).

6.12 Energy Radiation and Wave Resistance

The wave resistance of a three-dimensional ship can be related to the energy radiation downstream in the wake, as in the corresponding analysis of the two-dimensional problem. A control volume is considered, bounded downstream by a vertical control surface x = constant, which moves with the ship as in figure 6.12(b). The energy of the fluid in this control volume is constant, and the energy flux across the control surface can be equated to the work done to overcome the wave resistance.

In a fixed reference frame, the energy of each plane-wave component moves in the direction θ with the group velocity V_g. The velocity of energy transfer across the control surface, which moves through the fluid with velocity U, is given by $V_g \cos\theta - U$. Multiplying by the energy density $\frac{1}{2}\rho g A^2$ and integrating along the width of the control surface, we find that the total energy flux moving in the positive x-direction is

$$\frac{dE}{dt} = \frac{1}{2}\rho g \int_{-\infty}^{\infty} A^2 (V_g \cos\theta - U)\, dz. \tag{108}$$

This energy-flux integral can be extended to the case where the amplitude A and direction θ are functions of the position (x, z), if the changes in these quantities are small over distances comparable to a wavelength. The diverging and transverse components of the Kelvin ship-wave system are slowly varying in this sense, provided the distance downstream from the ship is large compared to the wavelength. With this caveat, (108) can be used to determine the energy flux of the diverging and transverse waves.

Since the distance downstream is large, the stationary-phase approximation (107) can be used to determine the wave amplitude in the form

$$A = |A(\theta)|(2\pi/|RG''|)^{1/2}. \tag{109}$$

With (63) and (88), the group velocity is given by

$$V_g = \tfrac{1}{2}V_p = \tfrac{1}{2}U\cos\theta. \tag{110}$$

Combining these results and setting (108) equal to the negative of the work DU as in (81) yields

$$D = \pi\rho g \int_{-\infty}^{\infty} \frac{|A(\theta)|^2}{|RG''(\theta)|}\left(1 - \frac{1}{2}\cos^2\theta\right)dz. \tag{111}$$

For each lateral position z, the two corresponding wave angles θ of the Kelvin wave system must be determined from (95), and the contribution from each component must be included in (111).

The coordinate z can be eliminated in (111) by transforming the variable of integration from z to the wave angle θ. Using the relation (94) with x held fixed, we can derive $dz/d\theta$ from

$$x\frac{d}{d\theta}(\sec^2\theta\sin\theta) + z\frac{d}{d\theta}[\sec^3\theta(1+\sin^2\theta)] = -\frac{dz}{d\theta}[\sec^3\theta(1+\sin^2\theta)]. \tag{112}$$

On the other hand, (106) is to be evaluated with the position (x, z) fixed, and this gives the relation

$$RG''(\theta) = (g/U^2)\left\{x\frac{d}{d\theta}[\sec^2\theta\sin\theta] + z\frac{d}{d\theta}[\sec^3\theta(1+\sin^2\theta)]\right\}. \tag{113}$$

Equations (112) and (113) are combined, so that the derivative relating the two variables of integration can be written as

$$\left|\frac{dz}{d\theta}\right| = \frac{|RG''|}{(g/U^2)\sec^3\theta(1+\sin^2\theta)} = \frac{U^2}{g}\frac{\cos^3\theta|RG''|}{(2-\cos^2\theta)}. \tag{114}$$

Using the last result, we can recast (111) in the relatively simple form[3]

$$D = \frac{1}{2}\pi\rho U^2 \int_{-\pi/2}^{\pi/2} |A(\theta)|^2\cos^3\theta\, d\theta. \tag{115}$$

Equation (115) expresses the wave resistance of a moving vessel as the weighted integral of the square of the wave amplitude. The factor $\cos^3\theta$

implies that the dominant portion of the resistance will be associated with the transverse waves.

An alternative derivation of (115) can be based on the energy flux passing through a pair of longitudinal planes z = constant on both sides of the ship. The integrand of (108) is then replaced by A^2 $(V_g \sin\theta)dx$. An approach independent of the stationary-phase approximation results from substitution of the free-wave potential corresponding to (89) in the energy-flux integral (74). The derivation for a transverse plane x = constant is outlined by Havelock (1963, pp. 391–393).

As a result of the positive-definite nature of·(115), the effects of interference are limited, and one cannot expect to find dramatic results of the type illustrated in the two-dimensional case by figure 6.13. Waves generated at different locations along the ship may still interfere, but in a manner that depends on θ and on the point of observation in the wake. By judicious design or chance, the wave amplitude may vanish at discrete positions in the wake. This is illustrated in figure 6.16 where the nodal values of the diverging wave system are indicated by the 180° phase shift along a few radial lines. Additional nodes may be expected for the shorter wavelengths as $\theta \to 90°$, but these are obscured and of little significance to (115). In any event, these nodes occur only for discrete values of θ. In general, $A(\theta)$ will be nonzero, contributing to a positive and nonvanishing value of the drag (115).

To utilize (115), the wave amplitude function $A(\theta)$ must be predicted from theory or measured in a suitable experiment. These possibilities are described separately in sections 6.13 and 6.14.

6.13 Thin-Ship Theory of Wave Resistance

The thin-ship theory of wave resistance was introduced by J.H. Michell in 1898 as a purely analytic approach for predicting the wave resistance of ships. The essential assumption is that the hull is *thin*, that is, the beam is small compared to all other characteristic lengths of the problem. The resulting solution can be expressed in terms of a distribution of sources on the centerplane of the hull, with the local source strength proportional to the longitudinal slope of the hull. This solution is analogous to the thickness problem of thin-wing theory, as discussed in chapter 5. However, the ship-wave problem is complicated by the need to choose the source potential, in the form suggested by equation (4.78), to satisfy the free-surface boundary condition.

The necessary source potential that satisfies the free-surface boundary condition is a complicated function; its derivation may be found in Wehausen and Laitone (1960). We content ourselves here by noting that this potential must possess an asymptotic approximation far downstream of the same form as the Kelvin wave system (107), with the velocity potential locally a plane wave of wavenumber specified by (90). In addition, this source potential must depend in a suitably symmetrical[4] manner on the position of the source and of the field point where the potential is measured.

If the resulting expression for the source potential is integrated over the centerplane $z = 0$ of the ship, the wave amplitude can be expressed in the form

$$A(\theta) = \frac{2}{\pi}(g/U^2)\sec^3\theta \iint \frac{\partial \zeta}{\partial x} \exp[(g/U^2)\sec^2\theta(y - ix\cos\theta)]dx\,dy. \tag{116}$$

Here $z = \pm\,\zeta(x, y)$ defines the local half-beam of the hull surface.

While (116) is given here without derivation, its form is not unreasonable. The source strength is proportional to $\partial\zeta/\partial x$, as in the thickness problem of lifting-surface theory, and for the reasons noted the far-field form of the source potential is proportional to the exponential factor shown in (116). Therefore the only part of (116) that cannot be guessed judiciously is the multiplicative factor outside the integral!

If the amplitude function (116) is substituted in (115), the result is a particular form of *Michell's integral*

$$D = \frac{4\rho g^2}{\pi U^2} \int_0^{\pi/2} \sec^3\theta \left| \iint \frac{\partial \zeta}{\partial x} \exp[(g/U^2)\sec^2\theta(y - ix\cos\theta)]dx\,dy \right|^2 d\theta. \tag{117}$$

This multiple integral is not the sort of expression one expects to find in a table of integrals, particularly since in practical cases the longitudinal slope of the ship hull cannot be expressed in terms of simple mathematical functions. Nevertheless, a fairly large number of numerical computations have been carried out, based on this expression or other equivalent forms, both for practical ship geometries and for simplified mathematical bodies.

To determine the practical value of this theoretical approach to wave resistance, one may compare numerical computations based on Michell's integral with measurements from towing-tank experiments. However, this type of comparison is subject to the uncertainties of Froude's hypothesis (equation 2.24). Typical results are reproduced in figure 6.18 for a destroyer form, which must be classified as being closer to a thin ship than would be the case for a supertanker. At higher Froude numbers the agreement is reasonable, particularly in the context of experimental variations resulting from different turbulence stimulators and model lengths. At lower speeds, however, the theory appears to exaggerate the effects of interference by comparison with the experiments.

A common explanation for this discrepancy is that viscous effects suppress the interference effects in a real fluid. An alternative possibility results from the fundamental assumption of thin-ship theory that the beam is small compared to all other length scales. The wavelength is significant in this context and should be recognized as an additional length scale not present in the steady-state lifting-surface problem. Since $\lambda \propto U^2$, it is inevitable that the Michell approximation will break down as the Froude number becomes small. Theories which seek to overcome this limitation are discussed by Gadd (1976) and by Noblesse and Dagan (1976).

The Michell theory has been extended and applied to a broad range of problems in ship hydrodynamics. Several relevant papers and comprehensive bibliographies may be found in the proceedings edited by Inui and Kajitani (Society of Naval Architects of Japan, 1976), as well as in Wehausen (1973). The contributions to this field by Havelock (1963) are particularly notable, and a monograph by Kostyukov (1959) is devoted to the theory of ship waves and wave resistance.

6.14 Wave Pattern Analysis

The use of the wave-resistance integral (115) with experimental measurements of the amplitude function $A(\theta)$ to determine the total wave resistance provides a direct measurement of the wave resistance without recourse to Froude's hypothesis. This approach is known as *wave pattern analysis*. A significant feature of this technique is that the measured quantity $A(\theta)$ is linear in the ship's disturbance, as compared to the quadratic wave resistance. By relating changes in the wave amplitude to changes in hull form, a linear optimization of the hull shape can be achieved more directly than is possible by studying the effects of shape on the total wave resistance.

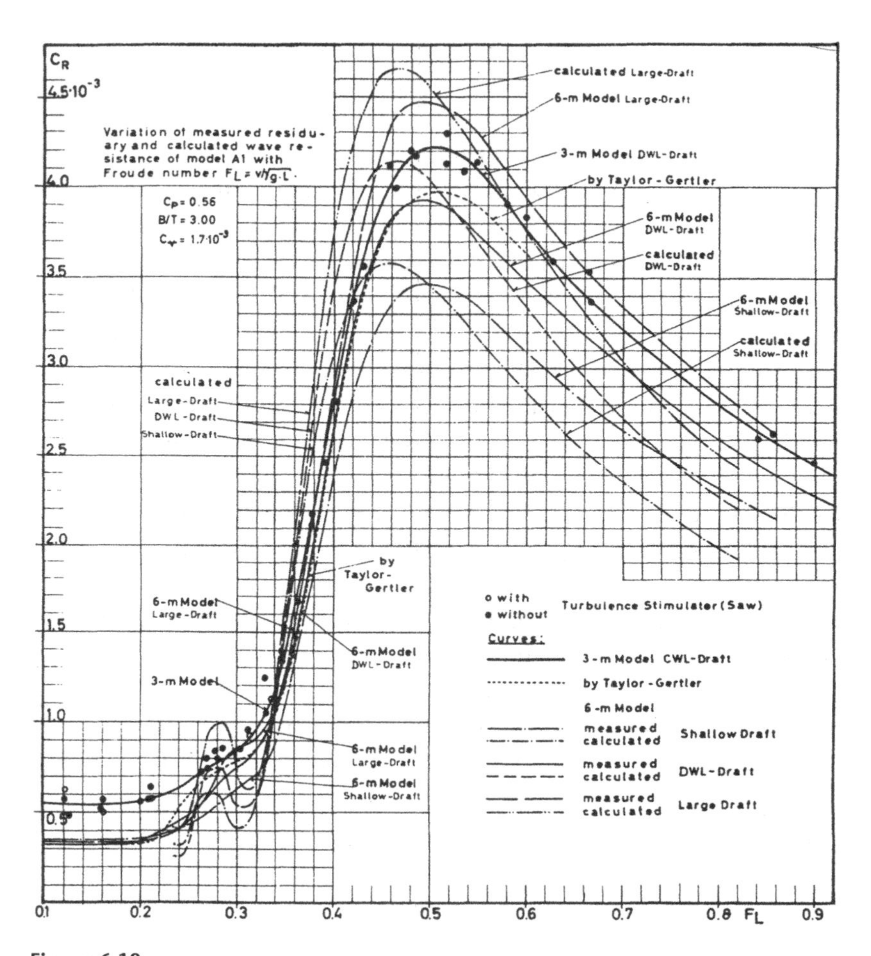

Figure 6.18

Comparison of the wave resistance coefficient calculated from Michell's integral (117) with the residual drag coefficient as measured from model tests. (From Graff, Kracht, and Weinblum 1964; reproduced by permission of the Society of Naval Architects and Marine Engineers)

The wave pattern analysis requires a relatively complex survey of the wake region to determine the amplitude function $A(\theta)$ for all relevant wave angles. By comparison, Froude's approach only requires a simple measurement of the total drag force on the model. Furthermore, the Froude approach accounts for the possibility of other contributions to the total resistance, such as the viscous form drag. Unless these additional contributions are known or can be neglected, the wave resistance component as given by (115) is of limited interest from the standpoint of predicting the total drag on a full-scale ship.

On the other hand, the wave-pattern analysis is a valuable diagnostic tool, particularly when used with a measurement of the momentum defect in the viscous-wake region. These techniques have led to the discovery of an additional drag component associated with wave *breaking*, and to a better understanding of the bulbous bows of supertankers and other ships with low Froude number.

In fine high-speed vessels such as destroyers and passenger liners, a bulbous bow promotes beneficial interference between the waves generated at different points along the length of the hull. Thus, for such vessels, the bow bulb reduces the wave resistance. Originally, bulbous bows of a similar form were fitted to supertankers on the basis of experimental measurements indicating significant reductions in the total drag, but these reductions often exceed the total estimated wave resistance.

This apparent paradox has been reconciled by careful experimental measurements of the wave-energy flux and of the momentum defect in the wake due to the viscous form drag. The latter measurement has revealed the existence of momentum associated with the breaking waves that are apparent in figures 6.16 and 6.17. Thus the wave-breaking resistance results from the breaking of waves near the ship, predominantly at the bow. The energy lost in this manner is convected downstream in the form of large-scale turbulence or eddies.

For supertankers and similar vessels, the bulbous bow is effective in reducing the magnitude of the bow wave and thereby in avoiding wave breaking. For predicting the total drag, one may argue that it makes little difference whether the energy is wasted in wave radiation or in wave breaking, but only through an understanding of the mechanisms involved can the total drag be reduced intelligently and systematically.

An extensive discussion of wave-pattern analysis is given by Eggers, Sharma, and Ward (1967). More recent references on this topic and on wave-breaking resistance are contained in the symposium volume issued by the Society of Naval Architects of Japan (1976).

6.15 Body Response in Regular Waves

A subject of great interest to ocean engineers and naval architects is the effect suffered by a floating or submerged vessel in the presence of ocean waves. The types of bodies of importance here include fixed structures and freely floating vessels, as well as the intermediate category of moored vessels. The most common *responses* of concern are the oscillatory motions of a free body or the structural loading imposed upon a fixed body.

In the simplest case it may be assumed that the waves incident upon the body are plane progressive waves of small amplitude, with sinusoidal time dependence. The results we will obtain about bodies in *regular* waves are not without physical interest, but their practical value might be questioned given the highly *irregular* nature of actual waves in the ocean. Fortunately, these two topics are connected by the description of irregular waves as a linear superposition of sinusoidal components. The actual ocean environment is most realistically described as a random process, which causes a random response of the body. These random motions will be described in sections 6.20 and 6.21, after we have dealt first with the problem of regular waves.

Plane progressive waves were analyzed in sections 6.2 and 6.3 on the assumption that the wave amplitude is sufficiently small to justify linearization. The same assumption is made here, not only to be consistent with the description of the incident waves, but also to permit superposition of different facets of the wave-body interaction in regular waves and ultimately to justify the superposition of these to represent the more realistic motions in irregular waves.

In the problem to be treated here, plane progressive waves of amplitude A and direction θ are incident upon the body, which moves in response to these waves with six degrees of freedom as illustrated in figure 6.19. Following the indicial notation used in chapter 4, we define three translational motions parallel to $(x, y, z) = (x_1, x_2, x_3)$ as *surge, heave,* and *sway,* and three rotational motions about the same axes as *roll, yaw,* and *pitch,* respectively. The corresponding velocities $U_j(t)$ will be sinusoidal in time, with the same frequency as the incident waves, and thus

$$U_j(t) = \mathrm{Re}(i\omega\xi_j e^{i\omega t}), \quad j = 1,2,\ldots,6. \tag{118}$$

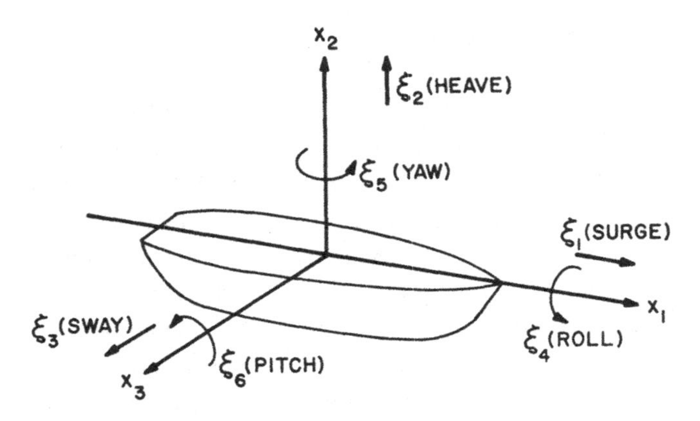

Figure 6.19
Definition sketch of body motions in six degrees of freedom.

Here the complex amplitude is employed in preference to the body velocity, since the hydrostatic restoring forces that will be encountered are proportional to the amplitudes of heave, roll, and pitch. These restoring forces are analogous to the familiar "spring constant" of a mechanical oscillator.

If the incident wave is sufficiently small in amplitude and the body is stable, the resulting motions will be proportionally small. The velocity potential ϕ can then be written in the form

$$\phi(x,y,z,t) = \mathrm{Re}\left\{\left(\sum_{j=1}^{6} \xi_j \phi_j(x,y,z) + A\phi_A(x,y,z)\right)e^{i\omega t}\right\}. \tag{119}$$

This decomposition of the total potential may be compared with equation (4.101), for the motions of a rigid body in an infinite unbounded fluid. The most important difference here is the presence of an additional term due to the incident waves. A more subtle difference is that in chapter 4 the coordinates (x_i) were fixed with respect to the body, and velocity potentials ϕ_j for each mode of motion were expressed without approximation in terms of that moving reference frame. Here, because of the free surface, we adopt a fixed coordinate system. Under the assumption that (119) is a small first-order quantity, the distinction between the inertial coordinate system and one fixed in the body will be a source of second-order effects that can be neglected.

In (119), the function ϕ_j represents the velocity potential of a rigid-body motion with unit amplitude, in the absence of the incident waves. For example, if the body is forced by an external mechanism to oscillate in heave with unit amplitude, in otherwise calm water, the resulting fluid motion will be represented by the potential ϕ_2. A similar interpretation applies for each mode of rigid-body motion. The appropriate boundary conditions to be imposed on the body surface S_B are obtained by equating the normal derivative of (119) to the normal component of the body velocity, as in equations (4.102–4.103). Since the amplitudes ξ_j are independent parameters, it follows that

$$\frac{\partial \phi_j}{\partial n} = i\omega n_j, \qquad j = 1,2,3, \tag{120}$$

$$\frac{\partial \phi_j}{\partial n} = i\omega(\mathbf{r} \times \mathbf{n})_{j-3}, \qquad j = 4,5,6, \tag{121}$$

on S_B. Here $\mathbf{n}$ is the unit normal vector on the body surface, directed into the body, and $\mathbf{r}$ is the position vector (x, y, z). The forced motion potentials ϕ_j, $j = 1, 2, \ldots, 6$, are generally known as solutions of the *radiation* problem.

The remaining potential represented by the last term in (119) is due to the incident waves and their interaction with the body. Given the assumption of linear superposition, this potential is independent of the body motions and may be defined with the body fixed in position. The appropriate boundary condition on the body surface is

$$\frac{\partial \phi_A}{\partial n} = 0, \quad \text{on} \quad S_B. \tag{122}$$

The problem so defined is known as a wave *diffraction* problem.

The presence of the body in the fluid results in diffraction of the incident wave system and the addition of a disturbance to the incident wave potential associated with the *scattering* effect of the body. This process can be emphasized by the additional decomposition

$$\phi_A = \phi_0 + \phi_7, \tag{123}$$

where ϕ_0 is the incident wave potential and ϕ_7 is the scattering potential that must be introduced to represent the disturbance of the incident waves by the fixed body. The incident wave potential may be regarded as known, from (16) for infinite depth or from (23) for finite depth, and the scattering potential must satisfy the boundary condition

$$\partial\phi_7 / \partial n = -\partial\phi_0 / \partial n, \quad \text{on } S_B. \tag{124}$$

In addition to the body boundary conditions (120–124), each potential ϕ_j must satisfy Laplace's equation

$$\nabla^2\phi_j = 0, \quad j = 0, 1, \ldots, 7, \tag{125}$$

throughout the domain of the fluid, and the condition (21) on the bottom for finite depth, or alternatively $\phi_j \to 0$ as $y \to -\infty$ for infinite depth. On the free surface, the linearized boundary condition follows from (6) and (119) in the form

$$-\frac{\omega^2}{g}\phi_j + \frac{\partial\phi_j}{\partial y} = 0, \quad \text{on} \quad y = 0, \quad j = 0, 1, \ldots, 7. \tag{126}$$

At this stage, boundary conditions have been prescribed on the body surface, free surface, and bottom; but the boundary-value problem is not unique. In particular, an arbitrary constant times the diffraction potential (123) can be added to any one of the potentials ϕ_j, $j = 1, 2, \ldots, 7$, without violating these boundary conditions. To overcome this problem it is necessary to impose a *radiation condition* at infinity, which states that the waves on the free surface, other than those due to the incident potential itself, are due to the presence of the body. Thus, the waves associated with the potentials ϕ_j, $j = 1, 2, \ldots, 7$, must be radiating *away* from the body. The appropriate condition in two dimensions is

$$\phi_j \propto e^{\mp ikx}, \quad \text{as } x \to \pm\infty, \quad j = 1, 2, \ldots, 7. \tag{127}$$

In three dimensions, with energy conservation or the stationary-phase approximation as a guide, the waves at infinity must be of the general form

$$\phi_j \propto R^{-1/2} e^{-ikR}, \quad \text{as } R \to \infty, \quad j = 1, 2, \ldots, 7. \tag{128}$$

Here $R = (x^2 + z^2)^{1/2}$, and the constants of proportionality in (127) and (128) may depend on the remaining coordinates, but not on x or R, respectively. With the time dependence (119), these are the most general forms of outgoing waves due to the presence of the body. The incident wave potential ϕ_0 is, by definition, excluded from these radiation conditions.

While the radiation conditions may appear contrived or arbitrary, the need for them results from our assumption of sinusoidal time dependence for all previous time. A similar situation was noted in sections 3.13 and 5.16, in dealing with viscous diffusion and unsteady lifting surfaces. In those cases appropriate initial conditions were prescribed, to make the solutions unique. The same approach could be used here with the final results unchanged, but imposing the radiation conditions (127–128) is simpler. This subject is discussed at some length by Stoker (1957).

The general form of the oscillatory force and moment acting on the body can be inferred by substituting (119) in Bernoulli's equation. Retaining only the first-order linear terms in (3), the total pressure is given by

$$\begin{aligned} p &= -\rho\left(\frac{\partial\phi}{\partial t} + gy\right) \\ &= -\rho\,\mathrm{Re}\left\{\left(\sum_{j=1}^{6}\xi_j\phi_j + A(\phi_0+\phi_7)\right)i\omega e^{i\omega t}\right\} - \rho g y. \end{aligned} \tag{129}$$

The force F and moment **M** can be determined by integrating the fluid pressure (129) over the wetted surface S_B. Substituting (129) in (4.81–4.82) gives expressions of the form

$$\begin{aligned} \begin{pmatrix}\mathbf{F}\\ \mathbf{M}\end{pmatrix} &= -\rho g\iint_{S_B}\begin{pmatrix}\mathbf{n}\\ \mathbf{r}\times\mathbf{n}\end{pmatrix} y\,dS - \rho\,\mathrm{Re}\sum_{j=1}^{6} i\omega\xi_j e^{i\omega t}\iint_{S_B}\begin{pmatrix}\mathbf{n}\\ \mathbf{r}\times\mathbf{n}\end{pmatrix}\phi_j\,dS \\ &\quad - \rho\,\mathrm{Re}\,i\omega A e^{i\omega t}\iint_{S_B}\begin{pmatrix}\mathbf{n}\\ \mathbf{r}\times\mathbf{n}\end{pmatrix}(\phi_0+\phi_7)\,dS. \end{aligned} \tag{130}$$

The three integrals in (130) represent distinctly different contributions to the total force and moment. The first is the hydrostatic component, which is both elementary and important. The second integral in (130) is reminiscent of the added-mass coefficients studied in chapter 4, except that in the present case the potentials ϕ_j are complex. The resulting force and moment are defined generally in terms of the *damping* and *added-mass* coefficients, corresponding to the real and imaginary parts of these integrals. Finally, the last term in (130) is the exciting force or moment, proportional to the incident wave amplitude. Each of these three contributions will be discussed separately, in sections 6.16–6.18.

6.16 Hydrostatics

Hydrostatics is the oldest and most elementary topic of naval architecture and fluid mechanics. Our task here is to evaluate the first term in (130), for the hydrostatic force

$$\mathbf{F} = -\rho g \iint_{S_B} \mathbf{n} y \, dS, \tag{131}$$

and moment

$$\mathbf{M} = -\rho g \iint_{S_B} (\mathbf{r} \times \mathbf{n}) y \, dS, \tag{132}$$

taking advantage of the linearization assumption. The coordinate y is fixed in space, whereas the wetted surface S_B oscillates with the body, and for a floating body the upper boundary of S_B is the free surface $y = \eta(x, z, t)$, as shown in figure 6.20.

Since the hydrostatic pressure is of order one, the distinction between the fixed coordinates ($\mathbf{x}$) and body coordinates ($\mathbf{x}'$) is significant. Assuming the body motions ξ_j are small and neglecting the second-order quantities, we can express the transformation between these coordinate systems as

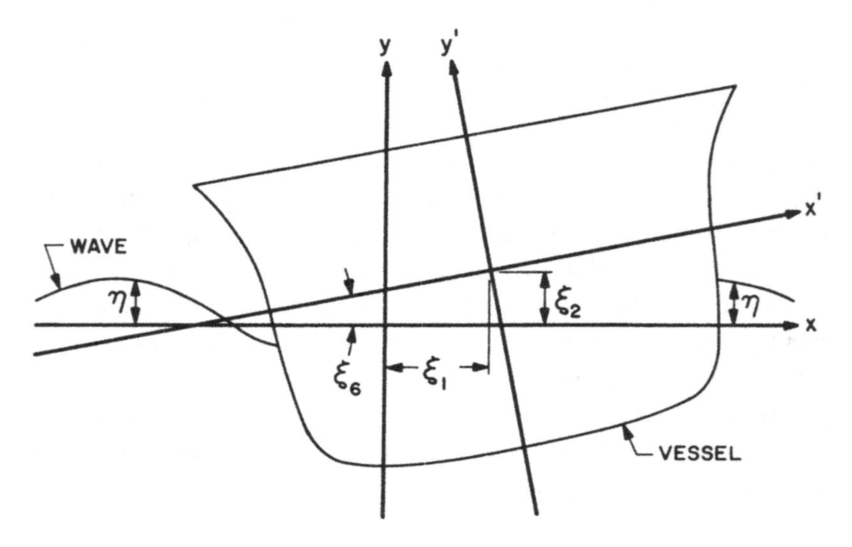

Figure 6.20
Two-dimensional sketch showing the fixed and moving coordinate system, wetted surface S_B of the vessel, and free surface $y = \eta$.

$$\mathbf{x} = \mathbf{x}' + \boldsymbol{\xi}_T + \boldsymbol{\xi}_R \times \mathbf{x}'. \tag{133}$$

Here $\boldsymbol{\xi}_T = (\xi_1, \xi_2, \xi_3)$ and $\boldsymbol{\xi}_R = (\xi_4, \xi_5, \xi_6)$ are vectors formed by the translation and rotation of the body; the latter are unique since the motions are small. Throughout this section, ξ_j will be used to denote the real time-dependent body motion. In the context of hydrostatics, ξ_j may be a constant or an arbitrary function of time.

The dominant portions of (131) and (132) are the static buoyancy force and moment, which are canceled for an unrestrained body by its weight. Since the center of gravity moves with the body, it is convenient to replace (132) by the moment $\mathbf{M}'$ about the origin of the body-fixed coordinates $\mathbf{x}'$. Thus

$$\mathbf{M}' = -\rho g \iint_{S_B} [(\mathbf{r} - \boldsymbol{\xi}_T) \times \mathbf{n}] y \, dS. \tag{134}$$

Before evaluating the force (131) and moment (134), we note that the integrands are proportional to y; therefore the contribution from the thin strip $0 < y < \eta$ will be of order η^2. This second-order quantity may be neglected in the linearized analysis, and the surface of integration S_B in these integrals can be redefined as the instantaneous body surface beneath the calm-water plane $y = 0$.

To evaluate these surface integrals, the surface S_B may be closed by adding to it the interior portion of the plane $y = 0$, where the integrand is zero. The volume $\mathcal{V}(t)$ within this closed surface is the instantaneous displaced volume of the body beneath the plane $y = 0$. Applying Gauss's theorem to (131) and the corresponding theorem (4.92) to (134) yields

$$\mathbf{F} = \rho g \iiint_{\mathcal{V}} (\nabla y) d\mathcal{V} = \mathbf{j} \rho g \iiint_{\mathcal{V}} d\mathcal{V}, \tag{135}$$

$$\begin{aligned} \mathbf{M}' &= -\rho g \iiint_{\mathcal{V}} [\nabla \times y(\mathbf{r} - \boldsymbol{\xi}_T)] d\mathcal{V} \\ &= \rho g \iiint_{\mathcal{V}} [\mathbf{k}(x - \xi_1) - \mathbf{i}(z - \xi_3)] d\mathcal{V}. \end{aligned} \tag{136}$$

The force (135) is obviously the buoyancy force proportional to the displaced volume $\mathcal{V}(t)$, and the moment (136) is the cross-product of this buoyancy force with the vector position of the center of buoyancy.

The volume integrals in (135–136) can be evaluated in terms of the body-fixed coordinates, after decomposing $\mathcal{V}$ into the static volume $\mathcal{V}_0$ beneath the plane $y' = 0$ and the thin layer bounded by the planes $y = 0$ and $y' = 0$. From (133) it follows that

$$\mathbf{F} = \mathbf{j}\rho g\left(\forall - \iint_{S_0} (\xi_2 + \xi_6 x' - \xi_4 z')\,dx'\,dz'\right), \tag{137}$$

$$\begin{aligned}\mathbf{M}' = \rho g\Bigg\{&\iiint_{\mathcal{V}_0} [\mathbf{k}(x' + \xi_5 z' - \xi_6 y') - \mathbf{i}(z' + \xi_4 y' - \xi_5 x')]\,d\mathcal{V} \\ &- \iint_{S_0} (\xi_2 + \xi_6 x' - \xi_4 z')(\mathbf{k}x' - \mathbf{i}z')\,dx'\,dz'\Bigg\},\end{aligned} \tag{138}$$

where $\mathcal{V}_0$ and S_0 denote the displaced volume and waterplane in the static condition.

Equations (137) and (138) can be expressed in the form

$$\mathbf{F} = \mathbf{j}\rho g[\forall - \xi_2 S + \xi_4 S_3 - \xi_6 S_1], \tag{139}$$

$$\begin{aligned}\mathbf{M}' = &-\mathbf{i}\rho g[\forall(z_B + \xi_4 y_B - \xi_5 x_B) - \xi_2 S_3 + \xi_4 S_{33} - \xi_6 S_{13}] \\ &+ \mathbf{k}\rho g[\forall(x_B + \xi_5 z_B - \xi_6 y_B) - \xi_2 S_1 + \xi_4 S_{13} - \xi_6 S_{11}].\end{aligned} \tag{140}$$

Here $\forall$ is the displaced volume, S the waterplane area, $\mathbf{x}_B$ the center of buoyancy

$$\mathbf{x}_B = \frac{1}{\forall}\iiint_{\mathcal{V}_0} \mathbf{x}\,d\mathcal{V}, \tag{141}$$

and the waterplane moments are defined as

$$S_j = \iint_{S_0} x_j\,dS, \quad j = 1,3, \tag{142}$$

$$S_{ij} = \iint_{S_0} x_i x_j\,dS. \tag{143}$$

Note that the integrals (141–143) are defined in terms of the static condition of the vessel, and the primes have been deleted with this understanding.

The horizontal components of the hydrostatic force and the vertical component of the moment are identically zero. For a submerged body the waterplane area S and moments (142–143) vanish, and (139–140) simplify accordingly.

In the conventional terminology of naval architecture, $(S_1/S, 0, S_3/S)$ are the coordinates of the *center of flotation*. This is the point of rotation for a freely floating body subject to an applied horizontal moment. The integrals (142) vanish if the origin is placed at the center of flotation, and with a suitable rotation of the horizontal coordinates it can be assumed that $S_{13} = 0$ as well.[5] This particular choice of the coordinate system will be assumed hereafter. Thus the vertical force (139) is independent of the roll and pitch motions, and the terms in (139–140), which involve S_1, S_3, and S_{13} can be deleted.

For subsequent use it is convenient to add to the hydrostatic force and moment the corresponding components due to the weight of the body. The force associated with the body weight is simply $(0, -mg, 0)$, where m is the body mass. The corresponding moment is given, to first order in the motions ξ_j, by the vector

$$(mgz_G + mgy_G\xi_4 - mgx_G\xi_5, 0, -mgx_G - mgz_G\xi_5 + mgy_G\xi_6).$$

Using an indicial notation for the components of the resulting *total static* force (F_1, F_2, F_3), and moment (F_4, F_5, F_6), it follows from (139–140) that

$$F_i = (\rho\forall - m)g\delta_{i2} + (mz_G - \rho\forall z_B)g\delta_{i4} - (mx_G - \rho\forall x_B)g\delta_{i6} - \sum_{j=1}^{6} c_{ij}\xi_j. \tag{144}$$

Here δ_{ij} is the Kroenecker delta function, equal to one if $i = j$ and zero otherwise, and c_{ij} is the matrix with nonzero elements

$$\begin{aligned} c_{22} &= \rho g S, \\ c_{44} &= \rho g S_{33} + \rho g \forall y_B - mgy_G, \\ c_{45} &= -g(\rho\forall x_B - mx_G), \\ c_{65} &= -g(\rho\forall z_B - mz_G), \\ c_{66} &= \rho g S_{11} + \rho g \forall y_B - mgy_G. \end{aligned} \tag{145}$$

For other values of (i, j), $c_{ij} = 0$.

The expression (144), for the total static force and moment, consists of terms of zero- and first-order in the body motions ξ_j, with second-order terms neglected. If the body is in equilibrium, for $\xi_j = 0$, the zero-order terms must vanish. For equilibrium of the vertical force, the mass of a freely floating body must equal the displaced mass,

$$m = \rho\forall, \tag{146}$$

in accordance with Archimedes' principle. With this equality, equilibrium of the zero-order moments in (144) requires that the center of gravity and center of buoyancy must lie on the same vertical line,

$$x_B = x_G, \tag{147}$$

$$z_B = z_G. \tag{148}$$

For a freely floating body, where (146–148) hold, the elements of the matrix coefficients (145) contained in parenthesis vanish, leaving as the only nonzero coefficients

$$\begin{aligned} c_{22} &= \rho g S, \\ c_{44} &= \rho g \forall [(S_{33} / \forall) + y_B - y_G], \\ c_{66} &= \rho g \forall [(S_{11} / \forall) + y_B - y_G]. \end{aligned} \tag{149}$$

The body is *statically* stable if the first-order components in (144) oppose small displacements ξ_j. Since there is no static restoring force or moment in surge, sway, and yaw, these are neutrally stable modes. In heave, roll, and pitch of a freely floating body, the necessary and sufficient condition for static stability is that the corresponding coefficients of (149) are positive. In heave this condition is always satisfied, provided the waterplane area is nonzero. A submerged body, with $S = 0$, is neutrally stable in heave, or *neutrally buoyant*, provided the condition (146) is satisfied.

Since static stability in roll and pitch requires that the coefficients c_{44} and c_{66} be positive, it follows from (149) that stability in these modes will depend on the height of the center of gravity. The factors in square brackets in these expressions are the vertical distances between the *metacenters* $(S_{jj} / \forall) + y_B$ and the center of gravity y_G. These differences in elevation are known as *metacentric heights* and often are denoted by the symbol GM. The metacenters of a submerged body coincide with the center of buoyancy, but for a floating body the metacenters are above y_B, since the moments of inertia S_{jj} about the center of flotation are positive. Thus the center of gravity of a stable floating body may be situated above the center of buoyancy.

Comprehensive treatments of static stability for ships are given by Moore (1967) and by Rawson and Tupper (1968).

6.17 Damping and Added Mass

In chapter 4 we showed that if a body moves in an infinite ideal fluid, hydrodynamic pressure forces and moments will result which can be expressed most simply in terms of the added-mass coefficients m_{ij}. A more complicated and less general situation follows for a body moving on or near a free surface; the corresponding force and moment are expressed by the second term on the right side of (130), with the restrictions that the body motions are small and sinusoidal in time.

The six components of this force and moment can be written in the matrix form

$$F_i = \mathrm{Re}\left\{\sum_{j=1}^{6} \xi_j e^{i\omega t} f_{ij}\right\}, \quad i = 1,2,\ldots,6, \tag{150}$$

where

$$f_{ij} = -\rho \iint_{S_B} \frac{\partial \phi_i}{\partial n} \phi_j dS. \tag{151}$$

The last result follows from (130) and the boundary conditions (120–121). The coefficient f_{ij} is the complex force in the direction i, due to a sinusoidal motion of unit amplitude in the direction j.

The coefficient defined by (151) is analogous to the added-mass coefficient m_{ij} defined by (4.114) for a body undergoing small oscillations in an unbounded fluid. Here, however, we shall find that the coefficients f_{ij} are complex as a result of the free surface, and the real and imaginary parts depend on the frequency ω. For this reason the coefficients (151) take the form

$$f_{ij} = \omega^2 a_{ij} - i\omega b_{ij}. \tag{152}$$

Equivalently, the force (150) can be expressed in the form

$$F_i = -\sum_{j=1}^{6} (a_{ij}\dot{U}_j + b_{ij}U_j), \tag{153}$$

which is a decomposition of the sinusoidal force, associated with each mode of motion, into components in phase with the velocity and acceleration of the corresponding modes.

The coefficient a_{ij} is known as the *added-mass* coefficient, since it represents the force component proportional to the acceleration. It should be emphasized that a_{ij} will differ from the corresponding added-mass coefficient m_{ij} for a body in the absence of a free surface, and one must not assume that all of the properties of m_{ij} apply to a_{ij}.

The coefficient b_{ij} in (153) gives a force proportional to the body velocity; for this reason b_{ij} is known as the *damping* coefficient. The presence of such a force results from the generation of waves on the free surface, due to the motions of the body. These waves radiate outward, with a corresponding energy flux that can be computed from (73–78). As in the steady-state wave-resistance problem, work must be done to oscillate the body and generate the radiating waves. For a single mode of body motion, the average work done to oppose the pressure force (153) over one cycle is

$$-\overline{F_i U_i} = b_{ii}\overline{U_i^2} = \tfrac{1}{2}\omega^2 b_{ii}\left|\xi_i\right|^2 . \tag{154}$$

Thus, there is a direct relation between the damping coefficients and the amplitude of the waves generated by the body. For any body motion where waves are generated on the free surface, the damping coefficient must be greater than zero, and therefore the force coefficient f_{ij} must be complex.

One simplification that the damping and added-mass coefficients share with the unbounded-fluid case is symmetry. The proof follows in an analogous manner to (4.117–4.118); however, Green's theorem must be applied not only on S_B but to the complete closed surface, including the free surface, bottom, and a control surface at infinity. Fortunately, a consequence of the boundary conditions on each of these additional surfaces is that none of them contribute to the integrand of (4.117). The two terms in this integrand vanish separately on the bottom and cancel on the free surface when (126) is imposed. The same cancellation occurs on the control surface at infinity if the two potentials satisfy the radiation conditions (127–128). Thus (4.117) applies here, in the form

$$\iint_{S_B}\left(\phi_i\frac{\partial\phi_j}{\partial n}-\phi_j\frac{\partial\phi_i}{\partial n}\right)dS=0, \quad i,j=1,2,\ldots,6; \tag{155}$$

from (151) it follows that

$$f_{ji}=f_{ij}, \quad i,j=1,2,\ldots,6. \tag{156}$$

Since the same result holds separately for the real and imaginary parts of f_{ij}, the damping and added-mass coefficients are symmetric.

The dependence of a_{ij} and b_{ij} on frequency can be verified by considering the two limits $\omega\to 0$ and $\omega\to\infty$. In these cases the free-surface condition (126) takes the form

$$\partial\phi_j / \partial y = 0 \quad \text{on } y = 0, \omega \to 0, \tag{157}$$

and

$$\phi_j = 0 \quad \text{on } y = 0, \omega \to \infty. \tag{158}$$

These can be regarded as the limits where gravitational forces dominate inertial forces, and vice versa. (Strictly, since ω and g are dimensional quantities, one should state that (157) and (158) are valid respectively for the two limiting values of the nondimensional ratio $\omega^2 l/g$, where l is the characteristic body length.)

Equation (157) is the *rigid-wall* boundary condition, where the free surface is replaced by a fixed horizontal plane. The resulting problem for each ϕ_j can be solved by the method of images, as outlined in section 4.18. The appropriate image above the plane $y = 0$ is a body of the same geometrical form, reflected about $y = 0$. The phase of the image body motion must be such that the potential is an even function of y, and the normal velocity takes the same value at corresponding points on the body and its image.

For low-frequency motions in the horizontal plane (surge, sway, and yaw), the image body must move with the same phase as the real body, as shown in figure 6.21. In this case the body plus its image form a rigid *double body*, and the potentials ϕ_j, $j = 1, 3, 5$, can be related directly to the corresponding solutions of chapter 4 for a rigid body in an unbounded fluid. The limiting values of the corresponding added-mass coefficients are

$$a_{ij} \to \tfrac{1}{2} m_{ij}, \tag{159}$$

for $i, j = 1, 3, 5$, as $\omega \to 0$, with m_{ij} the double-body added mass.

For low-frequency motions in the vertical plane (heave, roll, or pitch), symmetry requires the phase of the image body to be opposite that of the real body, as shown in figure 6.21. In this case the body plus its image can

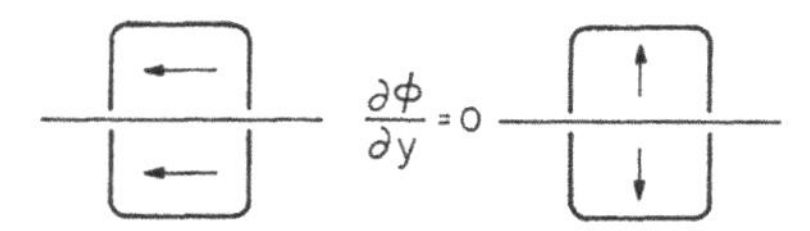

Figure 6.21
The low-frequency limit $\omega \to 0$ for surge, sway, or yaw (left), and for heave, roll, or pitch (right).

still be regarded as a double body in an unbounded flow, but the mode of motion corresponds to an oscillatory dilation of the body. The potentials ϕ_j must be derived separately, and the resulting added-mass coefficients a_{ij} are not related to the corresponding values of m_{ij}. As a particular example, the solution for heave will require a net oscillatory source strength to match the changing volume of the double body, with a completely different far-field behavior relative to the dipole-like potentials (4.123 and 4.129).

The opposite situation applies in the high-frequency limit (158), where the potential must be odd in y, and the resulting asymmetry requires the image phase to be as shown in figure 6.22. Here, the vertical modes of motion correspond to those of a rigid double body, and thus (159) will apply for $i, j = 2, 4, 6$, as $\omega \to \infty$. On the other hand, the horizontal modes correspond to shearing deflections of the double body that cannot be related to the potentials obtained in chapter 4 or to the resulting values of m_{ij}.

Comparing figures 6.21 and 6.22 shows that the added-mass coefficients a_{ij} must differ for $\omega \to 0$ and $\omega \to \infty$, since different types of double-body motions result in these two limits. It follows that the added-mass coefficients must depend on ω, in order to pass from one limit to the other as $0 < \omega < \infty$.

The damping coefficients vanish at the two limits, since the respective double-body flows are in an unbounded fluid without waves. Since the damping coefficients are greater than zero for intermediate frequencies, these too must depend on ω.

Many investigations have been carried out to compute the damping and added-mass coefficients of particular body forms, as functions of frequency. The majority of this work pertains to two-dimensional cylinders in deep water. Several references and examples are given by Wehausen (1971).

For the two-dimensional case, Vugts (1968) has made extensive calculations; some of these are reproduced by Wehausen (1971) and in a revised form in figure 6.23. In three dimensions systematic calculations have been

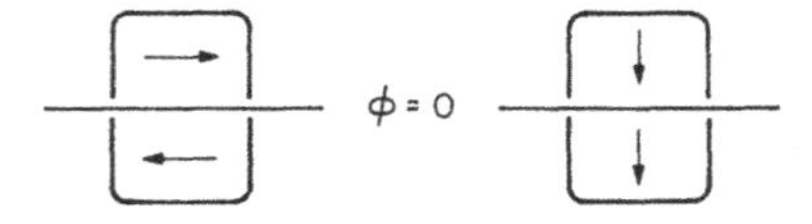

Figure 6.22
The high-frequency limit $\omega \to \infty$ for surge, sway, or yaw (left), and heave, roll, or pitch (right).

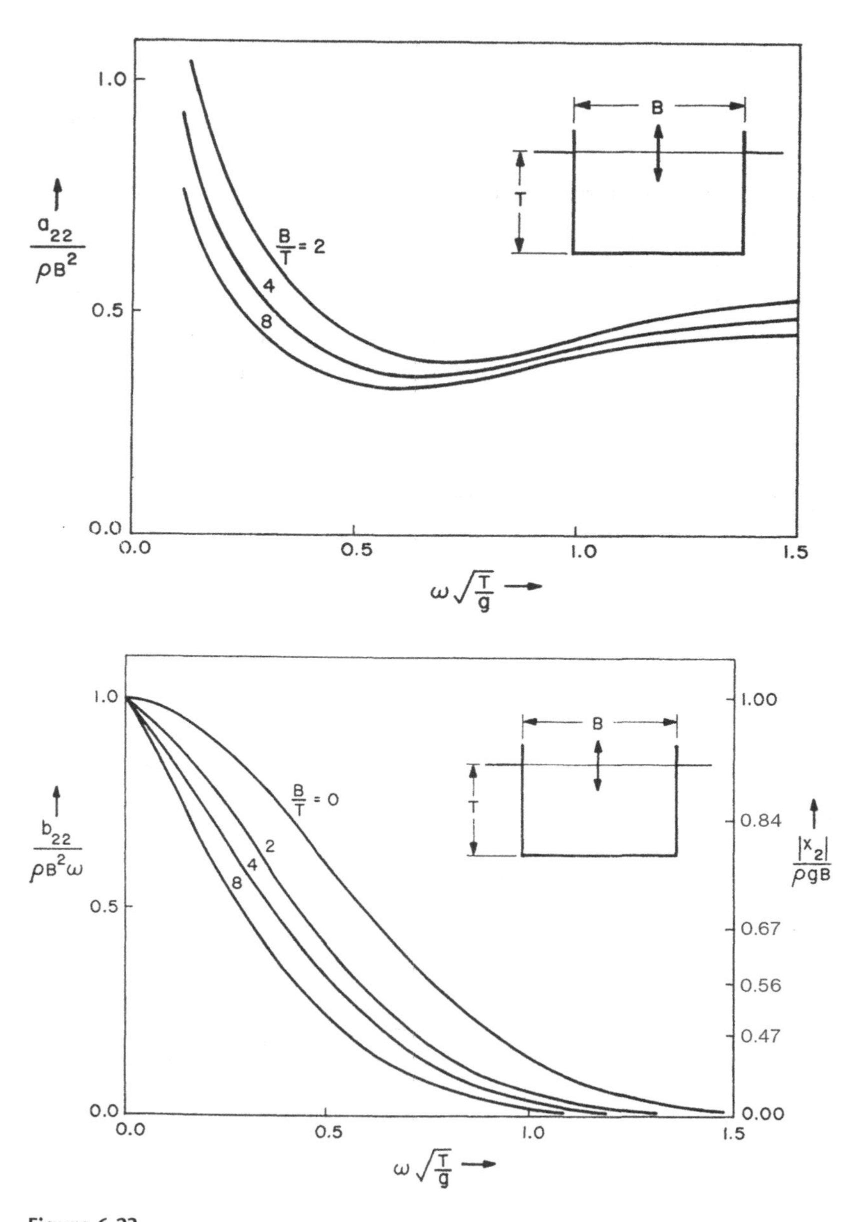

Figure 6.23
Added-mass and damping coefficients for a family of two-dimensional rectangular cylinders, heaving in deep water, based on the computations of Vugts (1968). Included below is the thin-ship approximation given by (179) and labeled $B/T = 0$. Also shown here by the scale on the right is the heave exciting-force coefficient, obtained in accordance with (174).

carried out by Kim (1966) for a series of ellipsoids, including the special case of a floating hemisphere. The latter was first considered by Havelock (1963), and the damping and added-mass coefficient for heave are reproduced in figure 6.24.

Since the complex force coefficient f_{ij} is an analytic function of ω in the lower half-plane, the damping and added-mass coefficients can be related by integral transforms known as the Kramers-Kronig relations. Further details may be found in the surveys of Ogilvie (1964) and Wehausen (1971). In addition, relations exist between the damping coefficients and exciting forces; these will be discussed in the next section and utilized to derive the damping coefficients for a thin ship.

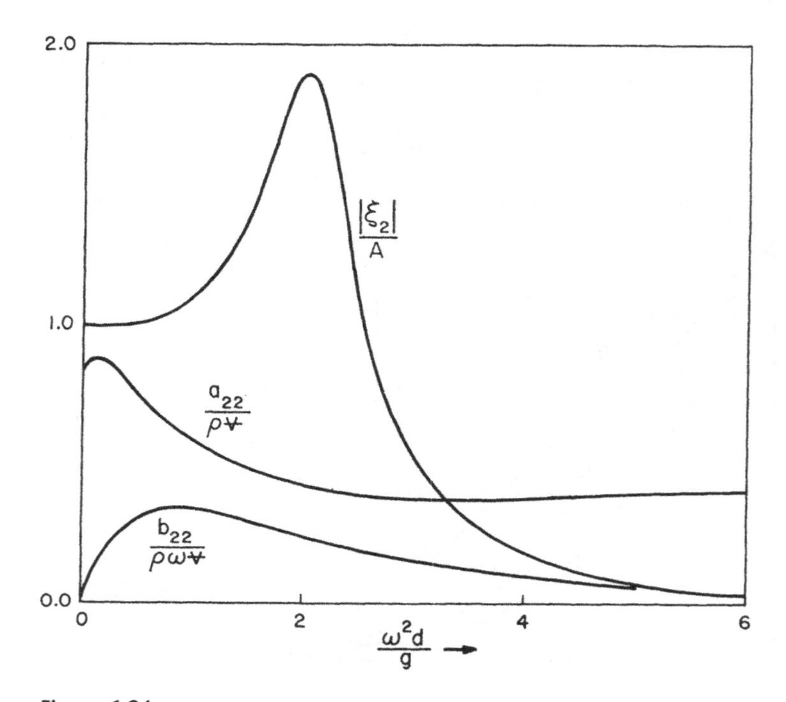

Figure 6.24
Added-mass and damping coefficients for a sphere of diameter d, half submerged in deep water. $\forall$ is the displaced volume $\pi d^3/12$. Also shown is the heave-response ratio (190).

6.18 Wave-Exciting Force and Moment

The last term in (130) gives the exciting force and moment proportional to the incident-wave amplitude,

$$\left.\begin{matrix}\mathbf{F}_{ex}\\ \mathbf{M}_{ex}\end{matrix}\right\} = -\rho\,\mathrm{Re}\, i\omega A e^{i\omega t}\iint_{S_B}(\phi_0+\phi_7)\begin{pmatrix}\mathbf{n}\\ \mathbf{r}\times\mathbf{n}\end{pmatrix}dS. \tag{160}$$

It will be convenient to define the six components of (160) in the form

$$F_{ex_i} = \mathrm{Re}\{Ae^{i\omega t}X_i\}, \quad i = 1,2,\ldots,6. \tag{161}$$

Thus X_i is the complex amplitude of the exciting force or moment, in the direction i, for an incident wave of unit amplitude. Using (160) and the boundary conditions (120–121) it follows that

$$X_i = -\rho\iint_{S_B}(\phi_0+\phi_7)\frac{\partial\phi_i}{\partial n}dS. \tag{162}$$

For a prescribed incident wave, the potential ϕ_0 can be found from (16) or (23) for infinite or finite depth, respectively. The diffraction potential ϕ_7 must be determined as the solution of the boundary-value problem (124–128).

Since the wave amplitude A is a first-order quantity, the surface integrals in (160–162) can be evaluated on the mean position of the body surface, with the neglected factor a second-order quantity proportional to A^2. Moreover, none of the boundary conditions governing the potentials ϕ_0 or ϕ_7 is affected by the body motions. Thus, the linearized exciting force and moment are independent of the body motions, just as the damping and added mass are independent of the incident-wave amplitude. In particular, (160–162) can be used to find the exciting force either for a fixed body or for one free to respond to the waves. The distinction between these is a second-order effect.

Since the boundary-value problem (124–128) differs from that for the potentials ϕ_j, j = 1, 2, ..., 6, only with regard to the boundary condition (124) on the body surface, the diffraction potential ϕ_7 can be considered the disturbance associated with a forced normal velocity on the body surface equal and opposite to that of the incident wave. This suggests the *long-wave approximation* where, if the body is small compared to the wavelength, the induced velocity field from the incident wave can be assumed constant over the body. Under this restriction the diffraction potential can be approximated in the form

$$\phi_7 \simeq \frac{i}{\omega}\left(\frac{\partial \phi_0}{\partial x}\phi_1 + \frac{\partial \phi_0}{\partial y}\phi_2 + \frac{\partial \phi_0}{\partial z}\phi_3\right), \tag{163}$$

where the derivatives of ϕ_0 are constants in this expression, evaluated at a suitable position on or within the body.

Using the long-wave approximation (163), the contribution to (162) from the diffraction potential ϕ_7 can be evaluated in terms of the coefficients f_{ij} defined by (151),

$$-\rho \iint_{S_B} \phi_7 \frac{\partial \phi_i}{\partial n} dS = \frac{i}{\omega} \sum_{j=1}^{3} f_{ij} \frac{\partial \phi_0}{\partial x_j}. \tag{164}$$

The integral in (162) involving ϕ_0 can be evaluated from Gauss's theorem. Consistent with (163) and (164), the potential ϕ_0 can be approximated by a Taylor series, and it follows that

$$\begin{aligned} \iint_{S_B} \phi_0 \, \mathbf{n} \, dS &= -\iiint_{\mathcal{V}_0} \nabla \phi_0 d\mathcal{V} - \iint_{S_0} \phi_0 \mathbf{n} \, dS \\ &\simeq -\forall \nabla \phi_0 + \mathbf{j}\left(\phi_0 S + \frac{\partial \phi_0}{\partial x} S_1 + \frac{\partial \phi_0}{\partial z} S_3\right). \end{aligned} \tag{165}$$

The corresponding integral for the moment follows from (4.92),

$$\begin{aligned} \iint_{S_B} \phi_0 (\mathbf{r} \times \mathbf{n}) dS &= \iiint_{\mathcal{V}_0} (\nabla \times \phi_0 \mathbf{r}) d\mathcal{V} - \iint_{S_0} \phi_0 (\mathbf{r} \times \mathbf{n}) dS \\ &\simeq \forall \nabla \phi_0 \times \mathbf{x}_B - \mathbf{i}\left(\phi_0 S_3 + \frac{\partial \phi_0}{\partial x} S_{13} + \frac{\partial \phi_0}{\partial z} S_{33}\right) \\ &\quad + \mathbf{k}\left(\phi_0 S_1 + \frac{\partial \phi_0}{\partial x} S_{11} + \frac{\partial \phi_0}{\partial z} S_{13}\right) \end{aligned} \tag{166}$$

where the hydrostatic parameters are defined in section 6.16. We shall avoid the step of combining (165) or (166) with (164) and substituting in (162), but we note that the derivatives of the incident wave potential will be proportional to $k\phi_0$. Since λ is large, the wavenumber k is small, and the leading-order force and moment will result from the terms in (165–166) proportional to ϕ_0. Substituting these in (160), and using (5) yields

$$\mathbf{F}_{ex} \simeq \mathbf{j} \rho g S \eta(t), \tag{167}$$

$$\mathbf{M}_{ex} \simeq (\mathbf{k} S_1 - \mathbf{i} S_3) \rho g \eta(t). \tag{168}$$

The leading-order contributions (167–168) are the hydrostatic force and moment associated with the elevation of the incident wave at the body.

The remaining contributions from (164–166) will be of order (kl), or proportional to l/λ, which is assumed small. The contribution from (164) can be related to the damping and added-mass coefficients, and the remaining contributions from (165–166) can be computed from the hydrostatic coefficients. A simple example is the vertical exciting force on a body symmetrical about the planes $x = 0$ and $z = 0$; this force can be obtained in the form

$$X_2 \simeq [\rho g S + i\omega b_{22} - \omega^2(a_{22} + \rho\forall)]. \tag{169}$$

Here the leading term is hydrostatic, and the *relative motion hypothesis* is suggested by the fact that the damping and added-mass coefficients in (169) multiply the velocity and acceleration of the incident wave. The term involving the displaced volume $\forall$ can be interpreted in a similar manner as the corresponding term in (4.150), for the force on a body in an unbounded fluid with a slowly-varying inhomogeneous velocity field. Equation (4.150) can be used to compute the exciting force on a submerged body, under the same assumptions as (169). The only difference that will result, apart from the absence of the waterplane area S, is that the complex force $-\omega^2 a_{22} + i\omega b_{22}$ is replaced in (4.150) by $-\omega^2 m_{22}$.

For $\lambda/l = O(1)$, the long-wave approximation is not valid, but the boundary conditions for ϕ_7 and the forced-motion potentials ϕ_j, $j = 1,2, \ldots, 6$, are similar except on the body surface. We shall take advantage of this similarity to derive the *Haskind relations*, which express the damping coefficients in terms of the exciting forces. No assumptions are required regarding the wavelength or body geometry.

Since ϕ_7 satisfies the same boundary conditions as ϕ_j on the free surface and bottom and the same radiation condition at infinity, Green's theorem in the form (155) can be applied to the diffraction potential. In particular,

$$\iint_{S_B} \left(\phi_i \frac{\partial \phi_7}{\partial n} - \phi_7 \frac{\partial \phi_i}{\partial n} \right) dS = 0, \quad i = 1,2,\ldots,6. \tag{170}$$

Substituting this result in (162), gives

$$X_i = -\rho \iint_{S_B} \left(\phi_0 \frac{\partial \phi_i}{\partial n} + \phi_i \frac{\partial \phi_7}{\partial n} \right) dS. \tag{171}$$

The diffraction potential can be eliminated from (171) by substituting the boundary condition (124), and the Haskind relations follow in the form

$$X_i = -\rho \iint_{S_B} \left(\phi_0 \frac{\partial \phi_i}{\partial n} - \phi_i \frac{\partial \phi_0}{\partial n} \right) dS. \tag{172}$$

The most significant feature of (172) is that the exciting force has been expressed, without approximation, in a form independent of the diffraction potential ϕ_7. Moreover, it follows from Green's theorem that the integral (172) can be evaluated on a control surface at infinity. This integral does not vanish, since the incident wave does not satisfy the radiation condition, but at infinity the forced-motion potential ϕ_i is proportional to the square root of the energy flux, associated in turn with the damping coefficient b_{ii}.

In three dimensions the method of stationary phase (105) can be used to show that

$$b_{ii} = \frac{k}{8\pi\rho g V_g} \int_0^{2\pi} |X_i(\theta)|^2 \, d\theta. \tag{173}$$

Here $X_i(\theta)$ is the amplitude of the exciting force, for waves of direction θ and unit amplitude. Equation (173) is particularly useful for axisymmetric bodies, with a vertical axis, where the heave exciting force is independent of the wave direction and the remaining forces and moments are proportional to $\sin\theta$ or $\cos\theta$. In this case the integral (173) can be evaluated, and a simple algebraic relation follows (see problem 12).

Similarly, in two dimensions, for a body symmetric about the plane $x = 0$, it can be shown that

$$b_{ii} = \frac{|X_i|^2}{2\rho g V_g}, \tag{174}$$

where X_i is the amplitude of the exciting force per unit length. Derivations and extensions of these relations are given by Wehausen (1971) and Newman (1976). Vugts (1968) has confirmed (174) from experimental measurements of the damping coefficients and exciting forces.

The character of the exciting force and moment as $\omega \to \infty$ or $\lambda \to 0$ can be inferred from a physical argument and confirmed from the Haskind relations. Physically, if the wavelength is short compared to the body length, several waves will exist simultaneously along the body and cancellation will occur. Moreover, the exponential decay of (16) with depth will result in a significant effect of the incident waves only in a thin "boundary-layer" region near the free surface. Alternatively, since the damping coefficients vanish in the high-frequency limit, it follows from (173) that the three-dimensional exciting force must tend to zero faster than $(V_g/k)^{1/2}$, or using (17) and (77),[6] gives

$$|X_i(\theta)| = o(\omega^{-3/2}), \tag{175}$$

as $\omega \to \infty$. In two dimensions, it follows from (174) that

$$|X_i| = o(\omega^{-1/2}), \tag{176}$$

as $\omega \to \infty$. These estimates are conservative, and more precise high-frequency limits for heave are given in the paper by Davis (1976), and references cited therein. For sway, see problem 15.

The simplest representation for the exciting force and moment is based on the assumption that the pressure field is not affected by the presence of the body and can be determined from the incident wave potential by itself. In other words, the diffraction potential ϕ_7 is neglected completely in the integrals (160–162). This approach was utilized in the earliest theories for ship motions in waves; it is known as the *Froude-Krylov hypothesis* in honor of the authors of those theories.

Historically, the Froude-Krylov approach was expedient, but in certain cases it can be justified. For a thin ship, with centerplane $z = 0$, Peters and Stoker (1957) have shown that the Froude-Krylov exciting force and moment are the leading-order contributions in the vertical plane, i.e., for surge, heave, and pitch ($j = 1, 2, 6$). A similar result is established for slender bodies in chapter 7. Both proofs are valid only if the wavelength is large compared to the beam.

To illustrate the Froude-Krylov approach for a thin ship, we start with (162) and restrict our attention to motions in the vertical plane ($i = 1, 2, 6$). The normal components n_x, and n_y can be approximated by the slopes $\partial\zeta/\partial x$ and $\partial\zeta/\partial y$, respectively, where $\zeta(x, y)$ is the local half-beam which is assumed small. From the boundary conditions (120–121), the normal derivatives $\partial\phi_i/\partial n$ will be first-order quantities, proportional to the beam. For *following or head* waves ($\theta = 0$ or $180°$), it follows from the boundary condition (124) that the diffraction potential ϕ_7 also will be a small first-order quantity proportional to n_x and n_y.

For oblique waves, the normal component n_z contributes to the boundary condition (124), and for this reason the effect of the diffraction potential on sway, roll, and yaw is not small. Nevertheless its contribution to the vertical modes remains small in oblique waves, as a consequence of symmetry.[7]

Since ϕ_0 is a function of order one, independent of the beam of the ship, the exciting force (172) is given to first order by

$$x_i \simeq -\rho \iint_{S_B} \phi_0 \frac{\partial \phi_i}{\partial n} dS, \quad i = 1,2,6. \tag{177}$$

Restricting our attention to the heave-exciting force in deep water, we obtain the potential ϕ_0 from (16),and using $\partial\phi_2 / \partial n \simeq i\omega \partial\zeta / \partial y$ as the thin-ship approximation of (120), we get

$$X_2 \simeq 2\rho g \iint \frac{\partial \zeta}{\partial y} e^{ky - ikx\cos\theta}\, dx\, dy. \tag{178}$$

Here the integral is over the centerplane $z = 0$, and the factor of two accounts for both sides of the body. The surge-exciting force and pitch moment are similar to (178), but with $\partial\zeta/\partial x$ and $(x\ \partial\zeta/\partial y - y\ \partial\zeta/\partial x)$ in place of $\partial\zeta/\partial y$. These expressions are analogous to (116) for the waves generated by a thin ship in steady forward motion.

The heave-exciting force acting on a thin two-dimensional cylinder can be obtained as a special case of (178), where ζ depends only on y, and the x-integration is from minus infinity to plus infinity. The exciting force cancels along the length unless the incident wave crests are parallel to the cylinder. In the latter case, the heave-exciting force, per unit length, is given by

$$X_2 \simeq 2\rho g \int_{-T}^{0} \frac{d\zeta}{dy} e^{ky}\, dy, \tag{179}$$

where T is the draft, and the integral is over the vertical extent of the body.

Substituting the exciting force (178) in (173) gives the heave-damping coefficient for a thin ship in the form

$$b_{22} = \frac{\rho g k}{2\pi V_g} \int_0^{2\pi} \left| \iint \frac{\partial \zeta}{\partial y} e^{ky - ikx\cos\theta} dx\, dy \right|^2 d\theta. \tag{180}$$

Similar expressions can be derived for surge and pitch, and the analogy with Michell's integral (117) should be noted.

For the two-dimensional cylinder, we use (174) and (179), to get

$$b_{22} = \frac{2\rho g}{V_g} \left| \int_{-T}^{0} \frac{\partial \zeta}{\partial y} e^{ky} dy \right|^2 . \tag{181}$$

From this result one can deduce the *wavemaker theory*, for the waves generated by a vertical wavemaker oscillating horizontally at one end of a wave tank (see problem 16). The wavemaker theory, as well as the damping coefficients (180) and (181), can be derived directly from solutions of the corresponding forced-motion potentials ϕ_i, as noted in Wehausen and Laitone (1960). However, the Haskind relations (172–174) can be used to deduce these results much more simply.

The proof of the Haskind relations is independent of the water depth, and (178–181) can be generalized to the case of finite depth simply by replacing the exponential factor e^{ky} by the ratio of hyperbolic cosines in (22).

6.19 Motion of Floating Bodies in Regular Waves

In sections 6.16–6.18 the various components of the total pressure force and moment (130) have been discussed, as well as the force and moment due to the weight of the body. We are now in a position to derive the equations of motion for free oscillations of the body in waves by equating the above forces to the inertial forces associated with acceleration of the body mass. Here we assume that the body is rigid, unrestrained, and in a state of stable equilibrium when in calm water.

The six components of the inertial force, associated with the body mass, can be obtained from section 4.16. With the assumption of linearized motions it follows that

$$F_i = \sum_{j=1}^{6} M_{ij} \dot{U}_j, \quad i = 1,2,\ldots,6, \tag{182}$$

where the matrix M_{ij} is defined by

$$M_{ij} = \begin{Bmatrix} m & 0 & 0 & 0 & 0 & -my_G \\ 0 & m & 0 & 0 & 0 & 0 \\ 0 & 0 & m & my_G & 0 & 0 \\ 0 & 0 & my_G & I_{11} & I_{12} & I_{13} \\ 0 & 0 & 0 & I_{21} & I_{22} & I_{23} \\ -my_G & 0 & 0 & I_{31} & I_{32} & I_{33} \end{Bmatrix}. \tag{183}$$

Here the body mass is

$$m = \iiint_{\mathcal{V}_B} \rho_B \, d\mathcal{V}, \tag{184}$$

the mass-density is ρ_B, and the moments of inertia can be defined in the form

$$I_{ij} = \iiint_{\mathcal{V}_B} \rho_B [\mathbf{x} \cdot \mathbf{x} \delta_{ij} - x_i x_j] d\mathcal{V}, \quad i,j = 1,2,3, \tag{185}$$

where δ_{ij} is the Kroenecker delta function, equal to one if $i = j$ and zero otherwise.

Equations of motion follow by equating the inertia forces (182) to the sum of the pressure forces (130) and the forces due to the body weight, which are incorporated in the total static restoring forces (144). Combining these relations, six simultaneous equations of motion are obtained in the form

$$\sum_{j=1}^{6} \xi_j(-c_{ij} + f_{ij}) + AX_i = -\omega^2 \sum_{j=1}^{6} M_{ij}\xi_j. \tag{186}$$

Rearranging this equation and replacing f_{ii} by the added-mass and damping coefficients from (152), we arrive at the result

$$\sum_{j=1}^{6} \xi_j[-\omega^2(M_{ij} + a_{ij}) + i\omega b_{ij} + c_{ij}] = AX_i. \tag{187}$$

These are six simultaneous linear equations, which can be solved for the body motions ξ_j by standard matrix-inversion techniques. Thus, in general, the body motion ξ_j will be given by an equation of the form

$$\xi_j = A\sum_{i=1}^{6} [C_{ij}]^{-1} X_i. \tag{188}$$

where C_{ij} denotes the total matrix in square brackets on the left side of (187). The ratio (ξ_j/A) is a quantity of fundamental significance, which we shall define separately by

$$Z_j(\omega,\theta) \equiv \xi_j/A = \sum_{i=1}^{6}[C_{ij}]^{-1}X_i. \tag{189}$$

Physically, this is the complex amplitude of body motion in the jth mode, in response to an incident wave of unit amplitude, frequency ω, and direction θ. This ratio is known as the *transfer function*, or the *response amplitude operator*. The transfer function can be calculated from (189) if the added-mass, damping, exciting, and hydrostatic forces are known.

To illustrate the nature of the transfer function (189), we return to the special case of heave for a body symmetrical about $x = 0$ and $z = 0$. From symmetry, there are no *cross-coupling* effects, except for possible coupling between surge and pitch or between sway and roll. Thus, in heave,

$$Z_2(\omega,\theta) = \frac{X_2}{C_{22}} = \frac{X_2}{-\omega^2(a_{22}+M_{22})+i\omega b_{22}+c_{22}}. \tag{190}$$

In the limit of low frequencies and long wavelengths, the exciting force in the numerator can be approximated by (169); then the limit is

$$Z_2(\omega,\theta) \to 1, \quad \text{as } \omega \to 0. \tag{191}$$

Physically, if the waves are sufficiently long, the body will simply ride up and down with the free surface.

At very high frequencies the denominator of (190) will be proportional to ω^2, while the exciting force in the numerator will tend to zero. Thus, as $\omega \to \infty$,

$$Z_2(\omega,\theta) = o(\omega^{-7/2},\omega^{-5/2}), \tag{192}$$

for three and two dimensions, respectively. More precise estimates can be developed from Davis (1976), and for horizontal motions as noted in problem 15.

Returning to the transfer function (190) for intermediate frequencies, the most significant feature will be a resonant response at the natural frequency where the virtual mass and restoring forces cancel, or where

$$\omega_n = \left(\frac{c_{22}}{a_{22}+M_{22}}\right)^{1/2}. \tag{193}$$

At or near this resonant frequency, the body will experience a response of large amplitude and a phase shift increasing from zero for low frequencies to 180 degrees for $\omega \gg \omega_n$.

As in the case of a mechanical oscillator, the resonant response will be inversely proportional to the damping coefficient, but here there is a connection between the exciting force and the damping coefficient from the Haskind relations. In particular, for two-dimensional or axisymmetric bodies the damping coefficient will be proportional to the square of the exciting force, and the resonant response will be *inversely* proportional to the exciting force. Thus, bodies deliberately designed with small exciting forces will experience a large resonant response, in a highly-tuned manner. Vertical spar buoys are important examples of this situation, as shown in figure 2.16 and in problem 17. By comparison, the floating hemisphere shown in figure 6.24 is typical of the situation where the exciting force, damping, and resonant response assume relatively moderate values.

The analogy with a mechanical oscillator can be emphasized by rewriting (187) in the time-dependent form

$$(M_{ij} + a_{ij})\ddot{\xi}_j(t) + b_{ij}\dot{\xi}_j(t) + c_{ij}\xi_j(t) = \mathrm{Re}(AX_i e^{i\omega t}). \tag{194}$$

However, the simplicity of this result is deceptive, and the analogy with a mechanical oscillator must not be pushed too far. Equation (194) has been derived assuming the motions are sinusoidal in time, as evidenced by the fact that the added-mass and damping coefficients depend on the frequency ω. Thus, one cannot extend (194) directly to the case where the time-dependence is more general, although one can do so by suitable superposition of the incident waves and body motions at each frequency. We shall return to this representation in section 6.21.

The distinction between the left side of (194) and a mechanical oscillator with constant coefficients can be further emphasized by noting that when the body is forced to oscillate, waves will be generated on the free surface. As time increases, these waves will propagate outward from the body, but they will continue to affect the fluid pressure and hence the body force for all subsequent times. Thus *memory effects* are introduced, in a manner precisely analogous to the unsteady lifting surface theory in chapter 5.

One consequence of the memory effects is that if $\xi_j(t)$ is an arbitrary function of time, and if the linearization assumption is invoked, the pressure force acting on the body must be of the general form

$$F_i(t) = \sum_{j=1}^{6} \int_{-\infty}^{t} K_{ij}(t-\tau)\dot{\xi}_j(\tau)d\tau. \tag{195}$$

This is a convolution integral, over the previous history of the fluid motion, and the kernel $K(t - \tau)$ can be interpreted as the force, at time t, due to a delta-function body velocity at an earlier time τ. This *impulse-response function* is discussed by Price and Bishop (1974). The analog in chapter 5 is the Wagner function shown in figure 5.35.

The restriction of (194) to sinusoidal motions and the significance of the convolution integral (195) have been emphasized by a number of authors including Haskind (1953), Cummins (1962), and Maskell and Ursell (1970). Maskell and Ursell point out that if a floating body in deep water is given an impulsive motion at time $t = 0$ and is unrestrained thereafter, the body will oscillate about its equilibrium position for a finite number of cycles, approaching equilibrium asymptotically as an inverse power of time. By comparison, the homogeneous solution of the differential equation (194), with constant coefficients, would decay exponentially with an infinite number of oscillations.

Throughout this discussion we have assumed that the body motions are oscillatory about a fixed mean position. An important exception to this situation is that of a ship, moving with forward velocity U while oscillating in response to ambient waves. The assumption of linearity permits this oscillatory problem to be superposed on the steady-state wave-resistance solution. Nevertheless there are fundamental effects of the forward speed on the oscillatory problem, the simplest and most important being the change in frequency due to the Doppler shift. The complete solution of this problem is not established to the same extent as the simpler problem with zero forward speed, but a strip-theory approach will be discussed in chapter 7.

6.20 Ocean Waves

The complexity of ambient wave motions in the ocean is evident to anyone who has observed them either from the beach or the deck of a ship. The wave patterns are ever-changing with time and space, in a manner that appears to defy analysis. This complexity is partly the result of dispersion, which already has been discussed in the context of deterministic wave motions generated by well-defined processes. Ship waves are an example where the uniformity of the resulting wave system is itself the most striking feature of all.

Ambient waves on the surface of the ocean are not only dispersive, but *random*. The generating mechanism is, predominantly, the effect upon the water surface of wind in the atmosphere. The wind is itself random, especially when viewed from the standpoint of the turbulent fluctuations and eddies which are important in generating waves. The randomness of ocean waves is subsequently enhanced by their propagation over large distances in space and time and their exposure to the random nonuniformities of the water and air. Thus ocean waves must be described in a probabilistic manner.

Despite their randomness, ocean waves can be modeled by a suitable distribution of plane progressive waves of all possible frequencies and directions, as in the double integral (59) where the wave motion is defined uniquely by the complex amplitude $A(\omega, \theta)$. In a deterministic problem with finite energy, this amplitude function can be found by Fourier techniques, for example, in terms of prescribed values of the wave elevation along the free surface, at some initial instant of time.

For random ocean waves this approach must be modified. First, the required initial data cannot be measured everywhere on the free surface. In addition, Fourier theory requires the function to be represented, that is, $\eta(x, z, t)$, to be square-integrable over the entire domain, in this case the free surface. Ambient ocean waves do not vanish anywhere on the free surface, and while their features may be noticeably different at widely separated points in the ocean, an integral such as $\iint_{-\infty}^{\infty} |\eta|^2 \, dx \, dz$ will be finite only in the sense that the area of the oceans is limited.

Faced with a dilemma of whether the area of the oceans is finite or infinite, we find it appropriate to consider the length-scale and time-scale of interest to us in our description of the ocean-wave environment. In the context of ocean engineering and naval architecture, our interest is to describe the response of vessels and structures with dimensions comparable to a few hundred meters, at the most, and with characteristic response times on the order of a few seconds or, in extreme cases, minutes.

For this purpose, it should suffice to describe the wave environment over a small portion of the oceans, say on the order of tens or hundreds of kilometers squared. We have no interest in predicting simultaneously the environment throughout the world, or even throughout the North Atlantic Ocean. The complexity of developing such an extensive description, assuming it to be possible, would outweigh completely the advantages of making the classical Fourier theory valid!

By the same argument, we are interested in typical wave environments that might exist in conjunction with storms of varying severity, but not with the transient development of these waves with time. Therefore it should suffice to describe the wave environment over a period of a few hours, and to assume that the wave motion is *stationary* during this interval of time. It corresponds to this assumption that the waves are *homogeneous* in space over the area in question. These statements have meaning only in the statistical sense, since for a random wave system it would be ridiculous to suggest that the precise wave motion is the same at different points in space or time.

Another consideration from the practical standpoint, which avoids the need or desire for a deterministic solution, is that our interest in the wave environment is confined not to any one instant of time, but to a typical situation that might be expected over a representative period on the order of several hundred wave encounters. For this reason, the precise phase of any one wave is of no interest, and it is reasonable to assume that the phase of each component is distributed randomly with equal probability between 0 and 2π.

This last assumption overcomes the problem of square-integrability in the Fourier integral theory, but strictly it is necessary to replace the distribution (59) by a Fourier-Stieltjes representation of the form

$$\eta(x,z,t) = \mathrm{Re}\iint dA(\omega,\theta)\exp[-ik(x\cos\theta + z\sin\theta) + i\omega t]. \tag{196}$$

Here the wavenumber k is defined in terms of the frequency ω by the dispersion relation (17) or (24). The differential wave amplitude dA is a complex quantity of random phase, with a magnitude proportional to the square root of the total energy in a differential element of the frequency-wave-angle space bounded by (ω, $\omega + d\omega$) and (θ, $\theta + d\theta$). This type of representation, described by Yaglom (1962), can be circumvented by using infinite series of discrete waves in place of the continuous distribution (196).

The average energy density can be computed by squaring (196) and taking its average value; it then takes the form

$$\overline{\eta^2} = \frac{1}{2}\overline{\iint dA(\omega,\theta)\exp[-ik(\omega)(x\cos\theta + z\sin\theta) + i\omega t]}$$
$$\overline{\times \iint dA^*(\omega',\theta')\exp[ik(\omega')(x\cos\theta' + z\sin\theta') - i\omega' t]}. \tag{197}$$

Here (*) denotes the complex conjugate. The only contribution to this average is from the combinations $(\omega, \theta) = (\omega', \theta')$, with the result

$$\overline{\eta^2} = \frac{1}{2}\iint dA(\omega,\theta)\,dA^*(\omega,\theta) \equiv \int_0^\infty \int_0^{2\pi} S(\omega,\theta)\,d\theta\,d\omega. \tag{198}$$

Unlike the wave amplitude A, the function $S(\omega, \theta)$ is regular and the last integral can be interpreted in the ordinary Riemann sense. Derivations of this result that avoid the subtleties of (196–197) can be found in Neumann and Pierson (1966) and Price and Bishop (1974).

This approach to random ocean waves has been motivated by the deterministic distribution of plane waves and the desire to extend that representation to the random case. The spectral density $S(\omega, \theta)$ is more commonly defined probabilistically as the Fourier transform of the correlation function for the free-surface elevation. This description is outlined in Phillips (1966) and in the statistical theory of turbulence by Monin and Yaglom (1971).

Comparing (198) and (69), we can infer that the total mean energy of the wave system per unit area of the free surface is equal to

$$\bar{E} = \rho g \int_0^\infty \int_0^{2\pi} S(\omega,\theta)\,d\theta\,d\omega. \tag{199}$$

It is customary to ignore the factor ρg and to refer to the function $S(\omega, \theta)$ as the *spectral energy density*, or simply the *energy spectrum*. More specifically, this is a *directional* energy spectrum; it can be integrated over all wave directions to give the *frequency* spectrum

$$S(\omega) = \int_0^{2\pi} S(\omega,\theta)\,d\theta. \tag{200}$$

If one attempts to find the ocean-wave spectrum from measurements of the free-surface elevation at a single point in space, for instance by recording the heave motion of a buoy, the directional characteristic of the waves will be lost. Only the frequency spectrum (200) can be determined from such a restricted set of data. A limited amount of directional information follows if one measures the slope of the free surface, for example by measuring the angular response of the buoy as well as its heave. A complete description of the directional wave spectrum requires an extensive array of measurements at several adjacent points in space, however, and there are practical difficulties associated with this task.

As a simpler alternative, one can assume that the waves are unidirectional, with the energy spectrum proportional to a delta function in θ. Wave spectra of this form are called *long crested,* since the fluid motion is two-dimensional and the wave crests are parallel. If one assumes that the spectrum is of this character, with the direction prescribed, the frequency spectrum (200) is sufficient to describe the wave environment.

If the waves are generated by a single storm, far removed from the point of observation, it might be presumed that these waves would come from the direction of the storm in a long-crested manner. The limitations of this assumption are apparent to anyone who has observed the ocean surface. While a preferred direction may exist, especially for long swell that has traveled large distances, even these long waves will be distributed in their direction, and for short steep waves the directional variation is particularly significant. Since the superposition of such waves from a range of different directions appears in space as a variation of the free-surface elevation in all directions, these waves are known as *short-crested* waves.

Information on directional spectra is limited, although the knowledge of this subject is likely to increase as sophisticated measurement techniques and data analysis are applied. More extensive discussion of this subject is given by Kinsman (1965), Neumann and Pierson (1966), Lewis (1967), and Price and Bishop (1974).

In part from necessity, but with the presumption that a conservative estimate of wave effects on vessels will result, it is customary to proceed in the fields of ocean engineering and naval architecture on the assumption that the waves are long-crested. With this simplification it is possible to use existing information for the frequency spectrum (200), which is based on a judicious combination of theory and full-scale observations.

Ocean-wave spectra depend on the velocity of the wind as well as its duration in time and the distance over which the wind is acting on the free surface. This distance is known as the *fetch*. Wave spectra that have reached a steady-state of equilibrium, independent of the duration and fetch are said to be *fully developed.* A semi-empirical expression for the frequency spectrum of fully developed waves is

$$S(\omega) = \frac{\alpha g^2}{\omega^5} \exp[-\beta(g/U\omega)^4]. \tag{201}$$

Here α and β are nondimensional parameters defining the spectrum, and U is the wind velocity at a standard height of 19.5 meters above the free surface. This two-parameter spectrum is sufficiently general to fit most observations and is consistent with theoretical predictions of the high-frequency limit.

The most common values for the parameters in (201) are

$$\alpha = 8.1 \times 10^{-3}, \tag{202}$$

$$\beta = 0.74; \tag{203}$$

with these values, (201) is known as the Pierson-Moskowitz spectrum. The resulting family of curves is shown in figure 6.25, for wind speeds of 10 to 50 knots.

If the spectrum is assumed to be of narrow bandwidth,[8] the wave amplitudes follow a *Rayleigh distribution.* Thus the probability that the wave amplitude A lies within the differential increment $(A, A + dA)$ is $p(A)\,dA$, where the normalized *probability density function* is given by

$$p(\zeta) = \zeta e^{-\zeta^2/2}. \tag{204}$$

Here, $\zeta = A/m_0^{1/2}$ is the normalized wave amplitude, and m_0 is the total energy of the spectrum, or the integral of $S(\omega)$ over all frequencies. More generally the *moments* m_j of the spectrum are defined as

$$m_j = \int_0^\infty \omega^j S(\omega)\,d\omega, \quad j = 0,1,2,\ldots. \tag{205}$$

Statistics for the wave height $H = 2A$ can be derived using the probability density function (204). Thus the average wave height $\bar{H}$ is given by

$$\bar{H} = 2\int_0^\infty A p(\zeta)\,d\zeta = (2\pi m_0)^{1/2}. \tag{206}$$

For most purposes, however, we are interested primarily in the larger waves. The most common parameter that takes this into account is the *significant*

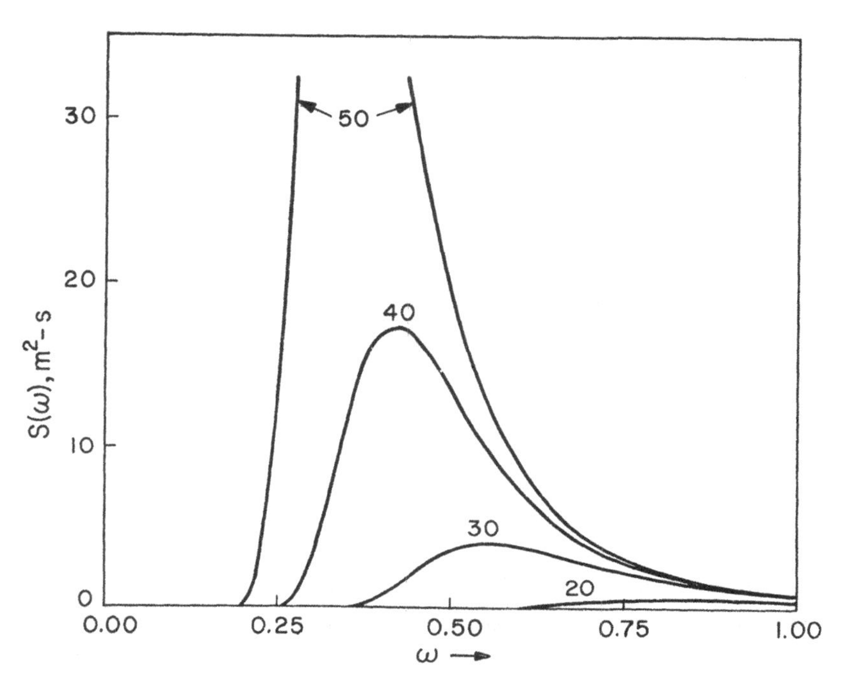

Figure 6.25
The Pierson-Moskowitz frequency spectrum (201–203) for the wind speeds in knots as listed in table 6.1. The peak value for 50 knots is at $S = 54$ m^2-s. The curve for 10 knots cannot be shown on this scale, its peak value being $S = 0.02$ m^2-s.

wave height $\bar{H}_{1/3}$, defined as the average of the highest one third of all the waves. This can be computed as the ratio

$$\bar{H}_{1/3} = 2\int_{\zeta_0}^{\infty} Ap(\zeta)\,d\zeta \Big/ \int_{\zeta_0}^{\infty} p(\zeta)\,d\zeta = 4.0(m_0)^{1/2}, \tag{207}$$

where the lower limit of the integrals is defined such that

$$\int_{\zeta_0}^{\infty} p(\zeta)\,d\zeta = 1/3. \tag{208}$$

Averages that focus upon less frequent maxima can be defined in a corresponding manner; for example, the average of the highest one tenth of all waves can be defined as

$$\bar{H}_{1/10} = 5.1(m_0)^{1/2}. \tag{209}$$

The average frequency of the spectrum can be defined as the expected number of zero upcrossings per unit time, that is, the number of times the wave elevation passes through zero with positive slope. The derivation is given by Price and Bishop (1974), and the final result is

$$\bar{\omega} \equiv \left(\int_0^\infty \omega^2 S(\omega)\, d\omega \Big/ \int_0^\infty S(\omega)\, d\omega \right)^{1/2} = \left(\frac{m_2}{m_0} \right)^{1/2}. \tag{210}$$

The moments m_0 and m_2 can be computed without difficulty for the two-parameter spectrum (201), and it follows that

$$\bar{H}_{1/3} = (2.0)\frac{U^2}{g}\left(\frac{\alpha}{\beta}\right)^{1/2}, \tag{211}$$

and

$$\bar{\omega} = (\pi\beta)^{1/4}(g/U). \tag{212}$$

Substituting the values (202–203) for the Pierson-Moskowitz spectrum gives the results shown in table 6.1. Also shown in this table is the approximate Beaufort scale of the sea state for each wind speed and the average value of the period, as derived by Neumann and Pierson (1966). The last column shows the average wavelength corresponding to each wave period, based on the dispersion relation for plane progressive waves.

The results shown in table 6.1 are based on the assumption of a fully developed sea. It is unusual for a storm of 50 knots' intensity to last long enough for the sea to develop fully. Thus, while wind speeds of this intensity are not uncommon, the corresponding wave height and length shown in table 6.1 are rare.

Table 6.1
Statistical Parameters and Approximate Sea State of the Pierson-Moskowitz Spectrum, for the Wind Speeds Shown in Figure 6.25

Wind Speed (knots)	Sea State (approx.)	Significant Wave Height (meters)	Average Period (seconds)	Average Wavelength (meters)
10	2	0.6	2.7	11
20	4	2.2	5.3	45
30	6	5.0	8.0	100
40	7	8.9	10.7	178
50	8	13.9	13.4	278

Much of the uncertainty regarding the process by which the wind speed is related to the waves can be avoided by identifying the Pierson Moskowitz spectrum in terms of the significant wave height rather than the wind speed or sea state. The situation is somewhat more complicated than this, however. If the sea is not fully developed, the corresponding wavelengths will be shorter since dispersion has less opportunity to occur. Thus, the Pierson-Moskowitz spectrum underestimates the peak frequency for the higher spectra and conversely for smaller waves. To account for this problem, the two-parameter spectrum (201) is used with full-scale observations of the significant wave height and period. The parameters α and β are then determined from (211–212).

Fetch-limited wave spectra measured during the Joint North Sea Wave Project (JONSWAP) show higher peaks compared to the Pierson-Moskowitz spectrum and the two-parameter family (201). The engineering implications of the JONSWAP results are discussed by Houmb and Overik (1976).

The significant wave height is relevant as an indication of the most likely waves to be encountered, but for design purposes a more conservative estimate is required. Thus we are led to consider the statistics of *extreme values*.

From the Rayleigh distribution (204), the cumulative probability of a single wave being less than $A = \zeta(m_0)^{1/2}$ is

$$P(\zeta) = \int_0^{\zeta} p(\zeta')d\zeta' = 1 - e^{-\zeta^2/2}. \tag{213}$$

The probability that the amplitudes of N statistically independent waves are all less than A is P^N, and the probability that at least one wave amplitude will exceed this value is

$$1 - P^N = 1 - (1 - e^{-\zeta/2^2})^N. \tag{214}$$

The negative derivative of (214) gives the probability density of the extreme value in N cycles, in the form

$$g(\zeta) = N_p(\zeta)[P(\zeta)]^{N-1}. \tag{215}$$

The wave amplitude for which this probability density function is a maximum can be found by setting the derivative of (215) equal to zero. Simplifying the resulting expression for $N \gg 1$, the most probable extreme value in N waves is given by

$$A = (2m_0 \log N)^{1/2}. \tag{216}$$

Ochi (1973) has shown that the probability of exceeding (216), for large N, is $1 - e^{-1} = 0.632$. Thus a more conservative criterion is required, and for this purpose we define the *design extreme value* as the wave amplitude that will be exceeded in N encounters with a probability of only 1 percent. From (214) it follows that

$$1 - P^N = 0.01, \tag{217}$$

and thus

$$P = (0.99)^{1/N} = e^{(1/N)\log(0.99)} \simeq e^{-0.01/N} \simeq 1 - 0.01/N. \tag{218}$$

Combining the last result with (213) and solving for the wave amplitude, gives

$$A = \left(2m_0 \log \frac{N}{0.01}\right)^{1/2}. \tag{219}$$

This is the design extreme value, or the maximum wave amplitude in N encounters that will not be exceeded with probability 0.99. Note from (216) that this result is equal to the most probable extreme value in $100N$ encounters.

Since the significant wave amplitude is half the wave height (207), the design extreme value exceeds the significant value by the ratio

$$\frac{\text{Design Extreme}}{\text{Significant}} = \left(\frac{1}{2}\log \frac{N}{0.01}\right)^{1/2}. \tag{220}$$

This factor is not sensitive to the value of N, nor to the arbitrary choice of 0.01 as the probability that the extreme wave will exceed this value. For example, with $N = 100$, (220) is equal to 2.15; for $N = 10^4$ this factor increases to 2.63. Since (219) is much more sensitive to the value of m_0 than to $(N/0.01)$, the most important factor to estimate correctly is the total wave energy.

In a comprehensive derivation of the extreme-value statistics, Ochi (1973) shows that the amplitude of the design extreme wave is reduced if the wave spectrum is not of narrow bandwidth. However, the net reduction is less than 10 percent in all cases.

6.21 Motions of Bodies in Irregular Waves

With the spectral description of ocean waves given in the preceding section, we can return to the problem of body motions and generalize the earlier treatment of regular waves with sinuosidal time-dependence. If the ocean waves are described by the random distribution (196), and if the response of the body to each component wave is defined by a linear transfer function $Z(\omega, \theta)$, the body response will be

$$\xi_j(t) = \mathrm{Re}\iint Z_j(\omega,\theta)e^{i\omega t}\, dA(\omega,\theta). \tag{221}$$

The principal assumption is that linear superposition applies, as it must in any event for the underlying development of the transfer function and spectrum.

Like the waves themselves, the response (221) is a random variable. The mean-square of this quantity can be computed in precisely the same manner as (197–198), giving the result

$$\overline{\xi_j^2} = \int_0^\infty \int_0^{2\pi} S(\omega,\theta)\left|Z_j(\omega,\theta)\right|^2 d\omega\, d\theta. \tag{222}$$

The distribution of response maxima is approximated by a Rayleigh distribution. Thus the statistics of the body response are identical to the wave statistics as analyzed in section 6.20, except that the wave-energy spectrum S is multiplied by the square of the transfer function.

To a large extent these relations provide the justification for our earlier analysis of regular-wave response. The transfer function $Z_j(\omega, \theta)$ is valid not only in regular waves, where it has been derived, but also in superpositions of regular waves, and ultimately in a spectrum of random waves. Generally speaking, a vessel with favorable response characteristics in regular waves will be good in irregular waves, and vice versa. This statement is oversimplified, however, and the relative "shape" of the energy spectrum and the transfer function is crucial to the value of integrals such as (222). For example, a large resonant response of the body will be of importance if the resonant frequency coincides with the peak of the wave-energy spectrum, and vice versa. Figure 2.16 shows the extreme resonant peak which results for a slender spar buoy, but the spar buoy FLIP, described by Rudnick (1967), is sufficiently large that the natural period in heave (27 seconds) is beyond the range of expected wave excitation. Similarly, a small unrestrained buoy will have a high natural frequency in heave; as such its resonant response will be of little significance. From the corresponding limits (192) and (191) for the transfer function in heave, it is obvious that FLIP will be stable with practically no heave motion, while the small buoy will follow the free-surface elevation. Resonance is not significant in either example for normal conditions.

As in the case of the ambient waves, the phase of the response (221) is a random variable of no importance. However the relative phase, between the response $\xi(t)$ and the wave $\eta(t)$, or between two different modes of response such as pitch and heave, may be a quantity of great significance. For example, the relative motion $(\xi_2 - \eta)$ between the body and the free surface determines the probability of submergence or emersion. Thus one may be concerned whether these random variables are in phase or out of phase. Here the argument of the transfer function is relevant, and can be used to determine the correlation between the different random variables noted above. For heave, we recall that for frequencies below the resonant frequency the argument of the transfer function is close to zero, whereas at high frequencies the body and waves move with a phase difference of 180°. For floating bodies, as for mechanical systems, coupling between two modes such as heave and pitch will introduce a pair of natural frequencies, both of which must be considered in this context.

Little has been said here about motions in the horizontal plane, where no resonance exists. It is essentially for the latter reason that these modes are of less importance, but their analysis can be carried through in much the same manner as has been outlined here for heave. Resonant horizontal motions may result from an elastic mooring, and second-order drift processes in the horizontal plane are of practical importance in some cases. Both subjects are beyond the scope of this text, but references can be found in Bishop and Price (1975). Additional references are noted in section 2.15.

Problems

1. For plane progressive waves of height 6 m and length 200 m in deep water, find the phase velocity, the maximum velocity of a fluid particle, and the position where this maximum occurs. How do these change if the fluid depth is 30 m?

2. Give the velocity potential for standing waves in finite depth, which reduces to (53) as $kh \rightarrow \infty$. Show that this corresponds to the solution of a problem where incident waves are totally reflected by a vertical wall. Using the linearized Bernoulli equation, find the horizontal wave force exerted on the wall. Compute the magnitude of this force if $A = \frac{1}{2}$ m, $h = 10$ m, and $\lambda = 50$ m.

3. The mass transport (49) can be computed to second order in finite depth by taking the time average of the horizontal flux across a vertical control surface, between $-h$ and η, after noting that $\phi_x = -(k/\omega)\phi_t$ for a progressive wave and that the time average of ϕ_t must be zero if the motion is periodic. In this manner show that the generalization of (49) in the form $Q = \frac{1}{2} g A^2 / V_p$ is valid for finite depths.

4. Derive expressions for the kinetic and potential energies of a standing wave in finite depth, and show that the total energy per wavelength is constant. Since the standing wave is equivalent to two identical plane progressive waves propagating in opposite directions, show that the energy of each plane wave is divided equally into kinetic and potential portions.

5. Compute the maximum electric power that can be generated by a device that converts all of the energy from plane progressive waves, of height 1 m and wavelength 100 m, over a width of 1 km parallel to the wave crests.

6. A wavemaker at $x = 0$ is turned on at time $t = 0$. At $t = 10$ s the maximum possible wave amplitude is desired at $x = 20$ m. Show that the appropriate frequency for the wavemaker is $\omega = C_1 - C_2 t$, assuming deep water. Find the constants C_1 and C_2. Describe qualitatively how this result can be changed to account for finite-depth effects.

7. A ship model of length 5 m is moved with a constant speed of 1 m/s for a distance of 100 m, in a wide towing tank. At the end of this run the model is stopped, and thereafter it is affected by the transverse waves coming upon it from astern. What is the period of these waves, and what is their wavelenth? Assuming the back edge of this wave system is sharp, separating a region of calm water from the region of the transverse waves, how many waves exist in the towing tank at the time the model is stopped? How many waves will ultimately move past the model after it has stopped?

8. Waves of period five seconds are incident upon a gradually shoaling beach in a two-dimensional manner. At what depth will the wave height

be a minimum? If the wave height $2A$ in deep water is 1 m and if the waves break when the depth is equal to the local wave height, what is the wave height when breaking commences? Explain why regular waves incident upon a beach from an oblique angle must appear steady-state to an observer moving along the beach at a suitable velocity, for all depths. Show from this that the waves are refracted toward the beach; that is, show that the wave crests become more nearly parallel to the beach as the depth decreases.

9. Show that a round homogeneous log is neutrally stable, with respect to static roll about its axis, for any density such that it will float. Show that the position of stable equilibrium of a square homogeneous log depends on its density.

10. A rectangular barge of length 100 m, beam 20 m, and draft 8 m is assumed to have an added mass equal to its displaced mass. Find the resonant wavelength for heave if the barge is moored with a slack mooring cable.

11. From energy conservation, show that the damping coefficient of a two-dimensional body symmetrical about $x = 0$ is related to the amplitude A_i of the far-field waves, generated by its motion in deep water, by the equation

$$b_{ii} = \frac{\rho g^2}{\omega^3}\left|\frac{A_i}{\xi_i}\right|^2 .$$

Substitute this result in (172) and carry out the vertical integration at infinity to derive the two-dimensional version of the Haskind relations in the form (174).

12. Using (173), show that the heave damping coefficient of a body of revolution with vertical axis is equal to

$$b_{22} = \frac{k}{4\rho g V_g}|X_2|^2 .$$

Show that for sway or surge a similar expression holds with a multiplicative factor of one half.

13. For a two-dimensional floating body, show that a suitable linear combination of heave and sway oscillations at the same frequency will result in outgoing waves only at $x = \infty$, with no waves at $x = -\infty$. By reversing the sign of time t, deduce that the corresponding body motions act to absorb an incident wave system with 100 percent efficiency.

14. Using the long-wave approximation (169) in conjunction with the exact result (173), obtain the first two nonvanishing terms in power-series expansions for the damping and exciting force, in powers of the frequency, for heaving motions in deep water. Compare these results with the exact calculations for a floating hemisphere shown in figure 6.24.

15. Consider the horizontal exciting force on a floating body in the short-wavelength limit $\lambda/l \ll 1$. If the body surface is vertical at the intersection with the free surface, explain the fact that the body can be replaced by a vertical cylinder with the same waterplane and large depth. Use the standing wave solution (53) to estimate the sway exciting force on a two-dimensional body for $\lambda/l \ll 1$.

16. Show from the respective boundary conditions that the wavemaker problem, of a prescribed horizontal velocity along a vertical wavemaker, is equivalent to a heaving thin ship of suitable shape. Using this result, (181), and the equation shown in problem 11, prove that the amplitude of the waves generated by a paddle-type wavemaker with horizontal velocity $U(y) \cos \omega t$ extending over the entire depth of deep water is given by

$$A = \frac{2\omega}{g} \int_{-\infty}^{0} U(y) e^{ky} \, dy.$$

Show from this result that a two-dimensional wavemaker, consisting of a paddle that rotates with oscillatory velocity about a submerged pivot point, will not generate waves at infinity if the depth of the pivot is suitably chosen in terms of the wavelength.

17. A vertical spar buoy of circular cylindrical form, draft T, and diameter d is freely floating. Compute the hydrostatic restoring forces and moments. Estimate the natural frequency in heave, assuming that the buoy is sufficiently slender that the added mass and damping coefficients can be neglected by comparison to the mass of the buoy. Estimate the exciting force from the Froude-Krylov approximation, the damping coefficient from the Haskind relations, and compute the heave response. Compare your answer with the results shown in figure 2.16.

18. A spherical floating platform of diameter 50 m is ballasted to float on its equatorial plane and is unrestrained in a sea state corresponding to a 40 knot Pierson-Moskowitz spectrum. On the same scale of frequency, plot the heave response per unit wave amplitude from figure 6.24, the

frequency spectrum of the incident waves, and the frequency spectrum of the heave response of the buoy in these waves. Using graphical integration, find the average of the one-third highest heave amplitudes, the most probable extreme value of the heave amplitude in 1000 wave encounters, and the design extreme value of the heave amplitude in the same number of waves.

References

Bishop, R. E. D., and W. G. Price eds. 1975. *Proceedings of the international symposium on the dynamics of marine vehicles and structures in waves.* London: Institution of Mechanical Engineering.

Cummins, W. E. 1962. The impulse response function and ship motions. *Schiffstechnik* 9:101–109.

Davis, A. M. J. 1976. On the short surface waves due to an oscillating, partially immersed body. *Journal of Fluid Mechanics* 75:791–807.

Eggers, K., S. Sharma, and L. Ward. 1967. An assessment of some experimental methods for determining the wavemaking characteristics of a ship form. *Society of Naval Architects and Marine Engineers Transactions* 75:112–157.

Gadd, G. 1976. A method of computing the flow and surface wave pattern around full forms. *Royal Institution of Naval Architects Transactions* 118:207–216.

Graff, W., A. Kracht, and G. Weinblum. 1964. Some extensions of D. W. Taylor's standard series. *Society of Naval Architects and Marine Engineers Transactions* 72:374–403.

Haskind, M. D. 1953. *Oscillation of a ship on a calm sea.* English translation, Society of Naval Architects and Marine Engineers. T & R Bulletin 1–12.

Havelock, T. H. 1963. *Collected papers. ONR/ACR-103.* Washington: U.S. Government Printing Office.

Houmb, O. G., and T. Overik. 1976. Parameterization of wave spectra and long term joint distribution of wave height and period. In *Proceedings of the conference on the behaviour of offshore structures* (BOSS '76). pp. 144–169. Trondheim: The Norwegian Institute of Technology.

Kim, W. D. 1966. On a freely floating ship in waves. *J. Ship Res.* 10:182–191+200.

Kinoshita, M. 1943. Wave resistance of a sphere in a shallow sea. *J. Society of Naval Architects of Japan,* 19–37.

Kinsman, B. 1965. *Wind waves—their generation and propagation on the ocean surface.* Englewood Cliffs, N. J.: Prentice-Hall.

Kostyukov, A. A. 1959. *Theory of ship waves and wave resistance*. Leningrad: Sudpromfiz. English translation 1968. Iowa City. E.C.I.

Lamb, H. 1932. *Hydrodynamics*. 6th ed. Cambridge: Cambridge University Press. Reprinted 1945, New York: Dover.

Le Méhauté, B. 1976. *An Introduction to hydrodynamics and water waves*. New York, Heidelberg, Berlin: Springer-Verlag.

Lewis, E. V. 1967. The motion of ships in waves. In *Principles of naval architecture*, ed. J. P. Comstock, 607–717. New York: Society of Naval Architects and Marine Engineers.

Lighthill, M. J. 1965. Group velocity. *J. Institute of Mathematics and its Applications* 1:1–28.

Lighthill, M. J. 1967. A discussion of nonlinear theory of wave propagation in dispersive systems. *Proceedings of the Royal Society*, Series A, 299: no. 1456.

Longuet-Higgins, M. S. 1953. Mass transport in water waves. *Philosophical Transactions of the Royal Society of London. Series A, Mathematical and Physical Sciences* 245:535–581.

Maskell, S. F., and F. Ursell. 1970. The transient motion of a floating body. *Journal of Fluid Mechanics* 44:303–314.

Monin, A. S., and A. M. Yaglom. 1971. *Statistical fluid mechanics. English translation.* Cambridge, Mass.: The MIT Press.

Moore, C. S. 1967. Intact stability. In *Principles of naval architecture*, ed. J. P. Comstock, 121–166. New York: Society of Naval Architects and Marine Engineers.

Neumann, G., and W. J. Pierson. 1966. *Principles of physical oceanography*. Englewood Cliffs, N. J.: Prentice-Hall.

Newman, J. N. 1970. Recent research on ship waves. In *Eighth symposium on naval hydrodynamics*, eds. M. S. Plesset, T. Y. Wu, and S. W. Doroff, pp. 519–545. Washington: U.S. Government Printing Office.

Newman, J. N. 1976. The interaction of stationary vessels with regular waves. In *Eleventh symposium on naval hydrodynamics*, eds. R. E. D. Bishop, A. G. Parkinson, and W. G. Price, pp. 491–501. London: The Institution of Mechanical Engineers.

Noblesse, F., and G. Dagan. 1976. Nonlinear ship-wave theories by continuous mapping. *Journal of Fluid Mechanics* 75:347–371.

Ochi, M. K. 1973. On prediction of extreme values. *J. Ship Res* 17:29–37.

Ogilvie, T. F. 1964. Recent progress toward the understanding and prediction of ship motions. In *Fifth symposium on naval hydrodynamics*, eds. J. K. Lunde and S. W. Doroff, pp. 3–128. Washington: U.S. Government Printing Office.

Peters, A. S., and J. J. Stoker. 1957. The motion of a ship as a floating rigid body in a seaway. *Communications on Pure and Applied Mathematics* 10:399–490.

Phillips, O. M. 1966. *The dynamics of the upper ocean*. Cambridge: Cambridge University Press.

Price, W. G., and R. E. D. Bishop. 1974. *Probabilistic theory of ship dynamics*. London: Halsted.

Rawson, K. J., and E. C. Tupper. 1968. *Basic ship theory*. London: Longmans.

Rudnick, P. 1967. Motion of a large spar buoy in sea waves. *J. Ship Res.* 11:257–267.

Ruellan, F., and A. Wallet. 1950. Trajectoires internes dans un clapotis partiel. *Houille Blanche* 5:483–489.

Schwartz, L. W. 1974. Computer extension and analytic continuation of Stokes' expansion for gravity waves. *Journal of Fluid Mechanics* 62:553–578.

Society of Naval Architects of Japan. 1976. *Proceedings of the international seminar on wave resistance*. Tokyo.

Stoker, J. J. 1957. *Water waves*. New York: Interscience.

Ünlüata, Ü., and C. C. Mei. 1970. Mass transport in water waves. *Journal of Geophysical Research* 75:7611–7617.

Vugts, J. H. 1968. The hydrodynamic coefficients for swaying, heaving, and rolling cylinders in a free surface. *International Shipbuilding Progress* 15:251–276.

Wehausen, J. V. 1971. The motion of floating bodies. *Annual Review of Fluid Mechanics* 3:237–268.

Wehausen, J. V. 1973. The wave resistance of ships. *Advances in Applied Mechanics* 13:93–245.

Wehausen, J. V., and E. V. Laitone. 1960. Surface waves. In *Encyclopedia of Physics*. vol. 9. Berlin, Gottingen, Heidelberg: Springer-Verlag; http://surfacewaves.berkeley.edu/.

Whitham, G. B. 1974. *Linear and nonlinear waves*. New York: Wiley-Interscience.

Wiegel, R. 1964. *Oceanographical engineering*. Englewood Cliffs, N.J.: Prentice-Hall.

Yaglom, A. M. 1962. *An introduction to the theory of stationary random functions. English translation*. Englewood Cliffs, N.J.: Prentice-Hall.

7 Hydrodynamics of Slender Bodies

Many vessels of interest in marine hydrodynamics are slender, with one length-dimension exceeding the others by an order of magnitude. For ships, submarines, sailboats, and fish, this shape is generally a consequence of the advantage of a streamlined body form, with the longitudinal length scale substantially greater than the beam and depth. Slenderness also occurs for various ocean platforms such as spar buoys and for bottom-mounted structures of similar form which are elongated vertically. For all of these vessels it is logical to simplify the hydrodynamic analysis by suitable approximations based on the slenderness of the body.

Slender-body theory originated in the field of aerodynamics, first as a technique for predicting the stability characteristics of dirigibles. In ship hydrodynamics, slender-body approximations have been utilized for a variety of problems, most of which have required substantial extensions or revisions of the aerodynamic theory. The strip theory of ship motions in waves is an important example, where the pioneering efforts of Korvin-Kroukovsky (1955) were motivated by parallel developments in aerodynamics, but the complexities of free-surface effects are such that the theory is still in a state of refinement.

The applications of slender-body theory have been less intensive in ocean engineering, although a strip theory can be used to analyze semisubmersible platforms supported by horizontal buoyancy hulls, as well as the wave forces on slender vertical structures. In a very simple form, slender-body theory has been applied successfully to spar buoys.

We shall begin our development of slender-body theory by discussing two relatively simple problems—the longitudinal and lateral motion of a slender body in an unbounded ideal fluid. These are problems that were first treated in aerodynamics, and our study of the lateral motion will lead to a derivation of the low-aspect-ratio theory for planar lifting surfaces. The results will be applied to the problem of ship maneuvering in the horizontal plane, where free-surface effects are generally small and the double-body approximation can be used to avoid the complications of waves. Subsequently, wave effects on the free surface will be considered, first for a slender body with no forward velocity and then for a moving ship.

Slender-body approximations can be applied to analyze the effects of shallow or restricted water on ships, as will be discussed in section 8, and to predict the interactions between adjacent ships. The latter problem is treated by Tuck and Newman (1974) and Yeung (1978). The applications of slender-body theory to ship hydrodynamics are reviewed by Newman (1970) and Ogilvie (1974, 1977). A survey of ship hydrodynamics in restricted waters is given by Tuck (1978).

7.1 Slender Body in an Unbounded Fluid

Let us consider the problem of a slender body, moving in an unbounded ideal fluid, with the fundamental assumption that the transverse dimensions of the body are small compared to its length. To formalize this assumption, we define the *slenderness parameter*

$$d/l = \varepsilon, \tag{1}$$

where d is the maximum lateral dimension of the body, and l is its length. The nondimensional parameter ε is assumed small, and on this basis an approximate solution is sought for the hydrodynamic quantities of interest.

The body motions are assumed to consist of a constant forward velocity U, parallel to the x_0-axis of a space-fixed coordinate system (x_0, y_0, z_0), and a small lateral motion in the z_0-direction. (Vertical motions in the y_0-direction can be treated in an identical manner.) A moving coordinate system (x, y, z) is defined by the transformations

$$x = x_0 - Ut, \tag{2}$$

$$y = y_0, \tag{3}$$

$$z = z_0. \tag{4}$$

The lateral motion of the body is described by its displacement $\zeta(x, t)$ from the x_0-axis, as shown in figure 7.1. For rigid-body motions ζ will be a constant or linear function of x. Undulatory motions of the body may also be considered where the dependence of ζ on x will be more general. We assume only that ζ is small relative to the body length and that this displacement is a slowly varying function of x. The latter restriction excludes flexural modes with a length-scale of order εl.

The body surface S_B is assumed to be elongated in the longitudinal direction and may include a tail fin as shown in figure 7.1. The intersection of S_B with a plane x = constant defines the cross-section profile $\Sigma_B(x)$, and the circumscribed area $S(x)$ is the sectional area of the body. We shall assume that the body ends are such that both $S(x)$ and $S'(x) \equiv dS/dx$ vanish at the ends.

The body surface can be described by the equation $F(x, y, z - \zeta) = 0$. For example, if the body is symmetrical about $z = \zeta$, then $F = z - \zeta \pm b(x, y)$ where b is the local half-beam. Using the substantial derivative (3.19), the boundary condition on S_B takes the form

$$\frac{D}{Dt}F(x_0 - Ut, y_0, z_0 - \zeta) = 0, \tag{5}$$

or

$$\left(\frac{\partial \phi}{\partial x_0} - U\right)\frac{\partial F}{\partial x} + \frac{\partial \phi}{\partial y_0}\frac{\partial F}{\partial y} + \left(\frac{\partial \phi}{\partial z_0} - \frac{\partial \zeta}{\partial x}\frac{\partial \phi}{\partial x_0} + U\frac{\partial \zeta}{\partial x} - \frac{\partial \zeta}{\partial t}\right)\frac{\partial F}{\partial z} = 0. \tag{6}$$

Here $\phi(x_0, y_0, z_0, t)$ is the velocity potential due to the presence of the body, which is defined initially with respect to the fixed coordinates (x_0, y_0, z_0).

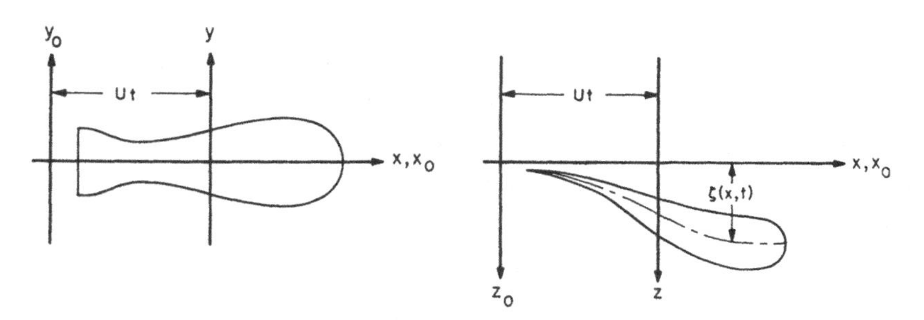

Figure 7.1
Definition sketch for slender body and coordinate systems.

Since ∇F is a vector normal to S_B, (6) can be rewritten in the form

$$\frac{\partial \phi}{\partial n} = Un_x + \left(\frac{\partial \zeta}{\partial t} - U\frac{\partial \zeta}{\partial x} + \frac{\partial \phi}{\partial x_0}\frac{\partial \zeta}{\partial x} \right) n_z, \quad \text{on } S_B, \tag{7}$$

where $\mathbf{n}$ is the unit normal vector into the body. The boundary-value problem is completed by requiring that the potential vanish at large distances from the body,

$$\phi \to 0 \quad \text{as } (x^2 + y^2 + z^2)^{1/2} \to \infty, \tag{8}$$

and by invoking Laplace's equation,

$$\frac{\partial^2 \phi}{\partial x_0^2} + \frac{\partial^2 \phi}{\partial y_0^2} + \frac{\partial^2 \phi}{\partial z_0^2} = 0, \tag{9}$$

throughout the fluid domain.

Before making the slenderness approximation, we note that the first term on the right side of (7) is associated with the boundary condition for the longitudinal-flow problem, and the remaining contribution to (7) is associated with the lateral motion. These two problems can be treated separately, since the lateral motion is assumed small, with a linear decomposition identical to the lifting and thickness problems treated in chapter 5. In particular, the boundary condition for the longitudinal-flow problem can be imposed on the "stretched-straight" position of the body with $\zeta = 0$.

The boundary-value problem (7–9) can be simplified if the body is slender. Geometrically, it follows from (1) that the unit normal vector $\mathbf{n}$ will coincide with the plane x = constant as $\varepsilon \to 0$, or

$$n_x = O(\varepsilon), \quad n_y = O(1), \quad n_z = O(1). \tag{10}$$

On this basis, the normal derivative in (7) can be approximated by the two-dimensional normal derivative in the y-z plane.

There is also a hydrodynamic consequence of slenderness. To see this, we adopt an *inner* reference frame, fixed with respect to the scale of the cross section Σ_B. From this viewpoint, $d = O(1)$ whereas $l = O(1/\varepsilon)$. As $\varepsilon \to 0$, the transverse scale remains fixed while the body length tends to infinity. The fluid velocity field then appears nearly constant in the longitudinal direction, with $\partial/\partial x_0 = O(\varepsilon)$. In this inner region it follows that

$$\frac{\partial \phi}{\partial x_0} \ll \left(\frac{\partial \phi}{\partial y_0}, \frac{\partial \phi}{\partial z_0} \right), \tag{11}$$

with a similar result for the second derivatives. Thus Laplace's equation reduces to the two-dimensional form

$$\frac{\partial^2 \phi}{\partial y_0^2} + \frac{\partial^2 \phi}{\partial z_0^2} = 0, \tag{12}$$

and ϕ can be replaced by a two-dimensional potential, say

$$\phi = \Phi(y_0, z_0; x_0). \tag{13}$$

Here the dependence on x_0 is included to emphasize that this potential will vary slowly along the body length, as a result of the change in the body geometry and lateral motion.

Since ζ and ε are both small, the body potential must be small in some sense, and if the forward velocity $U = O(1)$, then

$$\frac{\partial \phi}{\partial x_0} \ll U. \tag{14}$$

With these approximations the boundary condition (7) can be replaced by

$$\frac{\partial \Phi}{\partial N} = U n_x + \left(\frac{\partial \zeta}{\partial t} - U \frac{\partial \zeta}{\partial x} \right) n_z, \quad \text{on } \Sigma_B. \tag{15}$$

Here N denotes the two-dimensional unit vector normal to Σ_B in the y-z plane. Note that the longitudinal-flow term $U n_x$ has been retained, in spite of (10), since ε and ζ are independent small parameters.

From the boundary condition (15) and the linearity of the boundary-value problem, it is logical to express the solution as the sum of two potentials, one due to the longitudinal motion and the other to the lateral motion. With a notation suggested by (4.101–102), we shall write

$$\Phi = U\Phi_1 + W\Phi_3, \tag{16}$$

where, on Σ_B

$$\frac{\partial \Phi_1}{\partial N} = n_x, \tag{17}$$

$$\frac{\partial \Phi_3}{\partial N} = n_z, \tag{18}$$

and

$$W(x,t) = \frac{\partial \zeta}{\partial t} - U\frac{\partial \zeta}{\partial x}. \tag{19}$$

Physically, $W(x, t)$ is the lateral velocity of the body section, as observed in a fixed frame of reference. A similar form may be recalled for the boundary condition (5.193) on an unsteady lifting surface. The potentials Φ_1 and Φ_3 satisfy the two-dimensional Laplace equation and the boundary conditions (17–18), and they correspond to the solutions of two-dimensional flow problems at each section along the body length.

The lateral-flow problem for the potential Φ_3 can be identified precisely with the lateral translation of a two-dimensional body, with profile Σ_B, moving in the positive z-direction with velocity W. In the special but important case of a body of revolution, with local radius $R(x)$, the solution follows from table 4.2 in the form

$$\Phi_3 = -(R^2 / r)\cos\theta, \tag{20}$$

where $r = (y^2 + z^2)^{1/2}$ and $\theta = \tan^{-1}(y/z)$. This solution holds at each section along the body length, where $R(x)$ is equal to the local radius. For other body profiles the solution can be obtained by conformal mapping or other techniques. The lateral force acting on the body at each section is related to the two-dimensional added-mass coefficient in a manner which will be derived in section 7.3.

The longitudinal flow and the two-dimensional potential Φ_1 are more complicated. The boundary condition (17) does not correspond to a two-dimensional rigid-body motion but to a dilation of the body profile. Integrating (17) around the body section to obtain the local flux Q, we get

$$Q = -\oint_{\Sigma_B} \frac{\partial \Phi_1}{\partial N} dl = -\oint_{\Sigma_B} n_x \, dl = -S', \tag{21}$$

where S' denotes the derivative of the sectional area $S(x)$. Physically, (21) may be anticipated from the fact that longitudinal motion of the body displaces the surrounding fluid, with outward flux equal to the change in sectional area of the body.

For large radial distances from the body axis, the two-dimensional potentials are of the general form (4.70), and from flux considerations Φ_1 must include a source term with strength $-S'$. Thus,

$$\Phi_1 \simeq -\frac{S'}{2\pi}\log(r / l), \quad \text{for } r \gg d, \tag{22}$$

where the length l has been inserted for nondimensional purposes. On the forebody Φ_1 will be source-like, and on the afterbody sink-like, with corresponding radial outflow and inflow.

It is inevitable that as $r \to \infty$, the two-dimensional solution (22) will diverge and cannot be reconciled with (8). This incompatibility is associated with the assumption of a two-dimensional flow. In particular, the argument leading to (11) was made with the distance from the body fixed as its length tends to infinity. It is not appropriate to apply (8) in this inner region, nor to the inner solution Φ_1, since (8) holds only at large distances compared to all the body dimensions, including the length.

Since the condition (8) cannot be applied to the inner solution, the potential Φ_1 contains an arbitrary constant not determined either from Laplace's equation or from the boundary condition on the body. Thus, at this stage, the inner solution is not unique, nor is it well-behaved at infinity.

These deficiencies result from the assumption of two-dimensional flow, and thus a three-dimensional solution of the Laplace equation (9) is required to bridge the gap between the inner solution Φ_1 and the condition at infinity. The appropriate *outer* solution is a distribution of three-dimensional sources, along the body axis, which satisfy (8) at infinity and can be *matched* with (22) near the body. The latter condition serves to determine the strength of the three-dimensional source distribution; from this outer solution, the arbitrary constant of the inner solution is ultimately determined. The approach outlined here is essentially the method of matched asymptotic expansions referred to earlier in connection with the boundary-layer approximation and lifting-line theory. The common feature of these problems and slender-body theory is the disparity in length scales that characterize the flow. A more extensive discussion of such problems is given by Van Dyke (1975).

The method of matched asymptotic expansions will be employed in section 7.2 to solve the longitudinal-flow problem. Subsequently, in section 7.3, we shall return to the lateral-flow problem, which is in some respects simpler and more useful, to analyze the lateral force acting along the body length.

Before treating the longitudinal-flow problem in detail, it should be noted that the need for a three-dimensional outer solution does not arise for the lateral flow. Essentially this distinction occurs because no net flux is associated with the boundary condition (18) for lateral motion, and thus there is no source term in the local solution of the lateral flow problem. The potential Φ_3 behaves for large r as a two-dimensional dipole, as in (20) for a body of revolution or (4.129) for arbitrary body profiles. This dipole potential vanishes at infinity, and thus is consistent with (8). Strictly, this behavior is not identical to the more appropriate three-dimensional dipole (4.123), but the distinction does not affect the leading-order slender-body solution for the lateral-flow problem.

7.2 Longitudinal Motion

The longitudinal problem, associated with the steady forward motion of the body, can be solved most readily in the moving coordinate system (x, y, z). Here the body is in its stretched-straight position and is rigid. In accordance with (4.101), the potential due to the body disturbance will be of the form

$$\Phi = U\phi_1(x,y,z). \tag{23}$$

We shall seek a solution of this problem using the method of matched asymptotic expansions, where the inner and outer solutions are treated as complementary approximations, each of which is valid in its own domain. Thus the inner solution Φ_1 is governed by the two-dimensional Laplace equation (12) and the body boundary condition (17), but not by the condition (8) at infinity. Conversely, the outer solution φ_1 is governed by the three-dimensional Laplace equation and by the condition (8) at infinity, but need not satisfy the body boundary condition. These two solutions will be consistent provided they can be matched in an *overlap* region $\epsilon l \ll r \ll l$ far from the body in the inner region but very close to the body in the outer region. Thus the matching requirement

$$\varphi_1 \simeq \Phi_1, \qquad \epsilon l \ll r \ll l, \tag{24}$$

effectively replaces the missing boundary condition at infinity in the inner problem and at the body in the outer problem.

This method is useful because the two separate problems are simpler to solve, compared to the exact solution valid everywhere throughout the fluid. The inner solution is simplified because it is two dimensional, whereas the outer solution is simplified by replacing the body boundary condition with (24).

Combining (22) and (24) we get the inner condition for φ_1 in the form

$$\varphi_1 \simeq -\frac{1}{2\pi} S' \log(r/l), \quad r \ll l. \tag{25}$$

The appropriate outer solution is a source distribution, along the body length, with the same strength as (25). Using the three-dimensional source potential (4.31), and distributing these along the body axis gives

$$\varphi_1(x,r) = \frac{1}{4\pi}\int_l S'(\xi)[(x-\xi)^2 + r^2]^{-1/2} d\xi. \tag{26}$$

To confirm that (26) can be matched to the inner solution, it is necessary to develop the inner expansion of (26) for $r/l \ll 1$. For this purpose we shall use the relation

$$[(x-\xi)^2 + r^2]^{-1/2} = \frac{d}{d\xi}\log\left\{\frac{\xi - x}{r} + \left[\left(\frac{\xi - x}{r}\right)^2 + 1\right]^{1/2}\right\}, \tag{27}$$

which can be verified directly by differentiation. Substituting (27) in (26) and integrating by parts gives

$$\varphi_1 = -\frac{1}{4\pi}\int_l S''(\xi)\log\left\{\frac{\xi - x}{r} + \left[\left(\frac{\xi - x}{r}\right)^2 + 1\right]^{1/2}\right\} d\xi. \tag{28}$$

Here the contributions from the endpoints are deleted on the assumption that $S' = 0$ at the ends.

To approximate (28) for small r, we note that the limiting value of the argument of the logarithm is given by

$$\left\{\frac{\xi - x}{r} + \left[\left(\frac{\xi - x}{r}\right)^2 + 1\right]^{1/2}\right\} \simeq \frac{2(\xi - x)}{r}, \quad (\xi - x) > 0, \tag{29}$$

as $r \to 0$. Similarly, from a Taylor series expansion of the square-root function, or by noting from (27) that the logarithm must be an odd function of $(x - \xi)$, it follows that

$$\left\{\frac{\xi - x}{r} + \left[\left(\frac{\xi - x}{r}\right)^2 + 1\right]^{1/2}\right\} \simeq \frac{r}{2|\xi - x|}, \quad (\xi - x) < 0. \tag{30}$$

Substituting (29–30) in (28), the inner expansion of the outer solution for small r is obtained in the form

$$\varphi_1 \simeq \frac{1}{4\pi}\int_{x_T}^{x} S''(\xi)\log\left[\frac{2(x-\xi)}{r}\right]d\xi - \frac{1}{4\pi}\int_{x}^{x_N} S''(\xi)\log\left[\frac{2(x-\xi)}{r}\right]d\xi. \tag{31}$$

Here x_N and x_T denote the x-coordinates of the nose and tail. Separating the terms involving $\log r$ and integrating these, it follows that

$$\varphi_1 \simeq -\frac{1}{2\pi}S'\log(r/l) + f(x), \quad r \ll l. \tag{32}$$

where

$$f(x) = \frac{1}{4\pi}\int_{x_T}^{x} S''(\xi)\log[2(x-\xi)/l]d\xi - \frac{1}{4\pi}\int_{x}^{x_N} S''(\xi)\log[2(\xi-x)/l]d\xi. \tag{33}$$

Equation (32) is the inner expansion of the outer solution, which contains the desired two-dimensional source behavior, consistent with (25). However, a "constant" term (33) is left over; it is independent of the (y, z) coordinates but depends on the local value of x. Since the inner solution contains an arbitrary constant, the matching requirement dictates the value of this constant to be equal to (33), and the inner solution is uniquely determined by the requirement that

$$\Phi_1 \simeq -\frac{1}{2\pi}S'\log(r/l) + f(x), \quad r \gg \epsilon l. \tag{34}$$

Note that the body need not be axisymmetric, and the complexities of its geometry will be reflected in the inner solution Φ_1. The outer limit of this solution (34), and thus the outer solution φ_1, *are* axisymmetric, but that is a consequence of the source-like flow far from the body, where its detailed shape is unimportant. For a body of revolution, (34) is the inner solution for *all* values of r. For other body profiles, the inner solution must be found by conformal mapping or similar techniques.

This completes the solution of the longitudinal-flow problem for a slender body in an infinite fluid, and a brief review of the derivation may be useful. We began in section 1 with an assumed inner solution, valid near the body surface and satisfying the body boundary condition and the two-dimensional Laplace equation. However, this solution does not satisfy the condition at infinity in the outer field, and it contains an arbitrary additive constant. To provide a corresponding outer solution with the proper behavior at infinity, a three-dimensional source distribution is assumed, using flux conditions to guess the appropriate source strength. By examining the inner approximation of this outer solution, for small radius r, the outer solution can be matched to the inner solution provided the arbitrary constant of the latter is chosen as defined in (33). Thus, in the process of overcoming the deficiencies at infinity, the arbitrary constant of the inner solution is prescribed.

The "constant" (33) is neither disposable nor unimportant, since it affects the value of the pressure distribution along the body surface. To illustrate this, we note that the added-mass coefficient m_{11} may be computed from the inner solution, using (4.114). For a body of revolution it follows that

$$\begin{aligned} m_{11} &= \rho \iint_{S_B} \left[-\frac{1}{2\pi} S' \log(r/l) + f(x) \right] \frac{\partial \Phi_1}{\partial n} dS \\ &= \rho \iint_{l} \left[-\frac{1}{2\pi} S' \log(R/l) + f(x) \right] S'(x)\, dx. \end{aligned} \tag{35}$$

Since the body radius R is of order εl, the added-mass coefficient for longitudinal motion is proportional to $l^3 \varepsilon^4 \log \varepsilon$. By comparison, the body volume is of order $\varepsilon^2 l^3$. Thus, the added mass of a slender body in the longitudinal direction is negligible compared with the body mass. This conclusion is not restricted to bodies of revolution, as shown in problem 1.

7.3 The Lateral Force

To find the distribution of the lateral force along the body length, (4.90) may be applied to the thin slice of fluid of differential thickness dx_0, as shown in figure 7.2. This volume of fluid is bounded by the differential length of the body surface S_B, of profile $\Sigma_B(x)$ and width dx_0, as well as by a fixed control surface S_C. The surface S_C is composed of two lateral planes S_0 separated by a distance dx_0 exterior to the body, and by a closure surface S_∞, at large radial distance from the body. Applying (4.90) to this control surface, as well as to the differential portion of the body surface, the differential force is given by

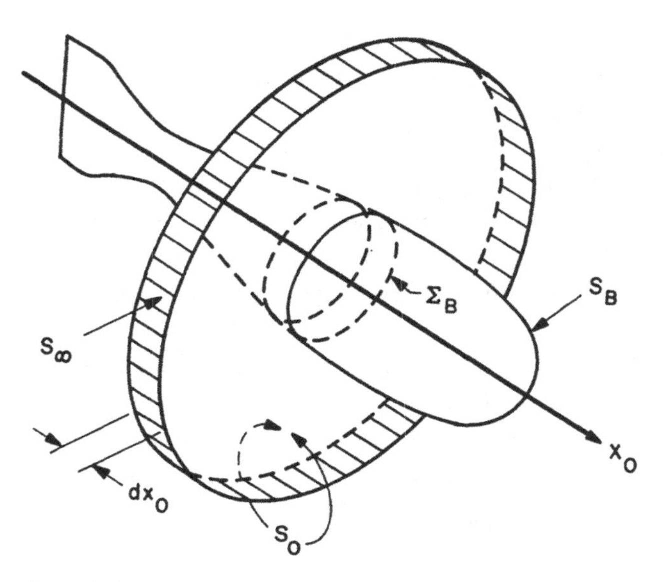

Figure 7.2
Control volume for the evaluation of the differential lateral force F_z'.

$$\begin{aligned} F_z' &= -\rho \frac{d}{dt} \oint_{\Sigma_B(x)} \Phi n_z \, dl - \rho \frac{\partial}{\partial x_0} \iint_{S_0} \frac{\partial \phi}{\partial x_0} \frac{\partial \phi}{\partial z_0} dy_0 dz_0 \\ &= -\rho \int_{\Sigma_\infty} \left(\frac{\partial \varphi}{\partial z_0} \frac{\partial \phi}{\partial n} - n_z \frac{1}{2} \nabla \phi \cdot \nabla \phi \right) dl. \end{aligned} \tag{36}$$

Here the differential operator d/dt is to be interpreted as the time-derivative in a fixed reference frame, with x_0 = constant, Σ_∞, is the profile defined by the limit of S_∞, as $dx_0 \to 0$, and S_0 is the plane bounded by $\Sigma_B\ (x)$ and Σ_∞. From the slenderness assumption (11) there is no contribution from the integral over S_0, and since the potential vanishes at infinity there is no contribution from the last integral in (36). Thus the differential lateral force acting along the body length is given by the expression

$$F_z' = -\rho \frac{d}{dt} \oint_{\Sigma_B(x)} \phi n_z dl. \tag{37}$$

On the body, the potential can be replaced by (16), and if the profile is symmetrical about $z = 0$, there is no contribution to the integral in (37) from Φ_1. Thus it follows that

$$F_z' = -\rho \frac{d}{dt}\left(W \oint_{\Sigma_B(x)} \Phi_3 n_z dl \right) = -\frac{d}{dt}(W(x,t)m_{33}(x)). \tag{38}$$

Here the two-dimensional added-mass coefficient $m_{33}(x)$ has been introduced, using the relation (4.114) and the boundary condition (18). In view of the coordinate transformation (2), the time-derivative in (36–38) can be replaced by partial derivatives in a moving reference frame, and (38) is equivalent to

$$F_z' = -\left(\frac{\partial}{\partial t} - U\frac{\partial}{\partial x}\right)[W(x,t)m_{33}(x)]. \tag{39}$$

Equation (39) has been derived by Lighthill (1960), in analyzing the swimming motion of slender fish. Various special cases were known earlier in the field of aerodynamics. One feature of (39) and of the derivation we have carried out thus far is that the time-dependence is relatively unimportant, and the distinction between steady and unsteady problems is trivial.

The total force acting on the body can be obtained in a strip-theory manner by integrating (39). The lateral added-mass coefficient, due to acceleration of a rigid slender body in the z-direction, is given by the integral

$$m_{33}^s = \int_l m_{33}(x)\,dx. \tag{40}$$

Similarly,

$$m_{35}^s = \int_l m_{33}(x)\,x dx, \tag{41}$$

and

$$m_{55}^s = \int_l m_{33}(x)x^2\,dx, \tag{42}$$

where m_{55}^s is the added moment of inertia for yaw acceleration, and m_{35}^s denotes the cross-coupling added mass between sway and yaw. The latter vanishes if the body is symmetrical about $x = 0$.

The superscript s in (42) is inserted to distinguish these strip-theory results. In this approximation, the local force at any section of the body is not affected by the shape of the body elsewhere; that is, there are no hydrodynamic interactions between adjacent sections of the body. The resulting simplification is significant. For example, the results for a slender spheroid can be deduced in terms of the added-mass coefficient for a circle without the complexity of solving the complicated three-dimensional problems represented by figure 4.8. Of course, a price must be paid for this simplification, and it is apparent from figure 4.8 that the sway added-mass coefficient will be overpredicted in the strip theory by a factor of 10 percent when the diameter-length ratio is 0.2, and by 100 percent in the extreme case of a sphere. A larger error results for the added moment of inertia, as emphasized by the fact that this coefficient vanishes for a sphere but is nonzero in the strip theory! (See problem 2.) It would be naive to suggest that a sphere is slender, and in general a diameter-length ratio of 0.1 to 0.2 is a reasonable upper limit for the slender-body approximation.

The total force acting on a slender body in steady motion can be derived by integrating the differential force (39),

$$F_z = U\int_l \frac{\partial}{\partial x}[W(x)m_{33}(x)]dx = U[W(x)m_{33}(x)]_{x_T}^{x_N}. \tag{43}$$

For a slender body, the nose is a point of zero transverse dimensions, and thus $m_{33}(x_N) = 0$. The same will be true at the tail for a pointed body, without a tail fin; in accordance with D'Alembert's paradox, the lateral force acting on such a body in steady motion is zero.

For a body with a tail fin, such as that shown in figure 7.1, the situation is fundamentally different. Using the flat-plate result from table 4.3, the added-mass coefficient of the tail is given by

$$m_{33}(x_T) = \frac{\pi}{4}\rho s^2, \tag{44}$$

where s is the tail span. Using (19) to evaluate W, and denoting the slope $\partial\zeta/\partial x$ at the tail by the local angle of attack α_T, it follows that there will be a transverse "lift" force

$$F_z = \frac{\pi}{4}\rho U^2 s^2 \alpha_T. \tag{45}$$

The yaw moment about the vertical axis of the body can be obtained from (39) in a manner similar to that used to obtain the force. For steady motion, we obtain the following expression in place of (43):

$$M_y = -U \int_l x \frac{\partial}{\partial x}[W(x)m_{33}(x)]\,dx$$
$$= U \int_l W(x)m_{33}(x)\,dx + Ux_T\, W(x_T)m_{33}(x_T).$$

Using (19) for W, and (45) in the last term,

$$M_y = -U^2 \int_l \frac{\partial \zeta}{\partial x} m_{33}(x)\,dx - x_T F_z. \tag{47}$$

The last term is associated with the moment due to the lift force on the tail fin and vanishes for a body with a pointed tail. The remaining integral in (47) gives a contribution for a body moving at an angle of attack, regardless of whether the tail is pointed. This *Munk moment* acts on a nonlifting body in steady translation, as noted in section 4.13.

The orders of magnitude of the velocity potential Φ_3 and lateral force F_3, in terms of the slenderness parameter ε, can be deduced from the above expressions. In terms of the inner solution of the problem, where the length scales are referred to the transverse dimensions of the body and where the body length is very large, the two-dimensional quantities Φ_3 and F_3' are quantities of order one.

On the other hand, if the body length $l = O(1)$ and the transverse dimensions are regarded as $O(\varepsilon)$, then the potential Φ_3 is $O(\varepsilon)$ and the lateral force is $O(\varepsilon^2)$. The latter results can be deduced from dimensional arguments, based on the fact that these are quantities of order one in the inner frame of reference. From either standpoint, the lateral force is of the same order as the body mass, and both will be of comparable importance in the equations of motion for an unrestrained body subject to a lateral disturbance. This situation contrasts with that of the longitudinal force.

This analysis of the lateral force on a slender body is based on the assumption that the flow is irrotational at each section along the body length. Thus, vortex sheets and separation are excluded in the fluid alongside the body. However, the flow downstream of the body does not affect this analysis, and a trailing vortex sheet may originate from the abrupt trailing edge at the tail. From the viewpoint of lifting-surface theory, this vortex sheet is associated directly with the lift force (45).

These results can be applied to planar lifting surfaces of small aspect ratio, where the body profile Σ_B is a flat plate of local span $s(x)$, provided no vorticity is shed upstream of the tail. This restriction requires the trailing edge to be abrupt and situated at the tail. In addition, the angle of attack must be sufficiently small to avoid leading-edge separation. Under these conditions, the results above reduce to the low aspect ratio lifting-surface theory. In the notation of chapter 5, the lift coefficient follows from (45) in the form

$$C_L = \tfrac{1}{2}\pi A\alpha, \tag{48}$$

where $A = s^2/S$ is the aspect ratio based on the maximum span and planform area S. As shown in figure 5.22, (48) gives precisely half the limiting value, for $A \ll 1$, of the lifting-line theory (5.148). This difference is not surprising, since the lifting-line theory assumes that $A \gg 1$.

The longitudinal distribution of the lift force on an uncambered lifting surface of small aspect ratio is proportional to the rate of change of added mass. For a "delta wing" with triangular planform, the lift force will be distributed uniformly along the length. If the span increases more rapidly, the center of pressure will move forward. In the limiting case of a rectangular planform, with constant span along the length, the lift force will be concentrated at the leading edge, with a differential lift given by

$$F_z' = \frac{\pi}{4}\rho U^2 \alpha s^2 \delta(x - x_N), \tag{49}$$

where $\delta(x - x_N)$ denotes a delta function. This limit is physically unreasonable, but experiments show that the center of pressure for a rectangular lifting surface of small aspect ratio is very close to the leading edge, particularly in the regime of small angles of attack where leading-edge separation does not occur.

For a planar lifting surface of small aspect ratio with the maximum span at the tail, the trailing vortex sheet downstream of the body must be independent of the coordinate x, with a constant downwash velocity equal to $U\alpha_T$, Thus, it makes no difference whether the body terminates at the position of maximum span or continues downstream with decreasing span, provided the local angle of attack is constant. For this reason, the low aspect ratio theory holds for uncambered lifting surfaces of more general planform, if the *maximum* span is used in (48). In this particular case, there is no lift force on the portion of the body downstream of the maximum span.

The extension of slender-body theory to account for the interaction of the afterbody with vortex sheets shed upstream has been carried out by Newman and Wu (1973) in the general case where the local lateral velocity of the body differs from the downwash of the trailing vortices. Vortex sheets that occur due to separation are analyzed by Fink and Soh (1974). In both cases, the vorticity along the body is convected with the free-stream velocity U, and "memory" effects will occur in the unsteady case as in the two-dimensional lifting-surface theory.

7.4 Ship Maneuvering: The Hydrodynamic Forces

The hydrodynamic forces of significance during a ship maneuver are primarily the sway force and yaw moment. Anticipating large excursions in the horizontal plane, as in the turning trajectory of a ship, it is customary to employ a coordinate system fixed with respect to the ship. In this frame of reference, the forward velocity is of order one, whereas the sway and yaw velocities are assumed small by comparison. The remaining three modes of rigid-body motion are ignored, on the presumption that changes in the roll angle, heave, and pitch of the ship will be unimportant.

We shall employ the same coordinates and notation here as in chapter 4. The body is assumed rigid, and the (x, y, z) axes are fixed with respect to the ship such that x is positive forward, y upward, and z to starboard. The three nonzero velocity components of the ship are U_1 (forward), U_3 (sway), and Ω_2 (yaw). These are indicated in figure 7.3, together with the sway force F_3 and yaw moment M_2, as well as the rudder angle δ_R defined to be positive for a turn to port.

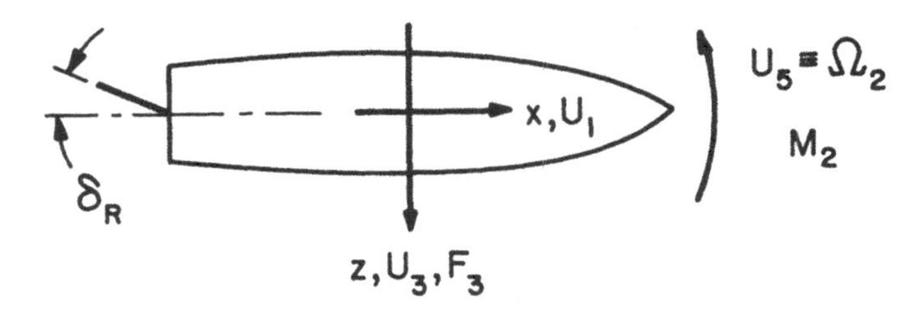

Figure 7.3
Horizontal motions of a ship as seen from above.

This convention differs from the standard throughout the field of ship maneuvering, where the y-axis is positive to starboard and the z-axis is directed vertically downward. In the standard notation, as described by Mandel (1967), the sway velocity v and force Y are identical to U_3 and F_3, but the yaw velocity r and moment N differ in sign from Ω_2 and M_2.

Ship maneuvers are relatively slow, compared to vertical motions in waves, due to the physical limitations of the control systems involved. Thus it is reasonable to assume that the sway velocity U_3 and yaw velocity Ω_2 are slowly changing functions of time and small by comparison to the scale of the forward speed. Neglecting wave effects associated with the steady forward velocity, the lateral flow associated with the sway and yaw motions can be analyzed on the basis of the low-frequency limit of the free-surface condition; that is, wave effects are neglected and the free surface is replaced by a rigid horizontal plane. The appropriate image of the ship hull can be inferred from figure 6.21, and for lateral motions it is appropriate to treat the wetted portion of the hull as the lower half of a rigid double body, as shown in figure 7.4.

In addition, it will be assumed that the fluid motion is ideal, and thus viscous effects can be ignored. Boundary layer phenomena and the related frictional drag are of no significance here, but if the angle of attack between the ship hull and the water becomes large, viscous separation will occur in the cross-flow past the hull. This is a nonlinear phenomenon, beyond the scope of our analysis, which must be regarded as a possible limitation of the theory.

With these preliminaries, we proceed to analyze the ship hull and its image as a rigid double body, of slender form, which can develop a lateral lift force in accordance with the results of section 3. Since U_3 and Ω_2 are the sway and yaw velocities defined with respect to the body fixed reference frame, the total lateral velocity is

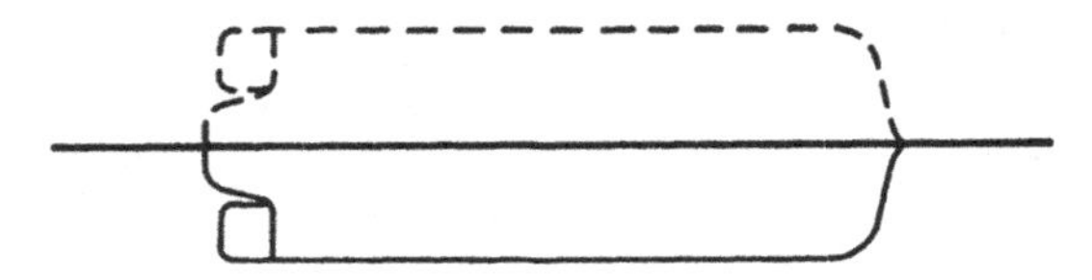

Figure 7.4
Elevation of the ship hull and its double-body image above the free surface.

$$W(x,t) = U_3(t) - x\Omega_2(t). \tag{50}$$

It follows from (39) that the local force acting along the hull is given by

$$\begin{aligned} F_z' &= -\left(\frac{\partial}{\partial t} - U_1 \frac{\partial}{\partial x}\right)[(U_3 - x\Omega_2)m_{33}(x)] \\ &= -\left(\dot{U}_3 - x\dot{\Omega}_2\right)m_{33}(x) + U_1 U_3 \frac{\partial}{\partial x}(m_{33}) - U_1\Omega_2 \frac{\partial}{\partial x}(xm_{33}). \end{aligned} \tag{51}$$

Integrating along the hull length gives the total sway force,

$$F_3 = -\dot{U}_3 m_{33}^s - \dot{\Omega}_2 m_{35}^s - U_1 U_3 m_{33}(x_T) + U_1 \Omega_2 x_T m_{33}(x_T). \tag{52}$$

Here the superscript s denotes the strip-theory added-mass coefficients (40–42), and x_T is the value of x at the effective trailing edge, nominally the stern.

Similarly, multiplying (51) by the moment arm $-x$ and integrating over the length gives the yaw moment

$$\begin{aligned} M_2 = &-\dot{U}_3 m_{35}^s - \dot{\Omega}_2 m_{55}^s + U_1 U_3[m_{33}^s + x_T m_{33}(x_T)] \\ &+ U_1\Omega_2[m_{35}^s - (x_T)^2 m_{33}(x_T)]. \end{aligned} \tag{53}$$

Equations (52–53) give the sway force and yaw moment acting on the hull in accordance with the slender-body results of section 7.3. In table 7.1 these expressions are compared with the corresponding equations which follow from (4.115–4.116), where no assumption of slenderness is made but where the body must be pointed at the stern and without lifting effects.

From a comparison of the results shown in table 7.1, the slender-body theory neglects the longitudinal added-mass coefficient m_{11}, which is consistent with our conclusion in section 7.2 that this coefficient is of higher order in the slenderness parameter ε. On the other hand, the slender-body theory includes the lifting force and moment associated with the action of the double body as a lifting surface of low aspect ratio. The physical relevance of these complementary descriptions depends on the nature of the ship's stern. Generally, the stern profile is as shown in figure 7.4, and it is not obvious whether this should be regarded as pointed or with a sharp trailing edge. However, the actions of the rudder and deadwood are significant in practice, and it is likely that additional vorticity is shed from the keel along the afterbody in a manner which coincides more nearly with the slender-body results.

Table 7.1
Hydrodynamic Coefficients in Equations (52–53), Based on the Slender-Body Theory with Lifting Effects Included

Present Notation	Standard Notation	Slender-Body Theory	Added-Mass Theory
$\partial F_3/\partial U_3$	Y_v	$-U_1 m_T$	0
$\partial F_3/\partial \Omega_2$	$-Y_r$	$U_1 x_T m_T$	$U_1 m_{11}$
$\partial F_3 / \partial \dot{U}_3$	$Y_{\dot{v}}$	$-m_{33}^S$	$-m_{33}$
$\partial F_3 / \partial \dot{\Omega}_2$	$-Y_{\dot{r}}$	$-m_{35}^S$	$-m_{35}$
$\partial M_2/\partial U_3$	$-N_v$	$U_1[m_{33}^S + x_T m_T]$	$U_1(m_{33} - m_{11})$
$\partial M_2/\partial \Omega_2$	N_r	$U_1[m_{35}^S + (x_T)^2 m_T]$	$U_1 m_{35}$
$\partial M_2 / \partial \dot{U}_3$	$-N_{\dot{v}}$	$-m_{35}^S$	$-m_{35}$
$\partial M_2 / \partial \dot{\Omega}_2$	$N_{\dot{r}}$	$-m_{55}^S$	$-m_{55}$

Note: The second column is the standard notation described by Mandel (1967), and the last column shows the coefficients that follow from equations 4.115–4.116). The superscript s denotes the strip-theory approximations, and m_T denotes the added-mass coefficient at the stern.

To determine the validity of the slender-body theory, rough estimates of the force coefficients shown in table 7.1 can be compared with experiments. To facilitate these estimates we shall assume that the draft T is constant along the length, and that the two-dimensional added-mass coefficient can be approximated by its value for a flat plate or ellipse. Utilizing the results from table 4.3, with a factor of 1/2 to include only the lower half of the double body, it follows that

$$m_{33}(x) = \tfrac{1}{2}\pi\rho T^2. \tag{54}$$

Taking the origin at the midship section, $x_T = -l/2$ and the results shown in table 7.2 follow from table 7.1 and equation (54).

The most significant discrepancy in table 7.2 occurs for the sway force due to a constant sway velocity, where the theory underpredicts the experiments by about 50 percent. Here the importance of viscous cross-flow drag is recalled. In this connection, Norrbin (1970) notes that the experimental value of this force coefficient is sensitive to nonlinear effects and the interpretation of the experimental results. Nor does the theory account for the cross-coupling yaw moment due to sway acceleration, due to our simplifying assumption concerning the planform and added-mass distribution.

Table 7.2
Theoretical and Experimental Values of the Coefficients Listed in Table 7.1, for a Mariner Ship Hull with Length-Draft Ratio $l/T = 21$

	Slender-Body Theory for Rectangular Planform	Theory for L/T of Mariner Hull ($\times 10^3$)	Experiments for Mariner Hull ($\times 10^3$)
$\partial F_3/\partial U_3$	$-\frac{\pi}{2}\rho U_1 T^2$	–7.1	-13.3 ± 3.6
$\partial F_3/\partial \Omega_2$	$-\frac{\pi}{4}\rho U_1 T^2 l$	–3.6	-2.62 ± 0.72
$\partial F_3/\partial \dot{U}_3$	$-\frac{\pi}{2}\rho T^2 l$	–7.1	-6.55 ± 0.93
$\partial F_3/\partial \dot{\Omega}_2$	0	0	0.25 ± 0.08
$\partial M_2/\partial U_3$	$\frac{\pi}{4}\rho U_1 T^2 l$	3.6	3.69 ± 0.78
$\partial M_2/\partial \Omega_2$	$-\frac{\pi}{8}\rho U_1 T^2 l^2$	–1.8	-2.40 ± 0.50
$\partial M_2/\partial \dot{U}_3$	0	0	0.22 ± 0.08
$\partial M_2/\partial \dot{\Omega}_2$	$-\frac{\pi}{24}\rho T^2 l^3$	–0.6	-0.36 ± 0.12

Note: The coefficients in the last two columns are nondimensionalized in terms of the quantities $\frac{1}{2}\rho$, U_1, and l. The theoretical values are based on the slender-body results shown in table 7.1, but with the simplifying assumptions that the draft T is constant along the length, and the added-mass coefficient can be approximated by (54). References to the experimental data are given by Mandel (1967), and Motora (1972).

Finally, the theory overpredicts the added moment of inertia in yaw by a substantial amount; again this may be attributed to our assumption of a rectangular planform with excessive added mass at the bow and stern.

Despite these differences, the slender-body theory gives a satisfactory qualitative prediction of the hydrodynamic force and moment. On this basis it will be used to study the maneuvers of a ship in the horizontal plane.

The rudder force can be predicted from the slender-body theory, on the premise that the rudder-hull combination is a single cambered lifting surface. With reference to figure 7.1, the action of the rudder can be regarded in the slender-body theory as a deflection of the hull equal to

$$\zeta(x,t) = \delta_R(t)(x - x_R), \tag{55}$$

for $x_T < x < x_R$. Here x_R denotes the longitudinal coordinate of the rudder axis. The contribution from the rudder to the lateral velocity (50) is

$$W(x,t) = \dot{\delta}_R(x - x_R) - U_1\delta_R. \tag{56}$$

Substituting (56) in (39) and integrating the resulting force from upstream of the rudder post to its trailing edge gives the rudder force

$$F_z = m_T\left[\ddot{\delta}_R l_R^2 / 2 + 2U_1 l_R \dot{\delta}_R + U_1^2\delta_R\right]. \tag{57}$$

Here m_T has been substituted on the premise that the two-dimensional added mass of the rudder is independent of position along its chord, which is denoted by l_R. Neglecting the small change in moment arm from the origin to different positions on the rudder, we may take the yaw moment as the product of (57) and $-x_T$.

The terms in (57) proportional to the angular velocity and acceleration of the rudder are unsteady effects that can be ignored in practice. These are significant only over time scales of order l_R/U_1, the time required for the ship to travel one chord length of the rudder, which is negligible compared to the response time of the ship.

Retaining only the steady term in (57) gives the rudder force and moment in the form

$$F_{3R} = \tfrac{1}{2}\pi\rho T^2 U_1^2\delta_R, \tag{58}$$

$$M_{2R} = -\tfrac{1}{2}\pi\rho T^2 U_1^2\delta_R x_T. \tag{59}$$

Here (54) has been used for the added-mass coefficient.

The results (58–59), for the rudder force and moment, overpredict the experimental values reported by Motora (1972) by a factor of two. This discrepancy may occur because the span of the actual rudder is significantly less than the draft T.

As an alternative to the use of (58–59), the rudder can be regarded as a separate appendage, with hull interactions neglected. The lifting surface theory or experimental data can be used to estimate the rudder force and moment, in a manner outlined by Mandel (1967). In practice there are significant interactions with the hull and with the slipstream from the propeller if this is upstream of the rudder. It is difficult to predict these interactions from model tests with Froude scaling, since the Reynolds number of the rudder is reduced significantly from its full-scale value.

7.5 Ship Maneuvering: The Equations of Motion

In the absence of external forces, the sway and yaw motions of a ship can be determined by equating the hydrodynamic force and moment derived in the preceding section to the body-mass force and moment. The body-mass contributions can be obtained by replacing the added-mass coefficients in table 7.1 by the body-mass coefficients M_{ij} defined in equation (4.141). Using the slender-body results of table 7.1 for the hydrodynamic forces on the hull, and using (58–59) to estimate the rudder force and moment, the equations of motion follow in the form

$$\begin{aligned}\tfrac{1}{2}\pi\rho T^2 U_1^2 \delta_R &= U_1 m_T U_3 - U_1(x_T m_T + m)\Omega_2 \\ &\quad + (m_{33}^S + m)\dot{U}_3 + (m_{35}^S + M_{35})\dot{\Omega}_2,\end{aligned} \tag{60}$$

$$\begin{aligned}-\tfrac{1}{2}\pi\rho T^2 U_1^2 \delta_R x_T &= -U_1(m_{33}^S + x_T m_T)U_3 \\ &\quad - U_1[m_{35}^S + M_{35} - (x_T)^2 m_T]\Omega_2, \\ &\quad + (m_{35}^S + M_{35})\dot{U}_3 + (m_{55}^S + M_{55})\dot{\Omega}_2.\end{aligned} \tag{61}$$

Equations (60–61) are a pair of coupled linear differential equations for the sway and yaw velocities; they can be solved for prescribed rudder angles if the hydrodynamic and body-mass coefficients are specified. Generally a digital or analog computer must be used for this purpose, but a few qualitative conclusions can be drawn here without elaborate analysis. To simplify the discussion as much as possible, we shall assume that the longitudinal position of the origin is chosen such that

$$M_{35}^S + M_{35} = 0. \tag{62}$$

For a ship in steady-state motion, with the initial conditions $U_3 = \Omega_2 = 0$, the response to a sudden change of rudder angle can be approximated by considering only the acceleration terms in (60–61). Thus for the initial period of time before the sway and yaw velocities develop substantial values, it follows that

$$\dot{U}_3 = \left(\frac{\tfrac{1}{2}\pi\rho T^2 U_1^2}{m_{33}^S + m}\right)\delta_R(t), \tag{63}$$

$$\dot{\Omega}_2 = \left(\frac{-\tfrac{1}{2}\pi\rho T^2 U_1^2 x_T}{m_{53}^S + M_{55}}\right)\delta_R(t). \tag{64}$$

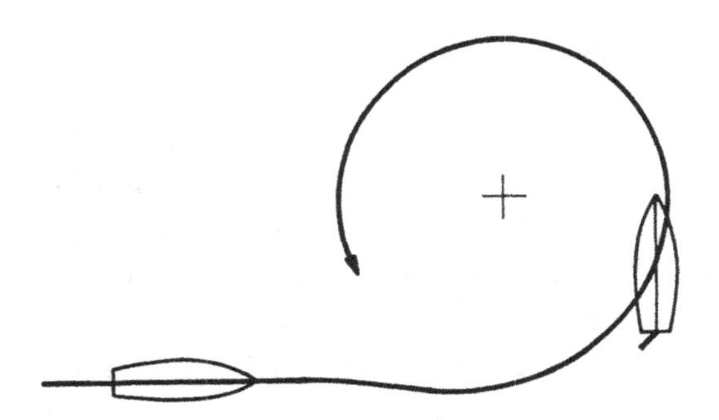

Figure 7.5
Trajectory of a ship in a steady-state turning maneuver.

Since x_T is negative, both of the factors in parentheses are positive, and a vessel commencing a turn to port will be accelerated initially to starboard, as shown in figure 7.5.

The ultimate, steady-state turning motions can be analyzed by assuming that U_3 and Ω_2 are constant. Thus, from (60) and (61),

$$\tfrac{1}{2}\pi\rho T^2 U_1 \delta_R = m_T U_3 - (x_T m_T + m)\Omega_2, \tag{65}$$

$$-\tfrac{1}{2}\pi\rho T^2 U_1 \delta_R x_T = -(m_{33}^S + x_T m_T)U_3 + (x_T)^2 m_T \Omega_2. \tag{66}$$

Using Cramer's rule, the solutions of these coupled linear equations are given by

$$U_3 / \delta_R = \frac{-\tfrac{1}{2}\pi\rho T^2 U_1 m x_T}{(m_T x_T)^2 - (m + x_T m_T)(m_{33}^S + x_T m_T)}, \tag{67}$$

$$\Omega_2 / \delta_R = \frac{\tfrac{1}{2}\pi\rho T^2 U_1 m_{33}^S}{(m_T x_T)^2 - (m + x_T m_T)(m_{33}^S + x_T m_T)}. \tag{68}$$

Since x_T is negative, U_3 has the same sign as Ω_2, and a ship in a steady-state turn is oriented with the bow pointed into the turn, as shown in figure 7.5. This gives an effective angle of attack and a resulting inward force, to maintain the centripetal acceleration. The radius of the steady-state turn is equal to U_1/Ω_2, to first order in the sway and yaw velocities.

The quantitative validity of (67–68) is limited because the denominator, or the determinant of (65–66), is close to zero for practical vessels. For example, using the approximations indicated in the second column of table 7.2, this determinant is equal to

$$\left(\frac{1}{2}\pi\rho l T^2\right)^2\left(\frac{1}{2}-\frac{BC_B}{\pi T}\right),$$

where $C_B = \forall / lBT$ is the block coefficient and B is the beam. For a ship with block coefficient 0.7, this estimate of the determinant vanishes if $B/T = 2.2$. For wider ships the determinant will be negative, and (68) predicts a turn opposite to that associated with the normal action of the rudder. Under these circumstances, where small changes in the values of the hydrodynamic forces affect the value of the determinant substantially, more accurate theoretical or experimental methods must be employed.

The vanishing of the determinant of (65–66) is related to the *directional stability* of the ship, which can be analyzed in terms of the homogeneous solutions of (60–61) with the rudder angle set equal to zero. Since (60–61) are coupled first-order differential equations with constant coefficients, the solutions for the sway and yaw velocities must be of the general form

$$U_3 = \text{Re}\{c_1 e^{\sigma_1 t} + c_2 e^{\sigma_2 t}\}, \tag{69}$$

$$\Omega_2 = \text{Re}\{c_3 e^{\sigma_1 t} + c_4 e^{\sigma_2 t}\}, \tag{70}$$

where the constants c_j and *stability indices* σ_j are complex.

Dynamic stability requires that the homogeneous solutions (69–70) decay to zero exponentially with time, or that

$$\text{Re}(\sigma_{1,2}) < 0. \tag{71}$$

The resulting motion will then appear as shown in figures 7.6 (a—b). If either or both indices violate this inequality, the vessel will be unstable, as in figure 7.6 (c).

Quadratic equations for the stability indices can be obtained by substituting (69–70) in (60–61). With the rudder angle set equal to zero, the only nontrivial solutions of (60–61) will occur when the determinant of the coefficients on the right side of these equations is zero. With the origin chosen in accordance with (62), it follows that

$$\begin{vmatrix} U_1 m_T + (m + m_{33}^S)\sigma_j & -U_1(m + x_T m_T) \\ -U_1(m_{33}^S + x_T m_T) & U_1(x_T)^2 m_T + (m_{55}^S + M_{55})\sigma_j \end{vmatrix} = 0. \tag{72}$$

The resulting quadratic equation is of the form

$$A\sigma_j^2 + B\sigma_j + C = 0, \tag{73}$$

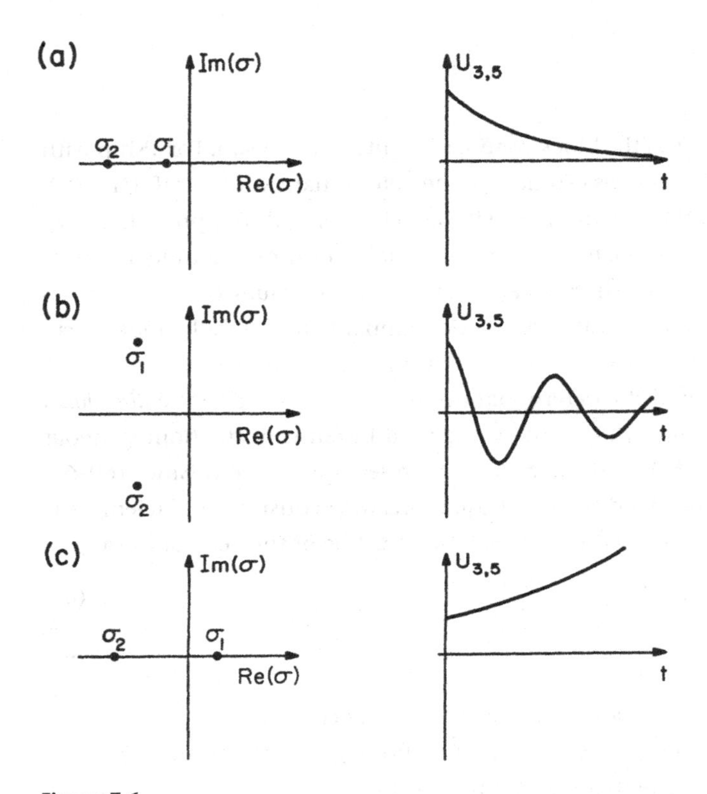

Figure 7.6
The three cases where the stability indices σ_i are real and negative (a), complex (b), or real and positive (c). The left figures show typical positions of the stability indices in the complex plane, and the right figures show the corresponding solutions for the ship motions, as functions of time, subject to an initial disturbance from equilibrium. All scales are arbitrary.

where

$$A = (m + m_{33}^S)(m_{55}^S + M_{55}), \tag{74}$$

$$B = U_1 m_T[(m_{55}^S + M_{55}) + (x_T)^2(m + m_{33}^S)], \tag{75}$$

$$C = -U_1{}^2[mm_{33}^S + m_T x_T(m_{33}^S + m)]. \tag{76}$$

The solutions of the quadratic equation (73) are given by

$$\begin{Bmatrix} \sigma_1 \\ \sigma_2 \end{Bmatrix} = \frac{1}{2A}[-B \pm (B^2 - 4AC)^{1/2}]. \tag{77}$$

It is clear that (74–75) will be positive quantities, whereas the sign of (76) is uncertain since x_T is negative. It follows that

$$\mathrm{Re}(\sigma^2) = \mathrm{Re}\left\{\frac{1}{2A}[-B-(B^2-4AC)^{1/2}]\right\} < 0. \tag{78}$$

However,

$$\begin{aligned}\mathrm{Re}(\sigma_1) = \mathrm{Re}\left\{\frac{1}{2A}[-B+(B^2-4AC)^{1/2}]\right\} &< 0, \quad \text{for } C > 0,\\ &> 0, \quad \text{for } C < 0,\end{aligned} \tag{79}$$

respectively. Thus C must be positive for stability, and the sign of (76) is crucial. Essentially the same factor occurs in connection with the steady-state turning motion, as the denominator of (67–68), and thus stability in a steady-state turn is necessary and sufficient to ensure dynamic stability in the more general sense.

For a nonlifting body with a pointed tail, such that $m_T = 0$, C is negative and the vessel is always unstable. This situation results from the destabilizing effect of the Munk moment; in general an elongated nonlifting body will be stable only when moving broadside to the flow. Directional stability of a streamlined body in the longitudinal direction depends on a tail fin, as in the case of an arrow or a wind vane. Specifically, directional stability of a ship requires that

$$m_T\,|x_T| > \frac{m m_{33}^S}{m + m_{33}^S} \tag{80}$$

For given values of the body mass and added-mass coefficients, the moment arm x_T must be sufficiently large for (80) to be satisfied. Noting our choice of the origin in accordance with (62), we can achieve this objective by moving the center of gravity forward, as well as in the more obvious manner by moving the tail fin aft. Thus the position of the center of gravity is of great significance for the dynamic stability of ships, and indeed for other vehicles as well[1].

From the hydrodynamic standpoint, the stability criterion (80) can be enhanced by increasing the added mass at the stern or by decreasing it elsewhere. Thus a ship's stability will increase as the vessel is trimmed down at the stern (see problem 7).

Excessive stability is undesirable in ship design because it impairs the turning ability. Thus a small positive value for the parameter C is optimum.

However, a limited degree of instability is often tolerated, particularly for supertankers, with the premise that this can be controlled by a suitable automatic pilot.

Since the sign of (76) is independent of the magnitude of U_1, the forward velocity is not a factor in determining the directional stability of a ship unless wave effects become significant, requiring a modification of the hydrodynamic force coefficients in table 7.1. The stability of a submarine in the vertical plane is speed-dependent, as noted in problems 8 and 9. This distinction results because the submarine has a hydrostatic restoring moment, independent of the forward velocity, and the relative importance of this and the hydrodynamic forces varies with the forward velocity.

The characteristic time-scale for maneuvering is inversely proportional to the stability indices; thus it is directly proportional to A/B or l/U_1. For constant Froude number, the time scale is proportional to $l^{1/2}$. This result is fundamental to the simulation of ship maneuvers for training purposes. Human operators can sense the response of a 10 m model with much greater ease than that of a 300m vessel. For this reason attempts to train operators of supertankers with small free-running models have generally been abandoned in favor of computer simulation, involving visual displays and mock-up of the bridge controls, operated in the time-scale of the full-scale vessel.

Our analysis of ship maneuvering has been based on the slender-body hydrodynamic force coefficients derived in section 7.4. This approach permits us to derive the most common form of the equations of motion from rational fluid mechanics and to draw qualitatively correct conclusions. For design purposes, greater accuracy is required; it is customary to assume the same equations of motion, but to measure the force coefficients from experiments where a captive model is forced to oscillate sinusoidally.

The latter approach to this subject is discussed by Mandel (1967), Abkowitz (1969), and by several authors in Bishop, Parkinson, and Eatock Taylor (1972). The survey by Norrbin (1970) deals with ship maneuvering in open and restricted water. Frank et al. (1976) discuss the more general equations of motion in convolution form, similar to (6.195), emphasizing the relation between these and the simpler equations (52–53).

7.6 Slender Bodies in Waves

The oscillatory motions of a slender body on the free surface can be described in accordance with the linear theory developed in chapter 6, with the additional assumption here that the body is slender. We shall consider the separate cases of a vertical slender body, such as a spar buoy, and a horizontal slender body, such as a ship or submarine.

The body is defined by two disparate length scales, the length l and maximum lateral dimension d, with the slenderness parameter $\varepsilon = d/l$ assumed small. The presence of waves on the free surface introduces a third length scale, the wavelength λ, and it is necessary to stipulate the order of magnitude of λ in relation to l and d. We shall consider explicitly the case where the wavelength is comparable to the body length, $\lambda/l = O(1)$, and thus the body diameter is small compared to the wavelength. The converse problem of short waves, with $\lambda/l \ll 1$, will be treated in section 7.7.

The slender-body theory for a body in an infinite fluid can be generalized by imposing the linearized free-surface boundary condition (6.6). This approach requires that the potential for a source beneath the free surface be utilized. However, the leading-order forces acting on the body are independent of this complication and can be discussed in a relatively simple manner. Here we shall follow the latter course, guided by our earlier conclusions regarding the orders of magnitude of the longitudinal and lateral flows and of the resulting hydrodynamic forces on the body.

Proceeding as in section 6.15, we may express the velocity potential in the form

$$\phi = \mathrm{Re}\left\{ e^{i\omega t} \left[\sum_{j=1}^{6} \xi_j \phi_j + A(\phi_0 + \phi_7) \right] \right\}. \tag{81}$$

Here ξ_j denotes the complex amplitude of the six rigid-body motions, A is amplitude of the incident wave, $A\phi_0 e^{i\omega t}$ is the corresponding velocity potential, and $A\phi_7 e^{i\omega t}$ is the scattered potential due to the disturbance of the incident wave by the body.

The incident wave potential is given by (6.16) or (6.23) for infinite or finite water depth, respectively. Since this potential is independent of the body, its order of magnitude with respect to ε is $O(1)$. The remaining potentials ϕ_j ($j = 1, 2, \ldots, 7$) are *body potentials* that will depend on ε.

The body potentials can be analyzed in the inner region, near the body section $\Sigma_B(x)$, using a two-dimensional approximation. The relevant length scale is $d = O(\varepsilon l)$, and since $\omega^2 d/g = O(\varepsilon)$, the free-surface condition in the inner problem can be replaced by a rigid boundary condition as in figure 6.21. The orders of magnitude of the various components of the body potential, as well as the resulting force components acting on the body, can be estimated as if the body and its image were situated in an infinite fluid. Thus, in general, axial forces will be $O(\varepsilon^4 \log \varepsilon)$, whereas lateral forces will be $O(\varepsilon^2)$.

In addition to the hydrodynamic pressure forces associated with the body potential, we must consider the forces due to the body mass, the hydrostatic restoring force, and the Froude-Krylov component of the exciting force due to the pressure field of the incident wave. Thus, for each mode of motion, four different types of force components are estimated, as displayed in table 7.3.

For a vertical spar buoy, the estimates of the body-induced pressure forces are shown in the first column. Heave induces a hydrodynamic force of order $(\varepsilon^4 \log \varepsilon)$, and one can infer from table 4.3 that the corresponding yaw moment will be proportional to d^4, and hence $O(\varepsilon^4)$. The remaining motions are lateral with respect to the body axis, and induce forces of order ε^2.

The body-mass forces for a spar buoy can be estimated from (6.182–6.183), and are as shown in the second column of table 7.3. The third column shows the hydrostatic force and moment, with the heave force proportional to the waterplane area $O(\varepsilon^2)$, and the roll and pitch moments proportional to the moment of inertia of order ε^4, as indicated by (6.144–6.149).

Finally, the Froude-Krylov component of the exciting force can be estimated for a spar buoy from Gauss' theorem, in a manner similar to (6.165–6.166). The results are as shown in the fourth column of table 7.3. In this case the displaced volume $\forall$ and waterplane area S are both $O(\varepsilon^2)$. The contribution to the exciting force from the scattering potential ϕ_7 is included in the body-potential contributions listed in column one.

Similar estimates can be made for the forces on a horizontal body, as shown in table 7.3. In surge, the body-induced force is longitudinal, and hence $O(\varepsilon^4 \log \varepsilon)$, while the body-mass force and the Froude-Krylov force are $O(\varepsilon^2)$. For vertical motions in the inner region, associated with pitch and heave, the two-dimensional flow is source-like since the submerged

Table 7.3
Orders of Magnitude of the Different Types of Forces and Moments Acting on a Vertical or Horizontal Slender Body, in Waves of Wavelength Comparable to the Body Length

	Vertical Body Axis				Horizontal Body Axis			
Mode	F_B	F_M	F_{HS}	F_{FK}	F_B	F_M	F_{HS}	F_{FK}
1. Surge	$\boldsymbol{\varepsilon}^2$	$\boldsymbol{\varepsilon}^2$	0	$\boldsymbol{\varepsilon}^2$	$\varepsilon^4 \log \varepsilon$	$\boldsymbol{\varepsilon}^2$	0	$\boldsymbol{\varepsilon}^2$
2. Heave	$\varepsilon^4 \log \varepsilon$	$\boldsymbol{\varepsilon}^2$	$\boldsymbol{\varepsilon}^2$	$\boldsymbol{\varepsilon}^2$	$\varepsilon^2 \log \varepsilon$	ε^2	$\boldsymbol{\varepsilon}$	$\boldsymbol{\varepsilon}$
3. Sway	$\boldsymbol{\varepsilon}^2$	$\boldsymbol{\varepsilon}^2$	0	$\boldsymbol{\varepsilon}^2$	$\boldsymbol{\varepsilon}^2$	$\boldsymbol{\varepsilon}^2$	0	$\boldsymbol{\varepsilon}^2$
4. Roll	$\boldsymbol{\varepsilon}^2$	$\boldsymbol{\varepsilon}^2$	ε^4	$\boldsymbol{\varepsilon}^2$	ε^4	ε^4	$\boldsymbol{\varepsilon}^2$	$\boldsymbol{\varepsilon}^2$
5. Yaw	$\boldsymbol{\varepsilon}^4$	$\boldsymbol{\varepsilon}^4$	0	$\boldsymbol{\varepsilon}^4$	$\boldsymbol{\varepsilon}^2$	$\boldsymbol{\varepsilon}^2$	0	$\boldsymbol{\varepsilon}^2$
6. Pitch	$\boldsymbol{\varepsilon}^2$	$\boldsymbol{\varepsilon}^2$	ε^4	$\boldsymbol{\varepsilon}^2$	$\varepsilon^2 \log \varepsilon$	ε^2	$\boldsymbol{\varepsilon}$	$\boldsymbol{\varepsilon}$

Note: F_B denotes the body-induced pressure force, F_M the body-mass force, F_{HS} the hydrostatic force, and F_{FK} the Froude-Krylov force due to the pressure field of the undisturbed incident wave. The dominant terms in each mode are indicated by bold-face type ($\boldsymbol{\varepsilon}$). In all cases the body length is assumed to be $O(1)$.

sectional area of the body is changing in time. As a result the body potentials in these modes are $O(\varepsilon \log \varepsilon)$ and the corresponding force components are $O(\varepsilon^2 \log \varepsilon)$. The dominant vertical forces are the hydrostatic component, proportional to the waterplane area of order ε, and the vertical component of the Froude-Krylov force which, from (6.165), is of the same order. An exception occurs for a submerged body, where the waterplane area is zero and the remaining vertical force components are $O(\varepsilon^2)$.

Equations of motion for the six degrees of freedom can be derived from the forces listed in table 7.3. In each mode the bold-faced terms are dominant, and thus the leading-order equations of motion are simplified. We shall outline the details of this procedure for the vertical and horizontal forces acting on a spar buoy, and for the vertical force acting on a slender ship.

For a spar buoy situated at $x = z = 0$, the vertical Froude-Krylov force can be expressed in the form

$$F_{FK} = \iint_{S_B} p_0 n_y \, dS \simeq \int_{-T}^{0} S'(y) p_0(y) \, dy. \tag{82}$$

Here $S(y)$ is the sectional area of the body, T is the draft, and $p_0(y)$ is the pressure field of the incident wave system, at a depth y. Note that in analyzing the vertical force, the variation of the pressure p_0 in the horizontal plane can be neglected, since the lateral dimensions of the body are small compared to the wavelength. For deep water, using the incident wave potential (6.16) and the linear form of Bernoulli's equation (6.3), it follows that

$$F_{FK} = \rho g A \,\mathrm{Re}\left\{ e^{i\omega t} \int_{-T}^{0} S'(y) e^{ky} dy \right\}. \tag{83}$$

Assuming the spar buoy is unrestrained, an equation of motion for heave follows by equating the exciting force (83) to the body-mass and hydrostatic forces. Thus

$$[-\omega^2 m + \rho g S(0)]\xi_3 = \rho g A \int_{-T}^{0} S'(y) e^{ky} dy, \tag{84}$$

where

$$m = \rho \forall = \rho \int_{-T}^{0} S(y) dy \tag{85}$$

is the body mass, and ξ_3 is the complex heave amplitude. Equation (84) can be solved for ξ_3. An obvious feature of the result is an unbounded resonance at the natural frequency

$$\omega_n = [gS(0)/\forall]^{1/2}. \tag{86}$$

Except for the immediate vicinity of resonance, the simple solution of (84) is quite accurate, as shown in figure 2.16. To render this solution valid near resonance, a damping coefficient of order ε^4 can be derived from the Haskind relations (6.173). However for long cylindrical buoys, the resulting wave damping will be exponentially small in proportion to e^{-2kT}. Under this circumstance viscous damping may be significant. Another higher-order effect that manifests itself near resonance, as shown in figure 2.16, is the reduction of the natural frequency due to the added mass of the body.

A complementary analysis is required for the horizontal force on a vertical spar buoy and for the resulting motions of the buoy. The inner flow near a body section can be analyzed using the technique developed in section 4.17, where the slowly varying velocity field is the horizontal component of the incident wave. Thus, if the incident waves are moving in the

x-direction, the local force on a body section at a depth y is given from (4.151) in the form

$$F_x = -[\rho S(y) + m_{11}(y)]\mathrm{Re}\left(-i\omega\frac{\partial\phi_0}{\partial x}\right)_{x=0} - m_{11}(y)\dot{U}(y)$$
$$= [\rho S(y) + m_{11}(y)]\mathrm{Re}(\omega k\phi_0)_{x=0} - m_{11}(y)\dot{U}(y). \tag{87}$$

Here $m_{11}(y)$ is the local two-dimensional added-mass coefficient of the body section, and $U(y)$ is the body velocity in the horizontal direction. Substituting the incident wave potential (6.16) or (6.23) and integrating over the body depth gives the total hydrodynamic force, which may be equated to the body-mass force. A similar result follows for the pitch moment. For an axisymmetric body in deep water, the details are carried out by Newman (1963). In general the solutions for surge and pitch are coupled and they are resonant unless the body is neutrally stable in pitch. Damping can be inferred from the Haskind relations, noting that the term in (87) proportional to ϕ_0 is the exciting force.

If the buoy diameter is comparable to or smaller than the wave amplitude, cross-flow viscous drag effects will become important, as discussed in section 2.14. To account for this in a semiempirical manner, an additional force of the following form may be added to (87),

$$\frac{1}{2}\rho C_D l^2\left(\frac{\partial\phi_0}{\partial x} - U\right)\left|\frac{\partial\phi_0}{\partial x} - U\right|,$$

as in equation (2.49). Here C_D is a viscous drag coefficient, for lateral flow across the body section, and l^2 is the area upon which the drag coefficient is based.

Turning our attention to the case of a horizontal floating body, the vertical force is dominated by the terms of order ε in table 7.3, that is, the hydrostatic restoring force and the Froude-Krylov exciting force. The Froude-Krylov force can be evaluated using (6.160) and (6.165), in the form

$$F'_{FK} = \rho g\,\mathrm{Re}\left\{AB(x)e^{i(\omega t - kx\cos\theta)}\right\}, \tag{88}$$

where $B(x)$ is the local beam of the waterplane and θ is the angle of incidence as shown in figure 6.6. Simple equations of motion follow for the coupled heave and pitch motions in the form

$$S\xi_2 + S_1\xi_6 = A\int_l B(x)e^{-ikx\cos\theta}\,dx, \tag{89}$$

$$S_1\xi_2 + S_{11}\xi_6 = A\int_l B(x)e^{-ikx\cos\theta}x\,dx. \tag{90}$$

Here the terms on the left-hand side result from the hydrostatic restoring force and moment (6.144–6.149), with S the waterplane area, and S_1 and S_{11} the first and second moments of this area.

Inertial and hydrodynamic effects are of higher order as indicated in table 7.3, and thus are absent from (89–90). As a result, the solutions of these equations are restricted to the range of frequencies substantially less than the resonant frequencies of the ship. The resonant frequencies ω_n can be estimated, as in (86), by equating the hydrostatic and virtual-mass forces which are of order ε and $\varepsilon^2\omega^2$ respectively. It follows that $\omega_n = O(\varepsilon^{-1/2})$, and in deep water the resonant wavelength will be of the order of the ship's beam. Thus resonant ship motions must be analyzed from the complementary short-wavelength theory where $\lambda/d = O(1)$.

Typical solutions of (89–90) are shown in figures 7.7–7.8 for an aircraft carrier in head seas ($\theta = \pi$). Also shown are experimental results for zero forward speed and a Froude number of 0.14. At zero speed the agreement is reasonable, except for an apparent resonance near $\lambda/l = 0.5$. The importance of this resonance is diminished because the exciting force and moment are small for $\lambda < l$. However if the ship proceeds with substantial forward velocity in head seas, the frequency of encounter is increased by the Doppler shift. Thus, the oscillation frequency is increased for a given wavelength, ultimately to the point where resonant motions occur in the range of substantial exciting forces and moments. This situation is apparent from the experiments shown in figures 7.7–7.8, and would become progressively more serious at higher speeds. In this case the practical value of the long-wavelength slender-body theory is diminished, and a complementary approach must be developed where the magnitude of the oscillation frequency is in the range of resonant motions, of order $\varepsilon^{-1/2}$.

7.7 Strip Theory for Ship Motions

The problem of a slender ship moving with forward velocity in a seaway is one of the most important topics in ship hydrodynamics. Here, by

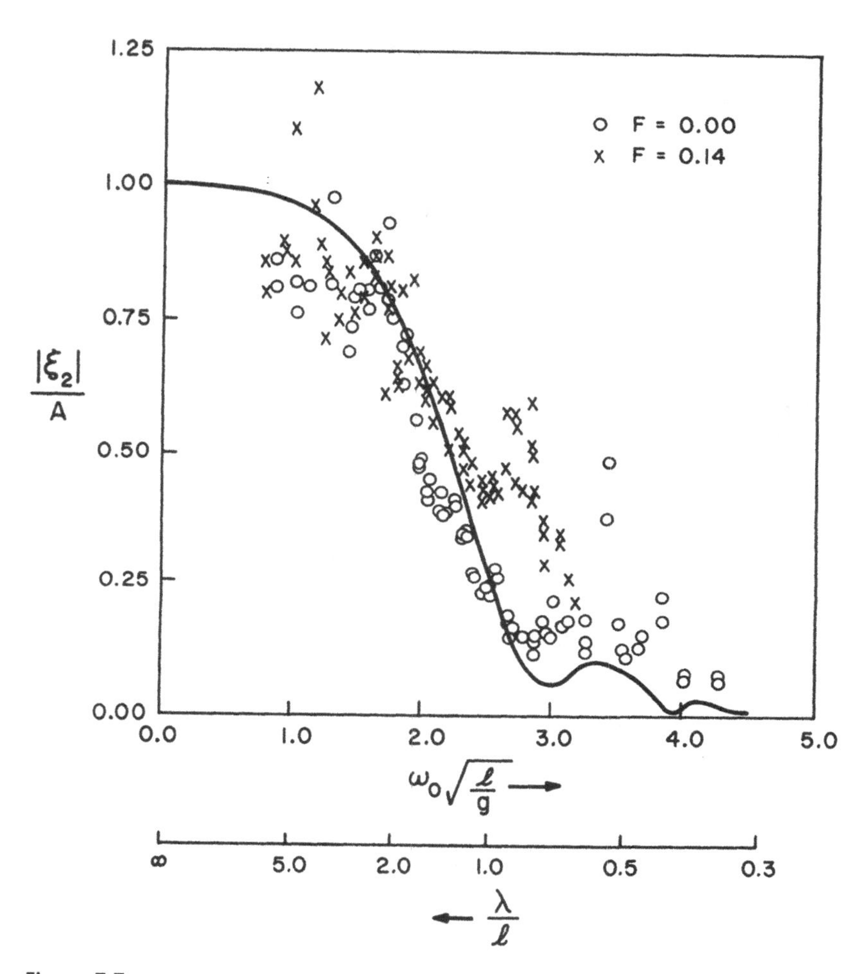

Figure 7.7
Heave motions predicted from (89–90) and compared with experiments. The wave frequency ω_0 is defined with reference to the fixed coordinate system, as in (91). (Adapted from Newman and Tuck 1964)

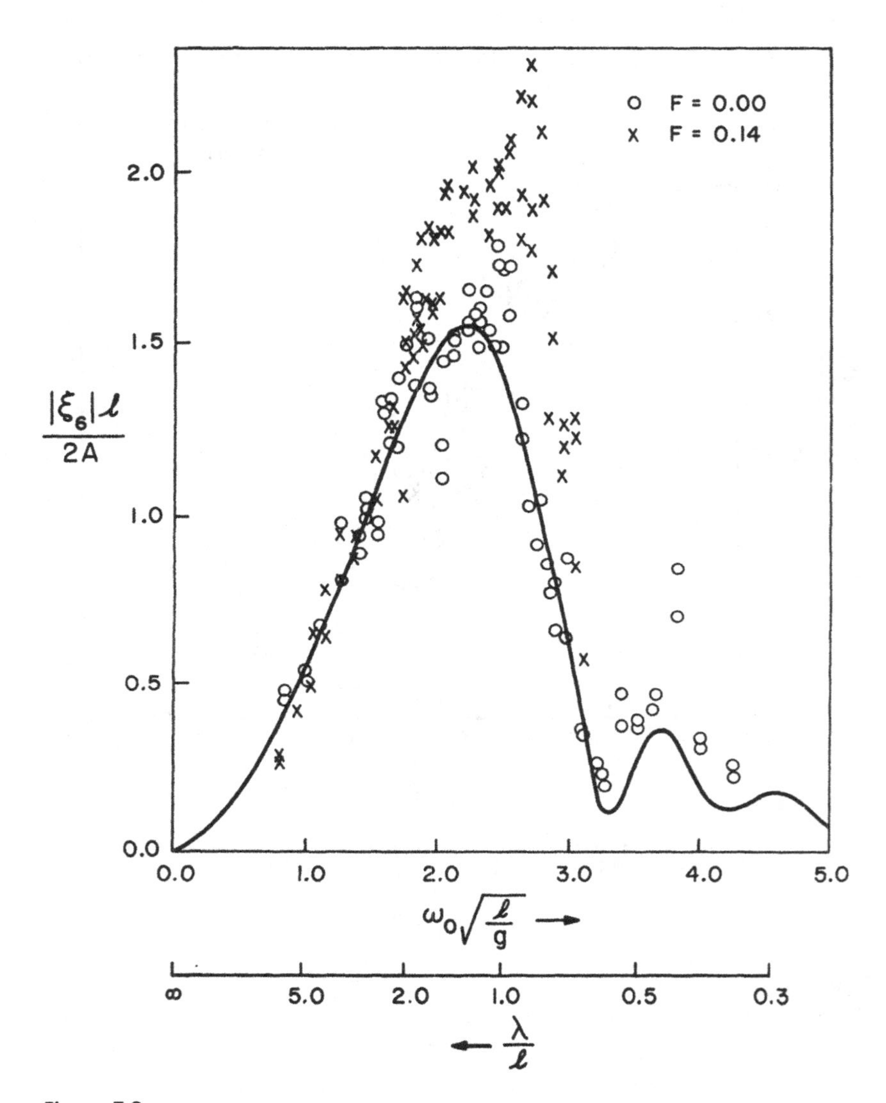

Figure 7.8
Pitch motions predicted from (89–90) and compared with experiments. (Adapted from Newman and Tuck 1964)

comparison to the simpler problem of a slender body in waves without forward speed, hydrodynamic interactions will exist between the steady-state flow field due to the ship's forward motions and the oscillatory flow associated with the unsteady motions. Since complete analysis of these interactions is beyond the present state of knowledge in this field, we must adopt a pragmatic approach, justified by experimental confirmation.

Two important effects of the ship's forward velocity can be analyzed without major difficulty. One is the change in the frequency of encounter, which is an elementary feature of the ship-motion problem but one with far-reaching consequences. The second effect is the longitudinal convection of momentum, which changes the oscillatory force distribution along the hull, as in the simpler case of a slender body in an unbounded fluid.

The effect of the forward speed on the frequency of encounter is identical to the Doppler shift in other fields of wave motion. Restricting our attention to harmonic motions in deep water, the velocity potential of the incident wave system given by (6.16) can be written in the complex form

$$\phi = (igA / \omega_0) \exp\left[ky_0 - ik(x_0 \cos\theta + z_0 \sin\theta) + i\omega_0 t\right], \tag{91}$$

where the real part is implied wherever an exponential time-dependence is displayed. Here A is the amplitude of the incident wave, θ is the direction of propagation relative to the x-axis (see figure 6.6), ω_0 is the wave frequency with respect to a fixed reference frame (x_0, y_0, z_0), and $k = \omega_0^2 / g$ is the wavenumber.

In a reference frame (x, y, z) moving with the steady forward motion of a ship, the transformations (2–4) apply if the ship moves in the positive x-direction. Thus, (91) can be expressed in the form

$$\phi = (igA / \omega_0) \exp[ky - ik(x \cos\theta + z \sin\theta) + i(\omega_0 - kU \cos\theta)t]. \tag{92}$$

The wave amplitude A and wavenumber k are unchanged from (91), but the frequency of encounter in the moving reference frame is given by

$$\omega = \omega_0 - kU \cos\theta. \tag{93}$$

For *following waves* ($\theta = 0$), $\omega < \omega_0$, but in *head waves* ($\theta = \pi$), the converse is true. The same conclusions apply to a lesser extent for *quartering* ($0 < \theta < \pi/2$) and *bow* ($\pi/2 < \theta < \pi$) waves, whereas for *beam* waves ($\theta = \pi/2$), the two frequencies coincide.

The pressure field of the incident wave system (91–92) can be derived in either reference frame. Using the linearized form of the Bernoulli equation (6.3), it follows from (91) that

$$p = -\rho \frac{\partial \phi}{\partial t} = -\rho(i\omega_0)\phi = \rho g A \exp[ky_0 - ik(x_0 \cos\theta + z_0 \sin\theta) + i\omega_0 t]. \tag{94}$$

Here the hydrostatic component of the pressure is neglected because this is not affected by the incident waves. In the moving reference frame, the time-derivative must be transformed, in a manner analogous to (19); thus

$$\begin{aligned} p &= -\rho\left(\frac{\partial \phi}{\partial t} - U\frac{\partial \phi}{\partial x}\right) \\ &= -\rho(i\omega + ikU\cos\theta)\phi \\ &= \rho g A \exp[ky - ik(x\cos\theta + z\sin\theta) + i\omega t]. \end{aligned} \tag{95}$$

The results of (95) are identical to (94), except for the change in frequency of encounter. Thus the incident-wave pressure field is not affected by the choice of reference frame.

We shall restrict our attention to the heave and pitch motions of a ship, in incident waves of the form (92). As noted at the end of section 7.6, the resonant frequencies in these modes are of order $\varepsilon^{-1/2}$, implying that the frequency of encounter ω is of the same order of magnitude in the regime where significant motions occur. For ω to be large relative to the wave frequency ω_0, it follows that the incident wave angles of greatest importance are head waves and bow waves. Thus we shall consider this regime, with the assumption that $\omega = O(\varepsilon^{-1/2})$ or, in terms of the ship's beam B, $\omega^2 B/g = O(1)$. As in section 6.15, the results are linearized in terms of the small oscillatory amplitudes of the incident wave and resulting body motions. On this basis, added-mass and damping coefficients can be derived for forced motions of the ship in otherwise calm water, and the exciting force and moment can be analyzed with the heave and pitch oscillations suppressed. In both problems the Froude number $U/(gl)^{1/2}$ is assumed to be $O(1)$.

First we consider the forced heave and pitch motions in calm water, in the frequency regime where $\omega^2 B/g = O(1)$. Under this circumstance, and unlike the low-frequency regime of section 7.6, both terms must be retained in the linearized free-surface condition (6.6). The motion is periodic in the moving reference frame, where (6.6) takes the form

$$\left(\frac{\partial}{\partial t}-U\frac{\partial}{\partial x}\right)^2\phi+g\frac{\partial\phi}{\partial y}=0 \tag{96}$$

or

$$-\omega^2\phi-2i\omega U\frac{\partial\phi}{\partial x}+U^2\frac{\partial^2\phi}{\partial x^2}+g\frac{\partial\phi}{\partial y}=0, \quad \text{on } y=0. \tag{97}$$

The total vertical displacement of the ship, at a longitudinal position x, is given by

$$\eta(x,t)=(\xi_2+x\xi_6)e^{i\omega t}, \tag{98}$$

where ξ_2 and ξ_6 are the complex amplitudes of heave and pitch, as defined in figure 6.19. This vertical motion, as well as the ship's geometry, are slowly varying in the x-direction. Thus a two-dimensional approximation equivalent to (16) can be adopted for the velocity potential in the inner region, near the ship hull, in the form

$$\phi=U\Phi_1+V\Phi_2. \tag{99}$$

Here U is the forward velocity, and

$$V(x,t)=\left(\frac{\partial}{\partial t}-U\frac{\partial}{\partial x}\right)[e^{i\omega t}(\xi_2+x\xi_6)] \tag{100}$$

is the local vertical velocity of the ship hull with respect to the fixed reference system.

Equation (100) is analogous to (19), for the lateral velocity of a slender body. Since ξ_2 and ξ_6 are constant, the only contribution from the term $U\partial/\partial x$ is an apparent vertical velocity equal to the product of the pitch angle and the forward speed.

The velocity potentials Φ_1 and Φ_2 in (99) satisfy the free-surface condition (6.6) with respect to the fixed reference frame, or (97) with respect to the moving reference frame, as well as the two-dimensional Laplace equation (12) and a suitable radiation condition.[2] On the hull surface, Φ_1 satisfies the boundary condition (17), whereas

$$\frac{\partial\Phi_2}{\partial N}=n_y, \qquad \text{on } \Sigma_B(x). \tag{101}$$

From the slenderness assumption, and the estimates (10), it can be anticipated that the potential Φ_1 will be of order ε^2, or possibly $O(\varepsilon^2\log\varepsilon)$, as in an infinite fluid; but the potential Φ_2 associated with the vertical motion will be of order ε. Thus, in terms of the slenderness parameter, the potential Φ_2 is dominant.

The differential vertical force acting on the ship can be derived as in (36), in a reference frame fixed with respect to the undisturbed fluid. The control contour surrounding the hull must include the portion of the free surface, Σ_F, between $\Sigma_B(x)$ and the contour Σ_∞ at large distance from the body. Thus

$$
\begin{aligned}
F_y' = &-\rho \frac{d}{dt} \int_{\Sigma_B(x)} \phi n_y dl - \rho \frac{\partial}{\partial x_0} \iint_{S_0} \frac{\partial \phi}{\partial x_0} \frac{\partial \phi}{\partial y_0} dy_0 dz_0 \\
&- \rho \int_{\Sigma_F + \Sigma_\infty} \left(\frac{\partial \phi}{\partial y_0} \frac{\partial \phi}{\partial n} - n_y \frac{1}{2} \nabla\phi \cdot \nabla\phi \right) dl,
\end{aligned}
\tag{102}
$$

where S_0 is the portion of the plane x_0 = constant interior to the closed contour $\Sigma_B(x) + \Sigma_F + \Sigma_\infty$.

If we restrict our attention to the linearized force, of first-order in the unsteady body motions and of leading order in the slenderness parameter ε, the nonlinear terms in (102) may be ignored. Thus, the local vertical force is given by the expression

$$
\begin{aligned}
F_y' &= -\rho \frac{d}{dt} \int_{\Sigma_B(x)} \phi n_y dl \\
&= -\rho \frac{d}{dt} [V(x,t) \int_{\Sigma_B(x)} \Phi_2 n_y dl].
\end{aligned}
\tag{103}
$$

The integral in (103) can be treated in a manner analogous to (38). Since the wave motion is periodic with respect to the moving coordinate system, the potential $\Phi_2(y, z; x)$ will be complex but independent of time. Thus the contour integral in (103) can be defined in terms of generalized added-mass and damping coefficients for the two-dimensional section,

$$
\rho \int_{\Sigma_B(x)} \Phi_2 n_y dl \equiv a_{22}(x,\omega) + \frac{1}{i\omega} b_{22}(x,\omega) \equiv Z(x,\omega). \tag{104}
$$

The desired expression for the differential vertical force follows by substituting (104) in (103),

$$
\begin{aligned}
F_y' &= -\frac{d}{dt}[V(x,t) Z(x,\omega)] \\
&= -\left(\frac{\partial}{\partial t} - U \frac{\partial}{\partial x} \right) [V(x,t) Z(x,\omega)].
\end{aligned}
\tag{105}
$$

This expression differs from (39) only with respect to the change in direction of the body motion, and the substitution of the complex coefficient (104) for the added-mass.

The total heave force is given by the integral of (105). With the assumption that (104) vanishes at the bow and stern, it follows that

$$F_y = \omega^2 e^{i\omega t} \int_l \left[\xi_2 + \left(x - \frac{U}{i\omega} \right) \xi_6 \right] Z(x,\omega) dx. \tag{106}$$

Similarly, after a partial integration of (105), the pitch moment is obtained in the form

$$M_z = \omega^2 e^{i\omega t} \int_l \left[\xi_2 \left(x + \frac{U}{i\omega} \right) + \xi_6 \left(x^2 + \frac{U^2}{\omega^2} \right) \right] Z(x,\omega)\, dx. \tag{107}$$

The factors of the pitch and heave amplitudes, in (106) and (107), are the complex force coefficients f_{ij} defined in (6.150). The three-dimensional added-mass and damping coefficients follow by taking the real and imaginary parts of these factors.

In this manner the three-dimensional force coefficients are derived from a strip-theory synthesis, as integrals of the two-dimensional force coefficient (104). However, the forward speed U leads to additional components of the force and moment (106–107) which are not evident from a simple integration of (104) along the length. These additional contributions result in part from momentum convection, as well as from the speed-dependent term in the vertical body-velocity (100).

The coefficients a_{22} and b_{22} in (104) are closely related to the two-dimensional added-mass and damping coefficients for forced motion of a two-dimensional body, with profile Σ_B, oscillating on the free surface with frequency ω. The only difference between these problems is in the free-surface boundary condition. The extra terms in (97), depending on U, distinguish the potential Φ_2 from the two-dimensional solutions described in chapter 6. Physically, these imply an interaction between adjacent sections of the ship hull, depending on the rate of change in the longitudinal direction of the inner solution.

The only role of the x-coordinate on Φ_2 is the parametric dependence on the hull profile $\Sigma_B(x)$. Thus Φ_2 will be purely two dimensional in those cases where Σ_B is locally independent of x, as along the parallel middle body of a ship hull. In this particular case the speed-dependent terms in (97) vanish, and the force coefficients a_{22} and b_{22} in (104) are identical to the added-mass and damping coefficients for the two-dimensional cylinder, with profile Σ_B. For this particular case results such as those illustrated in figure 6.23 can be substituted for the complex force coefficient Z and used to compute the differential force (105), as well as the total force and moment (106–107). Thus, for vertical heaving motions of a cylindrical body with constant profile, the differential force (105) is independent of the forward speed U. This is confirmed physically since a long cylindrical body can be translated in the axial direction, in an inviscid fluid, without affecting the surrounding flow except near the ends.

In the general case of a ship hull with varying profile Σ_B, the difficulties associated with the speed-dependent terms in (97) can be avoided by noting that $U/\omega = O(\varepsilon^{1/2})$. Thus, to leading order, (97) can be replaced by the linearized free-surface condition (6.126) for sinusoidal motion without forward speed. Once again, the zero-speed two-dimensional results can be used for the added-mass and damping coefficients in (104). Salvesen, Tuck, and Faltinsen (1970) proceed in this manner, using expressions equivalent to (106–107). Experimental confirmation is provided by Vugts (1970), for both the total force and moment (106–107) and the local force distribution (105).

The approach outlined is deficient from the theoretical standpoint, since some higher-order terms proportional to $U/\omega = O(\varepsilon^{1/2})$ have been retained, while others have been neglected. Indeed, one term retained in (107) is $U^2/\omega^2 = O(\varepsilon)$. In a consistent leading-order theory these terms should be discarded. Alternatively, a consistent higher-order theory should be developed, where all terms of the same order are retained in the free-surface condition (97) and in the integral over Σ_F in (102).

Ogilvie and Tuck (1969) have performed a consistent analysis, to order $\varepsilon^{1/2}$, including the two additional effects and a nonlinear contribution to the free-surface boundary condition. These complicated higher-order effects largely cancel each other, and to order $\varepsilon^{1/2}$ the total force due to heave and the moment due to pitch are correctly given by (106) and (107) respectively.

The analysis of Ogilvie and Tuck (1969) does reveal additional contributions to the cross-coupling coefficients, for the heave force due to pitch and the pitch moment due to heave, involving integrals of $(\Phi_2)^2$ over the free surface. These are re-derived from an energy analysis by Wang (1976).

Computations have been carried out by Faltinsen (1974), who shows improved agreement of the resulting cross-coupling coefficients with experiments. The net effect of these additional terms on the pitch and heave motions in waves is not well established.

In essence, the results (105–107) with the zero-speed two-dimensional added-mass and damping coefficients are relatively simple to use and give useful engineering predictions. The more complete results of Ogilvie and Tuck (1969) are superior from the mathematical standpoint, but their practical value is open to debate. The simple approach includes some forward-speed effects, which are identical to those associated with the lateral force acting on a slender body in an unbounded fluid. The additional forward-speed effects included by Ogilvie and Tuck (1969) result from the free surface. Apparently the latter are relatively unimportant, but this hypothesis lacks a rational justification.

The exciting force and moment can be analyzed with a similar approach, but under more restrictive assumptions. Assuming the forward speed $U = O(1)$, with respect to ε, and the frequency of encounter $\omega = O(\varepsilon^{-1/2})$, the only possible estimates for ω_0 and k compatible with (93) and the dispersion relation $k = \omega_0^2 / g$ are $k = O(\varepsilon^{-1/2})$ and $\omega_0 = O(\varepsilon^{-1/4})$. Thus the incident wavelength $2\pi/k$ must be short compared to the ship length, but long compared to the beam.

The exciting force and moment are determined from the solution of the diffraction problem, where the oscillatory potential consists of the incident wave ϕ_0, plus a scattering potential ϕ_7, as in (6.123). The contributions to the exciting force from these two potentials can be treated separately. The Froude-Krylov force, associated with the pressure field of the incident wave system, can be obtained directly from (95) in the form

$$F'_{y0} = \rho g A \exp(-ikx\cos\theta + i\omega t) \int_{\Sigma_B(x)} \exp(ky - ikz\sin\theta) n_y \, dl. \qquad (108)$$

This integral can be evaluated numerically, for a prescribed section $\Sigma_B(x)$, and no further approximation of the Froude-Krylov force is necessary from the computational standpoint. However, since $k = O(\varepsilon^{-1/2})$, it follows that $(ky, kz) = O(\varepsilon^{1/2})$ on Σ_B. Thus it is consistent with assumptions to be made in deriving the total exciting force to expand the integral of (108) in a Taylor series. Integrating the leading terms of this expansion gives

$$F'_{y0} \simeq \rho g A \exp(-ikx\cos\theta + i\omega t) \int_{\Sigma_B(x)} (1 + ky - ikz\sin\theta) n_y dl$$
$$= \rho g A \exp(-ikx\cos\theta + i\omega t)[B(x) - kS(x)], \tag{109}$$

where $B(x)$ and $S(x)$ are the local waterplane beam and sectional area, respectively, and symmetry about $z = 0$ has been assumed. The error incurred in approximating (108) by (109) can be computed for simple sectional shapes. For a rectangular section, (109) underestimates the Froude-Krylov exciting force by 30 percent when the wavelength/draft ratio is 10. For sections with less fullness, the error is smaller (see problem 11).

The scattering potential ϕ_7 can be analyzed in the inner region with the same two-dimensional approach used for the forced-motion problem. In a reference frame moving with the ship, the boundary condition for ϕ_7 is (6.124). Substituting the incident wave potential (92), it follows that

$$\frac{\partial \phi_7}{\partial n} = -\frac{\partial \phi_0}{\partial n}$$
$$= -i\omega_0 (n_y - in_x \cos\theta - in_z \sin\theta) \tag{110}$$
$$\times \exp[ky - ik(x\cos\theta + z\sin\theta) + i\omega t].$$

Neglecting the normal component $n_x = O(\varepsilon)$, and recalling that $(ky, kz) = O(\varepsilon^{1/2})$ on the body surface,

$$\frac{\partial \phi_7}{\partial n} \simeq -i\omega_0 (n_y - in_z \sin\theta)\exp(-ikx\cos\theta + i\omega t). \tag{111}$$

The form of this boundary condition suggests that the scattering potential can be expressed in the inner region by

$$\phi_7 = -i\omega_0 (\Psi_2 - i\sin\theta \Psi_3)\exp(-ikx\cos\theta + i\omega t), \tag{112}$$

where the functions Ψ_2 and Ψ_3 satisfy the boundary conditions

$$\frac{\partial \Psi_2}{\partial N} = n_y, \tag{113}$$

$$\frac{\partial \Psi_3}{\partial N} = n_z, \tag{114}$$

on the body contour Σ_B. From (113–114) it can be inferred that Ψ_2 and Ψ_3 are slowly varying functions of the longitudinal position along the ship length, in the same sense as the corresponding forced-motion potentials Φ_2 and Φ_3.

Operating with Laplace's equation on (112), and neglecting longitudinal derivatives of Ψ_2 and Ψ_3, it follows that these are solutions of the two-dimensional Laplace equation (12) to leading order in kB. However, application of the free-surface condition (96–97) to (112) gives the boundary condition

$$-\omega_0^2\Psi_j - 2i\omega_0 U\frac{\partial\Psi_j}{\partial x} + U^2\frac{\partial^2\Psi_j}{\partial x^2} + g\frac{\partial\Psi_j}{\partial y} = 0, \qquad \text{on } y = 0\ (j = 2,3). \tag{115}$$

The only difference between the boundary-value problems for the functions Ψ_j and the corresponding forced-motion potentials Φ_j is that the frequency of encounter ω appears in (97), whereas the incident wave frequency ω_0 appears in (115). *Thus the potentials Ψ_j can be regarded as forced-motion potentials, with oscillation occurring at the incident wave frequency ω_0, as measured in a fixed reference frame.*

Restricting our attention to the vertical exciting force and moment, for a ship with symmetry about the plane $z = 0$, the potential Ψ_3 in (112) can be neglected. Substituting (112) in (103), the contribution from ϕ_7 to the differential force is given by

$$\begin{aligned} F'_{y7} &= -\rho\left(\frac{\partial}{\partial t} - U\frac{\partial}{\partial x}\right)\int_{\Sigma_B(x)} \phi_7 n_y dl \\ &= \rho\omega_0 A\left(\frac{\partial}{\partial t} - U\frac{\partial}{\partial x}\right)\left[i\exp(-ikx\cos\theta + i\omega t)\int_{\Sigma_B(x)} \Psi_2 n_y dl\right] \\ &= \omega_0 A\left(\frac{\partial}{\partial t} - U\frac{\partial}{\partial x}\right)[i\exp(-ikx\cos\theta + i\omega t)Z(x,\omega_0)]. \end{aligned} \tag{116}$$

Here (104) has been used, along with the conclusion following (115).

Adding the forces (109) and (116), the differential exciting force is obtained in the form

$$\begin{aligned} F'_{yex} &= A\{\rho g[B(x) - kS(x)]\exp(-ikx\cos\theta + i\omega t) \\ &\quad + i\omega_0\left(\frac{\partial}{\partial t} - U\frac{\partial}{\partial x}\right)[z(x,\omega_0)\exp(-ikx\cos\theta) + i\omega t]\}. \end{aligned} \tag{117}$$

Before integrating (117) along the ship length, we note the special case of a cylindrical body with profile Σ_B independent of x, such as the parallel middle body of a ship, or a horizontal pipeline. In this case B, S, and Z are constants, and (117) reduces to the expression

$$F'_{yex} = gA\{\rho B - k[\rho S + Z(\omega_0)]\}\exp(-ikx\cos\theta + i\omega t). \tag{118}$$

Here the only effect of the forward speed is on the frequency of encounter, for fixed wave amplitude A and wavenumber k. This result is consistent with the fact, noted earlier, that axial motion of a slender cylindrical body in an inviscid fluid has no effect on the inner flow except near the ends.

The steady forward velocity of a cylindrical body through a fluid is identical to the flow of a current past a fixed body. Thus (118) can be applied in the case where a floating or submerged pipeline is fixed in position, in the presence of waves and a superposed axial current. From this relation it follows that such a current has no effect on the wave force, provided the wave amplitude and wavenumber are correctly measured in a frame of reference moving with the fluid.

The heave exciting force on a ship is given by the integral of (117) along the length, in the form

$$\begin{aligned} F_{yex} = {} & gAe^{i\omega t}\int_l e^{ikx\cos\theta}\{\rho B(x) - \rho k S(x) \\ & - k[1-(\omega_0 U / g)\cos\theta]Z(x,\omega_0)\}dx, \end{aligned} \tag{119}$$

where (93) has been used, and it is assumed that the factor in braces vanishes at the bow and stern. Here we note that the Froude-Krylov force, associated with the first two terms in braces, is independent of the forward velocity, whereas the remaining contribution due to the scattering potential is a linear function of the forward velocity.

The pitch exciting moment can be derived in a similar manner, after integrating by parts, with the result

$$\begin{aligned} M_{zex} = {} & gAe^{i\omega t}\int_l e^{-ikx\cos\theta}\{\rho[B(x) - kS(x)]x \\ & - [(1-(\omega_0 U / g)\cos\theta)kx - i(\omega_0 U / g)]Z(x,\omega_0)\}dx. \end{aligned} \tag{120}$$

Here again the Froude-Krylov component is independent of the forward speed, while the scattering component is linear in U.

A complementary derivation of the exciting force (119) and moment (120) is given by Salvesen, Tuck, and Faltinsen (1970), assuming that $k = O(\varepsilon^{-1})$ and using a generalization of the Haskind relations (6.172). In that approach the forced-motion potential Φ_2 enters as a consequence of the Haskind relations, *and as a function of the frequency of encounter instead of the wave frequency* ω_0. An alternative approach based on using the Haskind relations in a two-dimensional manner is outlined by Newman (1970), and is more closely related to the present results. The Haskind relations lead to an exponential factor in the scattering-force integral of (116), in a form similar to the Froude-Krylov integral (108). This additional factor makes the scattering force more difficult to compute, since the force due to the scattering potential cannot be related directly[3] to the added-mass and damping coefficients.

These ambiguities are not the only uncertainties in the exciting force problem. In general, oscillatory integrals such as (119) and (120) will tend to cancel along the length of the ship, if the wavelength is short compared to the ship length, as assumed here. Under these circumstances such integrals are dominated by local behavior near the ends. The importance of end effects is emphasized by Faltinsen (1973) in a study of the exciting force for the case where the incident wavelength is comparable to the ship's beam and the forward velocity is small.

The strip theory for predicting heave and pitch is completed by adding the body-mass and hydrostatic forces as given by (6.182–6.183) and (6.144–6.149), respectively. The result is a pair of coupled linear equations for the complex amplitudes of heave and pitch, analogous to (89–90). In solving these equations of motion the principal task is the computation of the two-dimensional added-mass and damping coefficients for each section of the ship.

Typical results are shown in figure 7.9 for a "Mariner" class ship. The calculations here are based on the exciting force and moment (119–120). Results using the complementary derivation of the exciting force and moment, given by Salvesen, Tuck, and Faltinsen (1970) and by Vugts (1970), show similar agreement with experiments.

It is apparent from figure 7.9 that the pitch and heave motions in head waves are predicted with sufficient accuracy for most practical purposes. Moreover, the regime of agreement with experiments is not limited to high frequencies in the vicinity of resonance, since the strip theory predicts the correct (hydrostatic) low-frequency limit of the exciting force and moment. Thus, as $\omega_0 \to 0$, $\xi_2/A \to 1$ and $|\xi_6|l/(2A) \simeq \omega_0^2 l/(2g)$, in agreement with the results shown in figures 7.7 and 7.8. As a result, the strip theory is useful over the complete range of frequencies of practical significance.

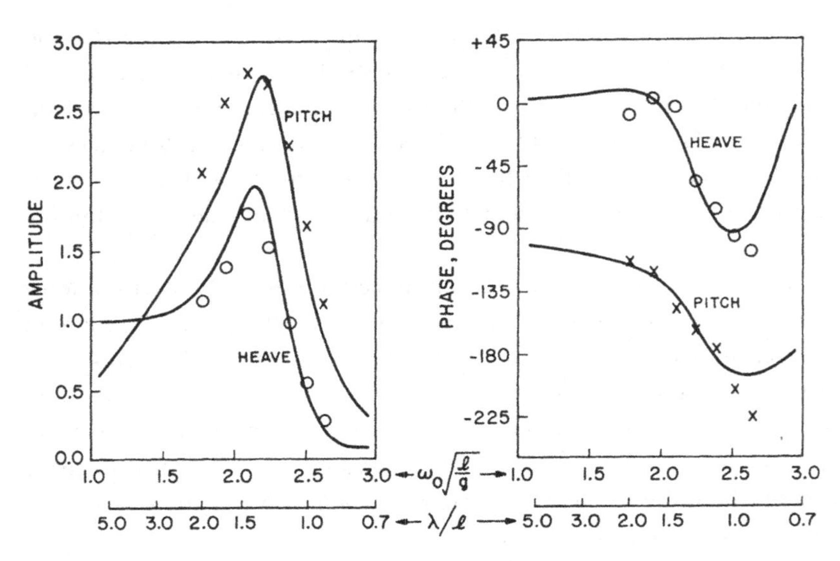

Figure 7.9
Amplitude and phase of the heave and pitch motions for a "Mariner" ship in head waves, at a Froude number of 0.3. The curves are strip-theory calculations. Experimental points are from Salvesen and Smith (1971). Amplitudes are nondimensionalized as in figures 7.7 and 7.8. The phase lead of the motion is shown, relative to the incident wave elevation at the midship section. (Calculations courtesy of C. M. Lee)

7.8 Slender Bodies in Shallow Water

Slender-body theory can be applied to the study of ship hydrodynamics in shallow water, and thus to the prediction of shallow-water effects on the hydrodynamic characteristics of ships. These predictions are of particular importance for large, deep-draft vessels during operations in coastal areas and harbors.

With the same assumptions regarding the body geometry as in earlier sections of this chapter, we shall assume in addition that the fluid is of constant depth $h = O(\varepsilon)$. This order of magnitude is assumed, so that the depth appears in the inner problem. In the converse case, where the depth is of the order of the body length, there are no effects of the bottom in the inner region.

The outer flow, far from the inner region near the body surface, is characterized by waves on the free surface. Since the depth is small, these waves will be nondispersive, as noted following (6.28). In order to describe this relatively simple wave motion, we shall assume that the characteristic wavelength is $O(1)$, and thus $kh = O(\varepsilon)$, where k is the wavenumber $2\pi/\lambda$. With this restriction, the three-dimensional velocity potential in the outer region can be expanded in a Taylor series about the bottom, of the form

$$\begin{aligned}\phi(x_0,y_0,z_0,t) = \phi(x_0,-h,z_0,t) + (y_0+h)\left(\frac{\partial\phi}{\partial y_0}\right)_{y_0=-h} \\ + \tfrac{1}{2}(y_0+h)^2\left(\frac{\partial^2\phi}{\partial y_0^2}\right)_{y_0=-h} + \cdots.\end{aligned} \tag{121}$$

The first derivative vanishes on the bottom, and hence, to second order, the potential can be expressed as

$$\phi \simeq \phi^{(0)}(x_0,z_0,t) + \tfrac{1}{2}(y_0+h)^2\phi^{(2)}(x_0,z_0,t). \tag{122}$$

Evaluating the first and second derivatives of (122) with respect to y_0, and substituting these in the free-surface condition and Laplace's equation, respectively,

$$\frac{\partial^2\phi^{(0)}}{\partial t^2} + gh\phi^{(2)} = 0, \tag{123}$$

$$\frac{\partial^2\phi^{(0)}}{\partial x_0^2} + \frac{\partial^2\phi^{(0)}}{\partial z_0^2} + \phi^{(2)} = 0. \tag{124}$$

Note that (123) holds on $y_0 = 0$, and (124) holds throughout the fluid domain ($-h < y_0 < 0$), but since the functions $\phi^{(0)}$ and $\phi^{(2)}$ are independent of y_0, that distinction is not significant. Thus (123–124) are valid throughout the x_0-z_0 plane in the domain of the outer solution.

Combining (123–124) to eliminate $\phi^{(2)}$ gives the linearized equation for shallow-water waves in the form

$$\frac{\partial^2\phi^{(0)}}{\partial x_0^2} + \frac{\partial^2\phi^{(0)}}{\partial z_0^2} - \frac{1}{gh}\frac{\partial^2\phi^{(0)}}{\partial t^2} = 0. \tag{125}$$

This is the *wave equation*, which governs two-dimensional acoustic waves with the speed of sound c replacing $(gh)^{1/2}$ in (125). Thus shallow-water waves are closely related to acoustic waves.

For steady-state motion in a moving reference frame, the time derivative $\partial/\partial t$ is replaced by $-U\partial/\partial x$, and (125) takes the form

$$(1-F_h^2)\frac{\partial^2\phi^{(0)}}{\partial x^2} + \frac{\partial^2\phi^{(0)}}{\partial z^2} = 0, \tag{126}$$

where F_h is the Froude number based on the fluid depth,

$$F_h = U/(gh)^{1/2}. \tag{127}$$

With this parameter replaced by the Mach number U/c, (126) is the equation governing linearized steady-state two-dimensional flow in aerodynamics, with compressibility effects represented by the speed of sound.

From the mathematical standpoint, (126) is an elliptic equation in the *subcritical* case ($F_h < 1$), and a hyperbolic equation in the *supercritical* case ($F_h > 1$). Physically these correspond respectively to the aerodynamic conditions of subsonic and supersonic flow. From either viewpoint the flow is expected to change markedly as one passes through the critical Froude number $F_h = 1$, and separate solutions of (126) must be developed for each case.

In the subcritical case, (126) can be reduced to Laplace's equation by the *Prandtl-Glauert transformation*. If the lateral coordinate is transformed by a factor

$$z' = z(1 - F_h^2)^{1/2}, \tag{128}$$

while the longitudinal coordinate $x = x'$ is unchanged, it follows that

$$\frac{\partial^2 \phi^{(0)}}{\partial x'^2} + \frac{\partial^2 \phi^{(0)}}{\partial z'^2} = 0. \tag{129}$$

For supercritical flow, $F_h > 1$, the method of characteristics can be used to show that the general solution of (126) is of the form[4]

$$\phi^{(0)}(x,z) = f[x \pm (F_h^2 - 1)^{1/2} z]. \tag{130}$$

For steady forward motion in the positive x-direction, the indeterminacy of the sign in (130) is removed by imposing a radiation condition. Thus only the characteristics that radiate downstream from the body are retained, and if the motion is symmetrical about $z = 0$ it follows that the velocity potential is of the form

$$\phi^{(0)}(x,z) = f[x + (F_h^2 - 1)^{1/2} |z|]. \tag{131}$$

In the inner region, near the body surface, motions that are slowly varying in the longitudinal direction can be approximated as two-dimensional flows. If we restrict our attention to the low-frequency regime, as in section 7.6, the inner solution is governed by the rigid free surface condition

$\partial\phi/\partial y = 0$ on $y = 0$, and the same condition on the bottom $y = h$. Thus the inner flow is constrained by two parallel and closely spaced rigid boundaries, as opposed to the unbounded case where the flow can spread radially in the y-z plane. This constraint is an essential feature of the inner flow and one of the principal effects of shallow water.

For lateral motions, the leading-order results for small ε can be constructed from a strip-theory synthesis, as in deep water (see problem 14). Thus, the sway added mass of a body in shallow water can be represented as in (40), but the relevant two-dimensional added-mass coefficient $m_{22}(x)$ is determined for the body section $\Sigma_B\ (x)$ and its rigid-body image, in a channel of total depth $2h$. Some examples of the resulting two-dimensional added-mass coefficient are discussed in section 4.18. Typically, as in (4.157), the lateral added-mass coefficient is increased by proximity to the bottom.

The lateral maneuvers of ships in shallow water can be described as in sections 7.4 and 7.5 provided the two-dimensional added-mass coefficient is modified to account for shallow-water effects. As a result, ships are generally more stable in shallow water and more difficult to maneuver.

Lateral flows that can be described in the inner region without influence from the outer flow are not affected to leading order by wave effects or by the Froude number F_h. However, the strip-theory approach for lateral flows breaks down when the "gap" between the bottom of the fluid and the body is very small, resulting in a "blockage" of the lateral flow past the body section. In the limit when this gap tends to zero (or the body runs aground!), the lateral flow must pass around the body ends, in a manner that is predominantly two-dimensional in the horizontal plane x-z, rather than in the lateral plane y-z.

The transition between these two regimes can be treated by a matching technique in which the outer flow is horizontal. Here, the extent to which the inner flow passes through the gap or around the body ends depends on the pressure jump between the two sides of the body, which depends in turn on the inner limit of the outer solution. This technique has been developed for the case $F_h = 0$ by Newman (1969) and extended to subcritical Froude numbers $0 < F_h < 1$ by Breslin (1972). The same technique is applied to horizontal motions in a seaway by Beck and Tuck (1972).

For longitudinal flows, as in the case of an unbounded fluid treated in section 7.2, it is essential to match the source-like flow in the inner region with an outer three-dimensional solution. This matching procedure determines an arbitrary function $f(x, t)$ in the inner solution, which affects the pressure distribution and resulting forces on the body. We shall illustrate this situation for the steady forward motion of a ship in shallow water, a problem that was solved first, in the sense described here, by Tuck (1966).

The problem to be solved is that of the steady forward motion of a slender ship moving on the free surface in shallow water. Since the motion is steady state in the moving reference frame (x, y, z), the inner solution near the hull surface can be expressed in the form

$$\phi(x,y,z) = U[\Phi_1(y,z;x) + f(x)]. \tag{132}$$

Here Φ_1 is a solution of the two-dimensional Laplace equation (12), and as in (17), the boundary condition

$$\frac{\partial \Phi_1}{\partial N} = n_x, \quad \text{on } \Sigma_B(x). \tag{133}$$

In addition, $\partial\Phi_1/\partial y = 0$ on the free surface $y = 0$ and on the bottom $y = h$.

The function $f(x)$ in (132) is a homogeneous solution of this boundary-value problem. As such, $f(x)$ is indeterminate from the inner solution, and must be found by matching (132) with the outer solution, where three-dimensional effects are significant but where the body geometry is unimportant.

Because of the boundary condition (133) and the relation (21), the solution for Φ_1 is source-like with a net flux Q equal to (minus) the rate of change of sectional area S'. In shallow water this flux is constrained by the free surface $y = 0$ and the bottom at $y = -h$. Thus, far from the body in the inner region, the flux must be a horizontal streaming flow, divided symmetrically in the two directions. The velocity of this streaming flow is $Q/2h$, and the outer expansion of the inner solution must be of the form

$$\frac{\partial \Phi_1}{\partial z} \simeq \mp S'/2h, \quad \text{for } z/\varepsilon \to \pm\infty. \tag{134}$$

The decomposition (132) can be made unique by integrating (134),and specifying the asymptotic form of Φ_1 as

$$\Phi_1 \simeq -(S'/2h)|z|, \quad \text{for } |z| \gg \varepsilon. \tag{135}$$

Equations (134–135) hold in the overlap region $\varepsilon \ll y \ll 1$, where they are to be matched to the inner expansion of the outer solution.

In the subcritical case, the outer solution is governed by Laplace's equation (129) in transformed coordinates. The solution corresponding to (134) near the body is a distribution of two-dimensional sources along the body axis. This outer solution is identical to the thickness problem of two-dimensional lifting-surface theory, and from (5.57) the appropriate source strength is given by

$$q = -US' / h. \tag{136}$$

The desired source distribution, of strength (136), satisfying Laplace's equation (129) in the transformed coordinates, or (126) in the physical coordinates (x, z), is of the form

$$\phi = -\frac{U}{4\pi h}(1 - F_h^2)^{-1/2} \int_l S'(\xi) \log[(\xi - x)^2 + z^2(1 - F_h^2)] d\xi. \tag{137}$$

Matching of this outer solution with (134–135) is assured by continuity, and in view of (135) the arbitrary function $f(x)$ in (132) can be obtained, from (137), simply by setting $z = 0$. Thus, in the subcritical case $F_h < 1$,

$$f(x) = -\frac{1}{2\pi h}(1 - F_h^2)^{-1/2} \int_l S'(\xi) \log|\xi - x| d\xi. \tag{138}$$

In the supercritical case, the outer solution can be deduced from (131) by noting that the velocity components are related linearly, in the form

$$\frac{\partial \phi^{(0)}}{\partial z} = \pm (F_h^2 - 1)^{1/2} \frac{\partial \phi^{(0)}}{\partial x}, \quad \text{for } z \gtrless 0. \tag{139}$$

Equating this result to (134) in the overlap region, it follows that

$$\frac{\partial \phi^{(0)}}{\partial x} = -(US' / 2h)(F_h^2 - 1)^{-1/2}, \tag{140}$$

for $|z| \ll 1$. Integrating both sides of this equation with respect to x, and requiring that $\phi^{(0)} \to 0$ as $x \to +\infty$, the outer solution is obtained in the form

$$\phi^{(0)} = -(U / 2h)(F_h^2 - 1)^{-1/2} S[x + (F_h^2 - 1)^{1/2} |z|]. \tag{141}$$

Here the sectional area $S(x)$ is defined to be zero, when the argument in (141) is outside the range of the body length. Differentiating (141) with respect to z confirms that this expression matches with the inner flux (134).

Moreover, the arbitrary function $f(x)$ of the inner solution (132) can be found by setting $z = 0$ in (141). Thus, in the supercritical case $F_h > 1$,

$$f(x) = -(1/2h)(F_h^2 - 1)^{-1/2} S(x), \tag{142}$$

with the understanding that (142) vanishes if x is outside the range of the body length.

With the function $f(x)$ determined by (138) or (142), the inner solution (132) is specified uniquely. The complete inner solution requires that the boundary-value problem for Φ_1 be solved. However, we shall show that the leading-order hydrodynamic pressure on the body is dominated by the function $f(x)$. Thus the pressure forces acting on the body can be found to leading order in ε, without solving for Φ_1.

The hydrodynamic pressure can be obtained in the inner region from Bernoulli's equation (6.3). In the moving frame of reference, it follows that

$$p = \rho\left(U\frac{\partial\phi}{\partial x} - \tfrac{1}{2}\nabla\phi\cdot\nabla\phi\right) \simeq \rho U^2[f'(x) - \tfrac{1}{2}\nabla\Phi_1\cdot\nabla\Phi_1]. \tag{143}$$

Here the last approximation holds in the inner region, and $f'(x) \equiv df/dx$.

From (138) or (142) the function $f(x)$ is of order $S/h = O(\varepsilon)$, whereas the boundary condition (133) gives the estimate $\nabla\Phi_1 = O(\varepsilon)$. Thus the dominant contribution to the pressure on the body surface is due to the first term in (143), or

$$p = \rho U^2 f'(x) + O(\varepsilon^2). \tag{144}$$

This approximation can be used to find the leading-order force and moment acting on the body. Note from symmetry that the only force components are the longitudinal drag force $-F_x$ and the vertical force F_y. The only non-zero moment is the pitch component M_z.

The drag force due to wave resistance can be obtained by pressure integration from the formula

$$D = -F_x = -\iint_{S_B} pn_x\, dS = -\rho U^2 \int_l f'(x)S'(x)\, dx. \tag{145}$$

Here we use (21) and the fact that the first-order pressure (144) is constant across the body section.

In the subcritical case, the wave resistance (145) vanishes. This can be confirmed directly, by substituting (138). The derivative of (138) is a

Cauchy principal-value integral, and the result of substituting this in (145) is a double integral that changes sign if the dummy variables are interchanged. Thus this double integral must vanish. More directly, since the outer solution in the subcritical case is governed by Laplace's equation, with transformed coordinates, it follows from D'Alembert's paradox that there can be no drag force. In essence, there can be no wave resistance, to leading order, since there are no waves in this outer solution.

In the supercritical case, the wave resistance is nonvanishing, essentially because the outer solution (141) propagates to infinity without attenuation downstream. Formally, substituting (142) in (145), it follows that, for $F_h > 1$,

$$D = \frac{\rho U^2}{2h(F_h^2 - 1)^{1/2}} \int_l [S'(x)]^2 \, dx. \tag{146}$$

Note that this drag force is positive definite.

The vertical force and pitch moment can be obtained from pressure integration in the forms

$$F_y = \iint_{S_B} p n_y \, dS \simeq \rho U^2 \int_l B(x) f'(x) \, dx, \tag{147}$$

$$M_z = \iint_{S_B} p n_y x \, dS \simeq \rho U^2 \int_l B(x) f'(x) x \, dx, \tag{148}$$

where $B(x)$ denotes the local waterplane beam.

In the subcritical case, it is convenient to integrate (147–148) by parts, before substituting (138). Assuming that the beam vanishes at the bow and stern, it follows that for $F_h < 1$,

$$F_y = \frac{\rho U^2}{2\pi h(1 - F_h^2)^{1/2}} \int_l B'(x) \int_l S'(\xi) \log|\xi - x| \, d\xi \, dx, \tag{149}$$

$$M_z = \frac{\rho U^2}{2\pi h(1 - F_h^2)^{1/2}} \int_l [xB(x)]' \int_l S'(\xi) \log|\xi - x| \, d\xi \, dx. \tag{150}$$

In the supercritical case, (142) can be substituted directly in (147–148), and thus for $F_h > 1$,

$$F_y = -\frac{\rho U^2}{2h(F_h^2 - 1)^{1/2}} \int_l B(x) S'(x) \, dx, \tag{151}$$

$$M_z = -\frac{\rho U^2}{2h(F_h^2-1)^{1/2}} \int_l B(x)S'(x)x\,dx. \tag{152}$$

The "sinkage" and "trim" of a ship moving in shallow water can be computed by equating the vertical force and pitch moment to the hydrostatic restoring force and moment (6.144) and (6.149). The resulting calculations are of practical importance in predicting the "squat" of a ship, and ultimately the occurrence of grounding due to increased draft. A comparison of these calculations with experiments is given by Tuck (1966) and reproduced in figure 7.10. The theory is invalid near the critical Froude number $F_h = 1$, but in other respects it agrees fairly well with the experiments. For typical ships the vertical "sinkage" is the dominant effect in the subcritical regime, whereas "trim" is dominant in the supercritical regime. These conclusions follow immediately if one assumes fore-and-aft symmetry (see problem 15).

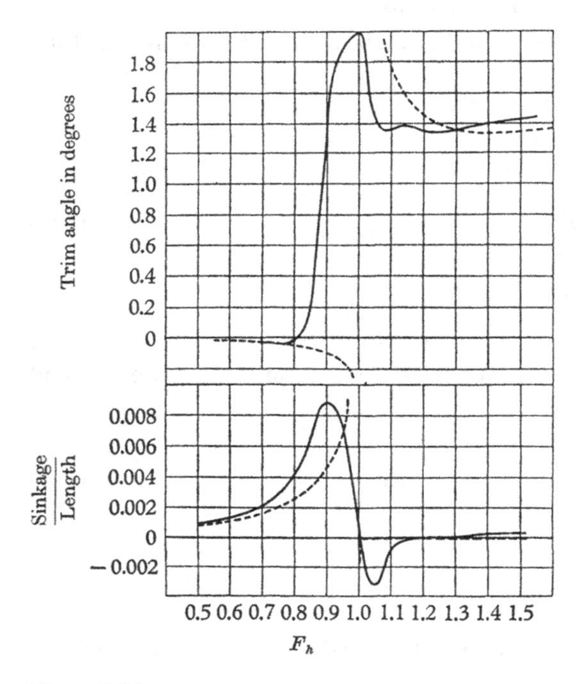

Figure 7.10
Sinkage and trim of a ship in shallow water vs. the Froude number (127). The solid lines are experimental values, and the dashed lines are theoretical predictions obtained from (149–152). (From Tuck 1966)

Various generalizations have been made of the original treatment by Tuck (1966). For example, Beck, Newman, and Tuck (1975) treat the case of a ship in a "dredged channel," bounded on the sides by shallow water. An interesting special case occurs when the flow in the channel is subcritical while the flow in the shallow regions is supercritical. Other related problems are discussed by Beck (1977) and Tuck (1978), and in references cited in these papers.

In the transcritical regime $F_h \simeq 1$, the linearized shallow-water approximation (125–126) is invalid. An appropriate nonlinear theory has been developed by Lea and Feldman (1972), and generalized by Mei (1976) to include dispersion as well as nonlinearities.

Problems

1. Derive from (26) the form of the outer potential at large distances from a body in longitudinal motion. Compare this result with (4.123) and show that the longitudinal added-mass coefficient is small compared to the displaced mass of the body.

2. Using the strip-theory results (40–42), compute the added mass and added moment of inertia for a slender spheroid of length $2a$ and maximum diameter $2b$. Compare the results with figure 4.8, and find the values of a/b where these approximations are in error by 10 percent of the exact value. Does a spheroid satisfy the restrictions on the behavior of the sectional-area curve at the two ends that have been assumed in deriving the slender-body theory?

3. Using (35), derive an integral representation for the longitudinal added mass of a slender spheroid. Show that as $b/a \to 0$, the dominant term in this expression is equal to

$$m_{11} = \frac{4}{3}\pi\rho\frac{b^4}{a}\left[\log\frac{a}{b} + O(1)\right].$$

Compare this approximation with the exact result shown in figure 4.8.

4. Following the approach outlined in problem 3, estimate the heave added mass of the spar buoy shown in figure 2.16 with (1) the rigid free-surface approximation and (2) the infinite-frequency approximation as illustrated in figure 6.22. Compare these results with each other and with the observed shift in resonance in figure 2.16.

5. Show that the pitch and roll motions of a slender spar buoy are resonant unless the centers of gravity and buoyancy coincide.

6. Explain the distinction between maneuverability and stability. In what sense do these conflict?

7. Assume that a ship has elliptical sections and that the planform is a quadrilateral with draft $T(x) = T_0 - \alpha x$ and vertical ends. Estimate the change in the derivatives in table 7.2 due to a small trim angle α and the effect of this trim angle on the stability parameter C. If $x_G = 0$, $C_B = 0.7$, $T/L = 0.05$ and $B/T = 2.5$, show that the ship will be stable if the bow-up trim angle is greater than 0.7 degrees.

8. Using the same assumptions as in the derivation of equations (60–61) derive the linearized equations of motion for a submarine in the vertical plane. Include a static stabilizing moment $k\theta$ where θ is the pitch angle. Show that the analogue of (72) for the stability roots is

$$A\sigma^3 + B\sigma^2 + [C + k(m - \partial F_2 / \partial \dot{U}_2)]\sigma - k(\partial F_2 / \partial U_2).$$

Show that the submarine will be stable at all speeds if $C > 0$, and at low speeds if $C < 0$.

9. Using the hydrodynamic derivatives for a streamlined body of revolution with fore-and-aft symmetry without lifting surfaces, show that the critical speed for a submarine to become unstable is

$$U_{crit} = \left[\frac{k(m+m_{22})}{(m_{22}-m_{11})(m+m_{11})}\right]^{1/2} \simeq [2g(y_B - y_G)]^{1/2}$$

where the approximation is valid if the body is slender.

10. Develop the linearized equations of motion for roll, pitch, surge, and sway for an oceanographic research buoy falling with constant vertical velocity U_2, assuming that it is symmetrical about the x, y, and z axes, except for the center of gravity which is at a position y_G below the centroid. What is the maximum terminal velocity for stability of the vessel in this orientation, assuming $m_{11} = m_{33} > m_{22}$?

11. Calculate the relative error incurred by the approximation (109) of the Froude-Krylov force for a rectangular and triangular section in head seas of various wavelengths. Confirm the statement following (109) regarding the magnitude of this error.

12. Rederive the strip theory results of section 7.7 for sway and yaw motions of a ship, noting the similarities and differences with respect to heave and pitch.

13. Show that the pitch and heave motions of a ship with fore-and-aft symmetry are uncoupled at zero forward speed. Is this true if $U \neq 0$?

14. Show that the expression (37) holds for the sway force on a ship in shallow water, with a rigid free-surface condition, but not for the heave force.

15. Compute the sinkage and trim for a simple ship with parabolic waterlines, constant draft, and vertical sides by balancing the hydrodynamic force and moment (149–152) with the hydrostatic restoring terms. Show that the trim angle of the ship is zero for subcritical Froude numbers and the sinkage (heave displacement) is zero for supercritical motion. Will a ship operating at excessive speed in shallow water ground at the bow or at the stern?

References

Abkowitz, M. A. 1969. *Stability and motion control of ocean vehicles.* Cambridge, Mass.: The MIT Press.

Beck, R. F. 1977. Forces and moments on a ship moving in a shallow channel. *J. Ship Res* 21:107–119.

Beck, R. F., J. N. Newman, and E. O. Tuck. 1975. Hydrodynamic forces on ships in dredged channels. *J. Ship Res.* 19:166–171.

Beck, R. F., and E. O. Tuck. 1972. Computation of shallow water ship motions. In *Ninth symposium on naval hydrodynamics,* eds. R. Brard and A. Castera, pp. 1543–1587. Washington: U.S. Government Printing Office.

Bishop, R. E. D., A. G. Parkinson, and R. Eatock Taylor eds. 1972. Proceedings of an international symposium on directional stability and control of bodies moving in water. [supplementary issue] *Journal of Mechanical Engineering Science* 14:7.

Breslin, J. P. 1972. Theory for the first-order gravitational effects on ship forces and moments in shallow water. *J. Hydronautics* 6:110–111.

Faltinsen, O. 1973. *Wave forces on a restrained ship in head-sea waves.* Oslo: Det Norske Veritas.

Faltinsen, O. 1974. A numerical investigation of the Ogilvie-Tuck formulas for added-mass and damping coefficients. *J. Ship Res.* 18:73–84.

Fink, P. T., and W. K. Soh. 1974. Calculation of vortex sheets in unsteady flow and applications in ship hydrodynamics. In *Tenth symposium on naval hydrodynamics,* eds. R. D. Cooper and S. W. Doroff, pp. 263–291. Washington: U.S. Government Printing Office.

Frank, T., D. J. Loeser, C. A. Scragg, O. J. Sibul, W. C. Webster, and J. V. Wehausen. 1976. Transient-maneuver testing and the equations of maneuvering. In *Eleventh symposium on naval hydrodynamics,* eds. R. E. D. Bishop, A. G. Parkinson, and W. G. Price, pp. 3–22. London: The Institution of Mechanical Engineers.

Gerritsma, J., and W. Beukelman. 1967. Analysis of the modified strip theory for the calculation of ship motions and wave bending moments. *International Shipbuilding Progress* 14:319–337.

Korvin-Kroukovsky, B. V. 1955. Investigation of ship motions in regular waves. *Society of Naval Architects and Marine Engineers Transactions* 63:386–435.

Lea, G. K., and J. P. Feldman. 1972. Transcritical flow past slender ships. In *Ninth symposium on naval hydrodynamics,* eds. R. Brard and A. Castera, pp. 1527–1541. Washington: U.S. Government Printing Office.

Lighthill, M. J. 1960. Note on the swimming of slender fish. *Journal of Fluid Mechanics* 9:305–317.

Mandel, P. 1967. Ship maneuvering and control. In *Principles of naval architecture,* ed. J. P. Comstock, 463–606. New York: Society of Naval Architects and Marine Engineers.

Mei, C. C. 1976. Flow around a thin body moving in shallow water. *Journal of Fluid Mechanics* 77:737–751.

Motora, S. 1972. Maneuverability, state of the art. In *Fortieth Anniversary Symposium of the Netherlands Ship Model Basin,* pp. 136–169.

Newman, J. N. 1963, *The motions of a spar buoy in regular waves.* David Taylor Model Basin report 1499.

Newman, J. N. 1969. Lateral motion of a slender body between two parallel walls. *Journal of Fluid Mechanics* 39:97–115.

Newman, J. N. 1970. Applications of slender-body theory in ship hydrodynamics. *Annual Review of Fluid Mechanics* 2:67–94.

Newman, J. N., and E. O. Tuck. 1964. Current progress in the slender-body theory of ship motions. In *Fifth symposium on naval hydrodynamics,* eds. J. K. Lunde and S. W. Doroff, pp. 129–166. Washington: U.S. Government Printing Office.

Newman, J. N., and T. Y. Wu. 1973. A generalized slender-body theory for fish-like forms. *Journal of Fluid Mechanics* 57:673–693.

Norrbin, N. H. 1970. Theory and observation of the use of a mathematical model for ship maneuvering in deep and confined waters. In *Eighth symposium on naval hydrodynamics*, eds. M. S. Plesset, T. Y. Wu, and S. W. Doroff, pp. 807–904. Washington: U.S. Government Printing Office.

Ogilvie, T. F. 1974. *Workshop on slender-body theory, Part 1: free surface effects*. University of Michigan, Department of Naval Architecture and Marine Engineering report 162.

Ogilvie, T. F. 1977. Singular perturbation problems in ship hydrodynamics. *Advances in Applied Mechanics* 17:92–187.

Ogilvie, T. F., and E. O. Tuck. 1969. *A rational strip theory of ship motions: Part 1*. University of Michigan, Department of Naval Architecture and Marine Engineering report 013.

Salvesen, N., and W. E. Smith. 1971. *Comparison of ship-motion theory and experiment for Mariner hull and a destroyer hull with bow modification*. David Taylor Naval Ship Research and Development Center report 3337.

Salvesen, N., E. O. Tuck, and O. Faltinsen. 1970. Ship motions and sea loads. *Society of Naval Architects and Marine Engineers Transactions* 78:250–287.

Tuck, E. O. 1966. Shallow-water flows past slender bodies. *Journal of Fluid Mechanics* 26:89–95.

Tuck, E. O. 1978. Hydrodynamic problems of ships in restricted waters. *Annual Review of Fluid Mechanics* 10:33–46.

Tuck, E. O., and J. N. Newman. 1974. Hydrodynamic interactions between ships. In *Tenth symposium on naval hydrodynamics*, eds. R. D. Cooper and S. W. Doroff, pp. 35–70. Washington: U.S. Government Printing Office.

Van Dyke, M. 1975. *Perturbation methods in fluid mechanics*. 2nd ed. Stanford: Parabolic Press.

Vugts, J. H. 1970. *The hydrodynamic forces and ship motions in waves*. Doctoral thesis, Delft Technological University.

Wang, S. 1976. Dynamical theory of potential flows with a free surface: A classical approach to strip theory of ship motions. *J. Ship Res.* 20:137–144.

Yeung, R. W. 1978. On the interactions of slender ships in shallow water. *Journal of Fluid Mechanics* 85:143–159.

Appendix: Units of Measurement and Physical Constants

The fundamental units of measurement in mechanics are mass (**M**), length (L), and time (T). In the SI system these are measured in the basic units of the kilogram (kg), meter (m), and second (s). Special names are given certain derived units. Thus the unit force of one Newton (N) is equal to one kilogram-meter per second-squared, the unit of work or energy is one Joule (J), equal to one Newton-meter, and the unit of power is one watt (W), equal to one Joule per second. Prefixes denote decade factors such as the kilo- (k) for 10^3, mega- (M) for 10^6, centi (c) for 10^{-2}, and milli- (m) for 10^{-3}.

Table A.1 lists various quantities in these and other units of measurement, and the relevant conversion factors. One nautical mile is the arc length equal to one minute of arc on the surface of the earth; one knot is the velocity unit equal to one nautical mile per hour. The (long) ton is equal to 2240 pounds and the metric tonne is equal to 1000 kg.

Various physical constants are required in marine hydrodynamics, notably the density and kinematic viscosity of water, which are listed for various temperatures in table A.2 with the corresponding properties of air at a pressure of one atmosphere. The standard acceleration of gravity (one g) is equal to 9.80665 m/s^2 or 32.174 ft/s^2. The standard atmospheric pressure at sea level is 1.01×10^5 N/m^2 or 14.7 pounds per square inch. The surface tension of the interface between air and water varies between 0.076 N/m and 0.071 N/m for the temperature range 0–30°C, with the value 0.074 N/m used commonly to describe capillary waves. The vapor pressure of water increases from 610 N/m^2 to 4230 N/m^2 over the same temperature range, or from 0.09 to 0.61 pounds per square inch.

Relative to these values for fresh water, the vapor pressure of salt water is slightly reduced and the surface tension increased. Both changes are insignificant compared to the effects of temperature and impurities.

Table A.1
Conversion Factors for Different Units of Measurement

Quantity	SI Unit	Other Unit	Inverse Factor
Length	1 m	3.281 feet (ft)	0.3048m
	1 km	0.540 nautical miles	1.852 km
Area	1 m^2	10.764 ft^2	0.0929 m^2
Volume	1 m^3	35.315 ft^3	0.0283 m^3
	1 m^3	264.2 gallon (US)	0.00379 m^3
	1 m^3	220.0 gallon (UK)	0.00455 m^3
Velocity	1 m/s	3.281 ft/s	0.305 m/s
	1 m/s	1.944 knot	0.514 m/s
Mass	1 kg	2.205 pound	0.454 kg
	1 Mg	0.984 ton (long)	1.016 Mg
	1 Mg	1 tonne (metric)	1 Mg
Force	1 N	0.225 pound force	4.448 N
	1 MN	100.4 ton force	9964 N
	1 MN	102.0 tonnef	9807 N
Pressure	1 N/m^2	0.000145 psi (pound per $inch^2$)	6895 N/m^2
Energy	1 J	0.738 foot pounds	1.356 J
Power	1 W	0.00134 horsepower	745.7 W

Table A.2
Density and Viscosity of Water and Air

	Density, ρ, kg/m^3			Kinematic Viscosity, ν, m^2/s		
Temperature, deg C	Fresh Water	Salt Water	Dry Air	Fresh Water	Salt Water	Dry Air
0	999.8	1028.0	1.293	1.79×10^{-6}	1.83×10^{-6}	1.32×10^{-5}
5	1000.0	1027.6	1.270	1.52	1.56	1.36
10	999.7	1026.9	1.247	1.31	1.35	1.41
15	999.1	1025.9	1.226	1.14	1.19	1.45
20	998.2	1024.7	1.205	1.00	1.05	1.50
25	997.0	1023.2	1.184	0.89	0.94	1.55
30	995.6	1021.7	1.165	0.80	0.85	1.60

Notes

Chapter 2: Model Testing

1. Advertisements for small toy racing cars typically claim a "speed of 760 scale miles per hour." One may question the basis for this scaling law, which certainly is not that suggested by Osborne Reynolds!

2. It may not be obvious at this stage that the drag force on a body in an unbounded fluid is independent of gravity. It will be shown in chapter 3 that the only role of gravity is to introduce a hydrostatic pressure and a corresponding buoyancy force on the body, which is additive to the hydrodynamic drag force.

Chapter 3: The Motion of a Viscous Fluid

1. Some authors prefer the symbol D/Dt, in equations such as (12) and (13) to emphasize that $\mathcal{V}(t)$ is a material volume, but since these integrals are functions only of time, we prefer the symbol d/dt in that context.

2. The roughness of the tube wall, ambient turbulence of the entering fluid, and degree of smoothness of the transition into the tube are all important. Reynolds observed the flow to remain at, or return to, the laminar state if the Reynolds number, based on the tube diameter and mean velocity across the tube, was less than about 2,000. With care, initially laminar fluid entering the tube would remain laminar up to Reynolds numbers on the order of 12,000. Subsequent investigations have attained much higher limits, as noted by Monin and Yaglom (1971).

3. This problem was first solved by Stokes, in 1851, as a means of estimating the drag on a pendulum.

4. This assumption is valid because the governing equation is parabolic. Thus $x < l$ is a sufficient condition, excluding the complicated region near $x = l$. Physically, the effects of viscosity are diffusive and the local solution in the boundary layer is independent of the flow downstream.

5. A rigorous justification for simultaneously changing both variables (U, x) to (u_τ, δ) requires that the Jacobian $\partial(U, x)/\partial(u_\tau, \delta)$ be nonzero. This can be verified from the empirical 1/7-th power relations to follow.

Chapter 4: The Motion of an Ideal Fluid

1. A strict interpretation of the material contour is essential to reconcile this statement with lifting-surface theory, and with other inviscid flows where vortices are shed into the fluid at a sharp corner.

2. The interpretation of $S + S_\varepsilon$ as a single closed surface can be justified by connecting these two separate surfaces with a "tube" of infinitesimal radius, as shown by the dashed lines in figure 4.6 (a); there is no contribution to the surface integrals from this tube in the limit where the tube radius shrinks to zero.

3. Since the body is rigid and of constant volume, the source term of order r^{-1} must vanish.

Chapter 5: Lifting Surfaces

1. Included here is a contribution proportional to $(\log A)/A^2$, which could not be anticipated from Prandtl's lifting-line equation (142).

2. This integral represents the downwash on the foil induced by the vortices in the wake.

Chapter 6: Waves and Wave Effects

1. There is a semantic problem here involving the number of dimensions. We shall refer to two- and three-dimensional motions of the fluid as in earlier chapters, corresponding respectively to one- and two-dimensionality of the free surface elevation $\eta(x, z, t)$.

2. A trochoid is the trajectory of any point on a circular disc rotating with its periphery in contact with a horizontal plane.

3. The derivative (114) and the resulting simplicity of (115) imply a fundamental relation between the factor G'' in the stationary-phase approximation and the spatial gradient of the wavenumber. This relationship is displayed explicitly in an alternative derivation of the stationary-phase approximation given by Lighthill (1965).

4. It is clear that if the horizontal coordinates of the source position and observation point are reversed, the flow will be unchanged provided the direction of the stream velocity U is also reversed. The dependence on the vertical coordinate and source position is less elementary, but in the asymptotic approximation used here both of these are in fact exponential.

5. The condition $S_{13} = 0$ is satisfied automatically for a body with a vertical plane of symmetry, such as a ship, provided one of the horizontal coordinates lies in this plane.

6. For finite depth h, these relations are still valid as $\omega \to \infty$, since for any fixed depth the wavelength will be short compared to h. The symbol "o" in (175–176) denotes that the order of magnitude of the exciting force is much smaller than that of the quantity in parentheses.

7. Since n_z is an odd function of z, the even part of ϕ_7, which contributes to the forces in the vertical plane, is a small first-order quantity.

8. Observed wave spectra, such as the Pierson-Moskowitz spectrum shown in figure 6.25, do not appear to satisfy this condition. Nevertheless the statistical distribution of observed wave amplitudes tends to follow the Rayleigh distribution with sufficient accuracy to justify the conclusions that follow.

Chapter 7: Hydrodynamics of Slender Bodies

1. The weighted head of an arrow and the forward placement of automobile engines are two common examples.

2. In the first-order solution, the appropriate radiation condition is that the waves far from the body section should be outgoing. The situation is more complicated in the higher-order solution, as discussed by Ogilvie and Tuck (1969).

3. This complication can be circumvented by introducing a mean effective depth for each section, where the scattering pressure acts, and evaluating the exponential factor for this particular depth, as outlined by, Gerritsma and Beukelman (1967).

4. Note that if (130) is extended to the subcritical case, one concludes that the solution of (126) is a function of the transformed complex variable $x' + iz'$ and its conjugate.

Index

(Italicized page numbers denote reference to material in figures. Italicized numbers in parentheses denote problem numbers.)

Printed in the United States
by Baker & Taylor Publisher Services